Books are to be returned on or before
the last date below.

Pelvic F **ISSUED IN BACKUP**

0 5 AUG 2009

DVD MISSING
MG 7/8/09

7-DAY LOAN

LIVERPOOL
JOHN MOORES UNIVERSITY
AVRIL ROBARTS LRC
TEL 0151 231 4022

D1615487

LIVERPOOL
JOHN MOORES UNIVERSITY
AVRIL ROBARTS LRC
TEL 0151 231 4022

WITHDRAWN

LIVERPOOL JMU LIBRARY

3 1111 01259 9302

Kaven Baessler, Bernhard Schüssler, Kathryn L. Burgio, Kate H. Moore,
Peggy A. Norton, and Stuart L. Stanton (Eds.)

Pelvic Floor Re-education

Principles and Practice

Second Edition

Springer

Kaven Baessler, MD
Consultant Gynaecologist
Department of Gynaecology
Charité University Hospital
Berlin, Germany

Kathryn L. Burgio, PhD
Professor of Medicine
Division of Gerontology, Geriatrics
  and Palliative Care
Department of Medicine
University of Alabama at Birmingham
Birmingham, AL, USA
and
Associate Director for
  Geriatric Research, Education, and
  Clinical Center
Birmingham Veterans Affairs Medical Center
Birmingham, AL, USA

Peggy A. Norton, MD
Professor
Chief of Urogynecology and Reconstructive
  Pelvic Surgery
Department of Obstetrics and Gynecology
University of Utah School of Medicine
Salt Lake City, UT, USA

Bernhard Schüssler, MD, PhD
Professor of Obstetrics and Gynaecology
Neue Frauenklinik
Kantonsspital Luzern
Luzern, Switzerland

Kate H. Moore, MBBS, MD, FRCOG,
  FRANZCOG, CU
Associate Professor
Department of Urogynaecology
University of New South Wales
St George Hospital
Sydney, NSW, Australia

Stuart L. Stanton, FRCS, FRCOG,
  FRANZCOG(Hon)
Professor of Pelvic Floor Reconstruction and
  Urogynaecology
Portland Hospital
London, UK

Library of Congress Control Number: 2006926873

British Library Cataloguing in Publication Data
Pelvic floor re-education : principles and practice. – 2nd
   ed.
   1. Pelvic floor   2. Urinary incontinence – Exercise therapy
   I. Baessler, K.
   616.6′2

ISBN-13: 9781852339685

ISBN: 978-1-85233-968-5 2nd edition   e-ISBN: 978-1-84628-505-9   Printed on acid-free paper
DOI: 10.1007/978-1-84628-505-9
ISBN: 3-540-19860-1/3-540-76145-4 1st edition

© Springer-Verlag London Limited 2008

First published 1994; Second edition 2008
The software disk accompanying this book and all material contained on it is supplied without any warranty of any kind. The publisher accepts no liability for personal injury incurred through use or misuse of the disk.
Apart from any fair dealing for the purposes of research or private study, or criticism or review, as permitted under the Copyright, Designs and Patents Act 1988, this publication may only be reproduced, stored or transmitted, in any form or by any means, with the prior permission in writing of the publishers, or in the case of reprographic reproduction in accordance with the terms of licences issued by the Copyright Licensing Agency. Enquiries concerning reproduction outside those terms should be sent to the publishers.
The use of registered names, trademarks, etc. in this publication does not imply, even in the absence of a specific statement, that such names are exempt from the relevant laws and regulations and therefore free for general use.
Product liability: The publisher can give no guarantee for information about drug dosage and application thereof contained in this book. In every individual case the respective user must check its accuracy by consulting other pharmaceutical literature.

9 8 7 6 5 4 3 2 1

springer.com

Preface

This book is designed to provide a clinically useful overview of our field – *Urogynecology and Reconstructive Pelvic Surgery*. Each chapter is meant to give a thorough yet concise amount of information. We assembled a world-class group of authors and asked each of them to focus on the information that they use in practice every day.

This text is appropriate for general gynecologists, physiotherapists, obstetrician-gynecologists, urologists, family practice and internal medicine physicians, nurse practitioners, physician assistants, and any other practitioners who regularly find themselves caring for women with pelvic floor dysfunction.

Kaven Baessler
Bernhard Schüssler
Kathryn L. Burgio
Kate H. Moore
Peggy A. Norton
Stuart L. Stanton

Contents

Preface v
Contributors xi

Part I Function and Dysfunction of the Pelvic Floor and Viscera

1.1 Functional Anatomy of the Pelvic Floor and Lower Urinary Tract 3
     Daniele Perucchini and John O.L. DeLancey

1.2 Neural Control of Pelvic Floor Muscles 22
     David B. Vodušek

1.3 The Effects of Pregnancy and Childbirth on the Pelvic Floor 36
     Kaven Baessler and Bernhard Schüssler

1.4 Muscle Function and Ageing 49
     Brenda Russell and Linda Brubaker

1.5 Urinary Incontinence and Voiding Dysfunction 62
     Annette Kuhn and Bernhard Schüssler

1.6 Pelvic Organ Prolapse 71
     Peggy A. Norton

1.7 Anal Incontinence, Constipation, and Obstructed Defecation 75
     Abdul H. Sultan and A. Muti Abulaffi

1.8 Overactive Pelvic Floor Muscles and Related Pain 83
     Wendy F. Bower

Part II Evaluation of the Pelvic Floor

2.1 Clinical Evaluation of the Pelvic Floor Muscles 91
     Diane K. Newman and Jo Laycock

2.2 Examination of Patients with Pelvic Organ Prolapse 105
     Ursula M. Peschers

2.3	Urodynamics	109
	Ursula M. Peschers	
2.4	Applying Urodynamic Findings to Clinical Practice	120
	Christopher K. Payne	
2.5	Anorectal Physiology	124
	David Z. Lubowski and Michael L. Kennedy	
2.6	Ultrasound Imaging	135
	Kaven Baessler and Heinz Kölbl	
2.7	Magnetic Resonance Imaging	144
	Thomas Treumann, Ralf Tunn, and Bernhard Schüssler	
2.8	Electrophysiology	155
	Clare J. Fowler and David B. Vodušek	
2.9	Outcome Measures in Pelvic Floor Rehabilitation	162
	Kate H. Moore and Emmanuel Karantanis	

Part III Techniques of Pelvic Floor Rehabilitation and Muscle Training

3.1	Concepts of Neuromuscular Rehabilitation and Pelvic Floor Muscle Training	177
	Jo Laycock	
3.2	Exercise, Feedback, and Biofeedback	184
	Pauline E. Chiarelli and Kate H. Moore	
3.3	Electrical Stimulation	190
	Wendy F. Bower	
3.4	Extracorporeal Magnetic Stimulation	196
	Alastair R. Morris and Kate H. Moore	
3.5	Devices	201
	Ingrid Nygaard and Peggy A. Norton	
3.6	Alternative Methods to Pelvic Floor Muscle Awareness and Training*	208
	Kaven Baessler and Barbara E. Bell	

* The authors gratefully acknowledge the assistance of Erna Alig in the preparation of this chapter.

Contents

Part IV   Treatment: Condition-Specific Assessment and Approaches

4.1   Behavioral Treatment 215
Kathryn L. Burgio

4.2   Stress Urinary Incontinence 221
Jo Laycock

4.3   Evidence for the Effectiveness of Pelvic Floor Muscle
Training in the Treatment and Antenatal Prevention of
Female Urinary Incontinence 228
E. Jean C. Hay-Smith and Kate H. Moore

4.4   Postpartum Management of the Pelvic Floor 235
Pauline E. Chiarelli

4.5   Role of a Perineal Clinic 242
Ranee Thakar

4.6   Overactive bladder 246
Kathryn L. Burgio, Dudley Robinson, and Linda Cardozo

4.7   Sexual Dysfunction and the Overactive Pelvic Floor 253
Wendy F. Bower

4.8   Anal Incontinence and Evacuation Difficulties 259
Christine Norton

4.9   Incontinence During Sports and Fitness Activities 267
Alain P. Bourcier

4.10 Pelvic Organ Prolapse – Pessary Treatment 271
Jane A. Schulz and Elena Kwon

Part V   What to Do if Physiotherapy Fails?

5.1   Stress Urinary Incontinence: Choice of Surgery 281
Stuart L. Stanton

5.2   Genital Prolapse: Surgery for Failed Conservative Treatment 285
Stuart L. Stanton

5.3   The Anal Sphincter 289
Klaus E. Matzel, Manuel Besendörfer, and Stefanie Kuschel

Index 293

Contributors

A. Muti Abulaffi, MS, FRCS
Colorectal Unit
Mayday University Hospital
Croydon, UK

Kaven Baessler, MD
Department of Gynaecology
Charité University Hospital
Berlin, Germany

Barbara E. Bell, BSc, BA
Pelvic Power
Chemmart Pharmacy
Acacia Ridge, QLD, Australia

Manuel Besendörfer, MD
Department of Surgery
University Hospital Erlangen
Erlangen, Germany

Alain P. Bourcier, PT
Pelvic Floor Rehabilitation Services
Department of Urology
Tenon Hospital
Paris, France

Wendy F. Bower, PhD
Department of Surgery
Division of Pediatric Surgery and Pedriatric
  Urology
The Chinese University of Hong Kong
Prince of Wales Hospital
Hong Kong, China

Linda Brubaker, MD, MS, FACOG, FACS
Department of Obstetrics and Gynecology
Female Pelvic Medicine and Reconstructive
  Surgery
Loyola University Medical Center
Maywood, IL, USA

Kathryn L. Burgio, PhD
Division of Gerontology, Geriatrics and Palliative
  Care
Department of Medicine
University of Alabama at Birmingham
Birmingham, AL, USA
and
Geriatric Research,
  Education, and Clinical Center
Birmingham VA Medical Center
Birmingham, AL, USA

Linda Cardozo, MD, FRCOG
Department of Urogynaecology
Kings College Hospital
London, UK

Pauline E. Chiarelli, PhD
Discipline of Physiotherapy
School of Health Sciences
University of Newcastle
Callaghan, NSW, Australia

John O.L. DeLancey, MD
Department of Obstetrics and Gynecology
Women's Hospital
Ann Arbor, MI, USA

Clare J. Fowler, FRCP
Department of Uro-Neurology
Institute of Neurology
London, UK

E. Jean C. Hay-Smith, PhD, MSc
Rehabilitation Teaching and Research Unit
Department of Medicine
Wellington School of Medicine and Health Sciences
University of Otago
Wellington, New Zealand

Emmanuel Karantanis, BMed, MBBS, UNSW, FRANZCOG
St George Hospital
Sydney, NSW, Australia

Michael L. Kennedy, BSc
Colorectal Research Fellow
St George Hospital Medical Centre
Sydney, NSW, Australia

Heinz Kölbl, MD
Gynecology and Obstetrics
Johannes-Gutenberg Universität
Mainz, Germany

Annette Kuhn, MD
Department of Urogynecology
Frauenklinik Inselspital
Bern, Switzerland

Stefanie Kuschel, MD
Department of Obstetrics and Gynecology
Kantonsspital Luzern
Luzern, Switzerland

Elena Kwon, BSc
Faculty of Medicine
University of Alberta
Edmonton, AB, Canada

Jo Laycock, PhD, FCSP
The Culgaith Clinic
Penrith, UK

David Z. Lubowski, MD, FRACS
Department of Colorectal Surgery
St George Hospital Medical Centre
Kogarah, NSW, Australia

Klaus E. Matzel, MD, PhD
Section of Coloproctology
Department of Surgery
University Hospital Erlangen
Erlangen, Germany

Kate H. Moore, MBBS, MD, FRCOG, FRANZCOG, CU
Department of Urogynaecology
University of New South Wales
St George Hospital
Sydney, NSW, Australia

Alastair R. Morris, MBChB, MRCOG, FRANZCOG, FRCSEd, MD
Department of Obstetrics and Gynaecology
Royal North Shore Hospital
Sydney, NSW, Australia

Diane K. Newman, RNC, MSN, CRNP, FAAN
Penn Center for Continence and Pelvic Health
Division of Urology
University of Pennsylvania Health System
Philadelphia, PA, USA

Christine Norton, PhD, MA, RN
Burdett Institute of Gastrointestinal Nursing
King's College London
St Mark's Hospital
Harrow, UK

Peggy A. Norton, MD
Urogynecology and Reconstructive Pelvic Surgery
Department of Obstetrics and Gynecology
University of Utah School of Medicine
Salt Lake City, UT, USA

Ingrid Nygaard, MD
Department of Urogynecology
University of Utah Health Science Center
Salt Lake City, UT, USA

Christopher K. Payne, MD, FACS
Section of Female Urology and NeuroUrology
Stanford University School of Medicine
Stanford, CA, USA

Daniele Perucchini, MD
Department of Obstetrics and Gynecology
University Hospital Zürich and Praxis am Stadelhoferplatz
Zürich, Switzerland

Contributors

Ursula M. Peschers, MD
Department of Obstetrics and Gynecology
Frauenklinik
Dachau, Germany

Dudley Robinson, MBBS, MRCOG
Department of Urogynaecology
Kings College Hospital
London, UK

Brenda Russell, PhD
Department of Physiology and Biophysics
The University of Illinois at Chicago
Chicago, IL, USA

Jane A. Schulz, MD, FRCSC
Urogynecology Unit
Department of Obstetrics and Gynecology
Royal Alexandra Hospital
University of Alberta
Community Services Centre
Edmonton, AB, Canada

Bernhard Schüssler, MD, PhD
Neue Frauenklinik
Kantonsspital Luzern
Luzern, Switzerland

Stuart L. Stanton, FRCS, FRCOG, FRANZCOG (Hon)
Pelvic Floor Reconstruction and Urogynaecology
Portland Hospital
London, UK

Abdul H. Sultan, MBChB, MD, FRCOG
Department of Obstetrics and Gynaecology
Mayday University Hospital
Croydon, UK

Ranee Thakar, MD, MRCOG
Department of Obstetrics and Gynaecology
Mayday University Hospital
Croydon, UK

Thomas Treumann, MD
Institute of Radiology
Kantonsspital Luzern
Luzern, Switzerland

Ralf Tunn, MD
German Pelvic Floor Center, Urogynecology
St Hedwig Hospitals
Berlin, Germany

David B. Vodušek, MD, PhD
Division of Neurology
University Medical Centre
Ljubljana, Slovenia

Video Material on the DVD

Stefanie Kuschel, MD
Department of Obstetrics and Gynecology
Kantonsspital Luzern
Luzern, Switzerland

Part I
Function and Dysfunction of the Pelvic Floor and Viscera

1.1
Functional Anatomy of the Pelvic Floor and Lower Urinary Tract

Daniele Perucchini and John O.L. DeLancey

Key Message

Pelvic floor rehabilitation is dependent on a meticulous insight into relevant anatomy. Therefore, this chapter describes not only the anatomy of the organs and muscles involved but also their topography and innervation. Predominantly its focus is on functional anatomy. Besides other issues, the following questions, which are necessary for understanding pelvic floor function, are extensively discussed:

- How is the pelvic floor muscle (PFM) able to empower the urethral closure mechanism?
- What are the anatomical deficiencies related to the prevention of successful pelvic floor re-education?
- How are the pelvic organs kept in place?
- What is the anatomical deficit when stress urinary incontinence (SUI) or prolapse occurs?
- What is the mechanism of the anal sphincter unit?

Introduction

The bony pelvis lies in the middle of the human body. It supports the spinal column, which attaches to it posteriorly, and provides the points of articulation for the femur and the lower extremities. It cradles the abdominopelvic organs that rest above and within it; however, because the bony pelvis is only a hollow ring, its contents would plummet to the ground unless it had a bottom. The "pelvic floor" is the bottom of this pelvic container and includes all of the structures that lie between the pelvic peritoneum and the vulvar skin (Fig. 1.1.1).

Pelvic floor disorders are common, affecting one in nine women.[1] They include pelvic organ prolapse, urinary incontinence, and anal incontinence. These are debilitating conditions that not only lead to medical problems and costs but are also associated with embarrassment that can lead to isolation, loss of independence, and diminished quality of life.

The levator ani muscles are the primary source of support for the pelvic organs. These muscles close off the pelvic floor, allowing structures that lie above them to rest on the upper surface of the muscle. This closure is usually remarkably effective; however, because of injuries and deterioration of the muscles, as well as of the nerves and connective tissue that support and control normal function, urinary incontinence, fecal incontinence, and pelvic organ prolapse can result.

Female incontinence is strongly associated with pregnancy, childbirth, and ageing. However, there is universal consensus that the pathophysiology of urinary incontinence and pelvic organ prolapse is multifactorial and incompletely understood. Incompetence of the sphincter mechanism, weakness of the muscles that support the urethra and bladder neck, overactive detrusor muscle, neurological disorders, injury during childbirth or other trauma, age-related changes in structural integrity, nervous control, hormone balance, and systemic disease have all been implicated as causative agents.

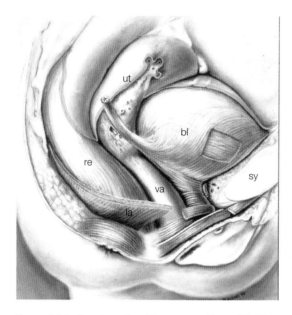

FIGURE 1.1.1. Overview of pelvic anatomy. The pelvic floor extends between the two red areas (peritoneum and levator ani muscle). re = rectum, ut = uterus, bl = bladder, sy = pubic symphysis, va = vagina, and la = levator ani muscles.

lead to the loss of continence early in life, even if an individual was born with a "good" continence mechanism (black line). Little is known about the individual risk factors for incontinence and prolapse; therefore, it is difficult to identify women who are at risk.

Nulliparous women without known damage to the pelvic floor may also leak urine. In a series of studies, approximately 30% of young, nulliparous, healthy, athletic women experienced problems with incontinence. The sports producing the highest percentage of incontinence occurrence were gymnastics (67%) and basketball (66%).[4] This suggests that there is a continence threshold that when exceeded can result in urine loss, even in the absence of known risk factors for incontinence.

A consensus also exists that the continence mechanism deteriorates over time. The prevalence of prolapse increases with age,[1] and vaginal birth confers a 4–11 fold increase in the risk of developing pelvic organ prolapse that increases with higher parity.[2] The deterioration of pelvic floor function may be acute, as with vaginal delivery. There may be recovery after that acute injury until another acute injury occurs, or there may be a gradual decline in function, especially with age.[3] As graphically depicted in Figure 1.1.2, despite a repeatedly damaged continence mechanism, a patient is able to retain continence and compensate for the damage (blue line). Compensation is possible because the damage to the pelvic floor and the continence control system remain above the continence threshold.

An individual may begin with far less reserve (red line), and although the number and magnitude of the insults she suffers over time are no greater than in a women who remains continent, her initial low reserve level leads to incontinence over her lifespan.

Pregnancy and devastating damage to the pelvic floor because of a difficult delivery can

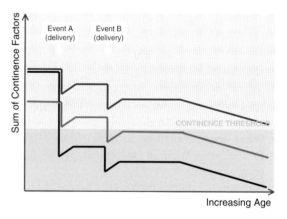

FIGURE 1.1.2. Deterioration of pelvic floor function over a lifetime. Continence is achieved if the sum of all continence factors remains over the (imaginary) continence threshold (blue area). Events A and B (pregnancy/delivery) and ageing contribute to the deterioration of the continence mechanism (pelvic floor). Various scenarios are possible:

- Blue line: Despite the fact that the continence mechanism is repeatedly damaged, a patient is able to remain continent and compensate the damage.
- Red line: An individual may start with far less reserve initially, even though the number and magnitude of the insults she suffers over time are no greater than in women who remain continent. She has less reserve and, therefore, becomes incontinent during her life.
- Black line: Devastating damage as consequence of an acute insult (event A) can lead to loss of continence early in life, even if an individual was born with good continence factors.

This chapter addresses the functional anatomy of the pelvic floor in women and, specifically, focuses on how the pelvic organs are supported by the surrounding muscle and fasciae. This chapter also addresses how pelvic visceral function relates to the clinical conditions of urinary incontinence and pelvic organ prolapse.

Anatomy of the Lower Urinary Tract

Clinicians have traditionally divided the lower urinary tract into the bladder and urethra (Fig. 1.1.1). At the junction of these two continuous, yet discrete, structures lies the vesical neck.

The Bladder

The bladder lies in the anterior (ventral) part of the pelvic cavity (Fig. 1.1.1). The proportion of the cavity that it occupies is dependent on the volume of fluid contained within the vesical lumen. Vesical filling results in direct contact between the bladder and the anterior abdominal wall above the pubic symphysis. The bladder is composed of an epithelium surrounded by layers of smooth muscle. The urothelium is much more than a classical barrier that separates urine from extracellular fluid. It is also an active absorptive epithelium and secretory tissue. The urothelium has specialized cell-surface proteins and ion pumps, plus proteglycans and glycoproteins, all of which function together to maintain the impermeability of the membrane.[5,6] These same mechanisms also provide an active defense against bacterial colonization.

Bladder Function

The primary function of the bladder is the storage of urine. The secondary function is the evacuation of urine (voiding). As a muscle, the detrusor must accommodate both processes using a "storage phase" and an "emptying phase." During the storage phase, the muscular layers relax to facilitate urine so that increasing physiologic volumes may be stored without any appreciable increase in intravesical pressure. When the bladder reaches its physiological capacity or when a woman voluntarily voids, cerebral tonic inhibition of the detrusor muscles' relaxation is released and a reflex voiding contraction is initiated. This is often referred to as the emptying phase, during which the tone of the urethral sphincter is relaxed and bladder detrusor muscles contract in a spiral fashion.

The Bladder Neck

The term "bladder neck" denotes the area at the base of the bladder where the urethral lumen passes through the thickened musculature of the bladder base (Fig. 1.1.3 A). It is a region where the detrusor musculature surrounds the trigonal ring and the urethral meatus[7]; therefore, it is sometimes considered part of the bladder musculature, but it also contains the urethral lumen. The bladder neck has come to be considered separately from the bladder and urethra because of its unique functional characteristics. The bladder neck plays an important role in the initiation of micturition. Its opening, coupled with the relaxation of the urethra, is required during bladder emptying. Urine entering into the proximal urethra through an open bladder neck can also contribute to the sensation of urgency and facilitate detrusor overactivity. In patients with SUI, increased bladder neck mobility is often present. Specifically, damage to this area results in its remaining open while at rest.[8]

Innervation of the Bladder

The bladder receives sympathetic innervation from the superior hypogastric plexus via hypogastric nerves into the inferior hypogastric plexus. Postganglionic fibers primarily innervate the bladder base and urethra. At the level of the inferior hypogastric plexus, contributions from the S2–S4 preganglionic fibers join the sympathetic nerves to form the pelvic plexus. These parasympathetic fibers lead to ganglia in the wall of the bladder, where the postganglionic fibers innervate the detrusor muscle. Afferent fibers toward the pelvic plexus and central nervous system travel with both the sympathetic and parasympathetic fibers. Overactivity of the detrusor muscle is attributed to increased activity of the parasympathetic components, and anticholinergic agents have become a mainstay of therapy.

FIGURE 1.1.3. **A.** Normal urethral anatomy of a nulliparous woman. The outermost layer (red) is composed of striated muscle that has three components: the sphincter urethra, the compressor urethra, and the urethrovaginal sphincter, which are known collectively as the striated urogenital sphincter muscle. The sphincter urethra encircles the midurethral wall, whereas the compressor urethra and sphincter urethrovaginalis arch over the distal urethra. The smooth muscle layers are highlighted in yellow. **B.** Illustration of striated muscle loss at the vulnerable zone of the urethra, at the bladder neck. Measurements of layer thickness in our specimens showed that there was localized disappearance of striated muscle, which contrasted with regions that seemed more resistant to damage. The pattern of striated muscle loss suggests that striated muscle in the proximal and the dorsal wall of the urethra might be more vulnerable (arrows) to one or more insults or processes than the distal urethra.

The Urethra

The adult female urethra is a complex 2–4-cm-long fibromuscular tube with a diameter of approximately 1 cm, and it extends from the bladder neck to the external urinary meatus (Fig. 1.1.3 A). The urethra lies on a supportive layer that is composed of the endopelvic fascia and the anterior vaginal wall (Fig. 1.1.4). This layer gains structural stability through its lateral attachment to the arcus tendineus fascia pelvis (ATFP) and arcus tendineus levator ani (ATLA) muscles[9,10] and through connections to the pelvic bones by the pubourethral ligaments, which contain dense connective tissue and smooth muscle.[11] The integrity of all of these connections is important for the transmission of PFM contraction to the closure function of the urethra. The female urethral wall contains concentric layers of muscle, connective tissue, and vasculature that contribute to urethral closure and are relevant for understanding lower urinary tract dysfunction.[12]

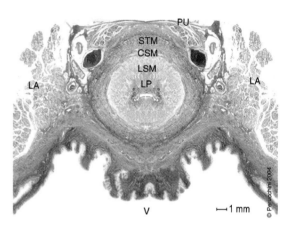

FIGURE 1.1.4. Midurethral cross-section with the levator ani muscle visible on both sides of the urethra. The outermost layer of the urethra is composed of striated muscle (STM). The female striated muscle of the urethra is predominantly slow twitch in nature. The striated muscle layer surrounds a two-layered smooth muscle component (CSM and LSM). Circularly arranged muscle cells occur in the outer aspect of the smooth muscle layer (CSM) and sometimes intermingle with the striated muscle. The innermost layer is longitudinally arranged (LSM). The urethral lamina propria (LP) extends from the longitudinal smooth muscle layer to the urothelium and fills the lumen of the urethra. The submucosa constitutes a relatively thick layer of loosely woven connective tissue with a rich vascular supply. The urethral mucosa consists of a transitiocellular epithelium in the proximal third. This epithelium fades out in a regular squamous mucosal epithelium in the distal two thirds of the urethra. Collagen is the major structural component of the connective tissue in the female urethra, whereas elastic fibers are exceedingly rare. LA = levator ani, V = anterior vaginal wall, PU = pubourethral ligaments.

Urethral closure function, as measured by urethral resting pressure, has been shown to decrease with age,[13] and groups of women with incontinence had statistically proven lower urethral closure pressure than continent women.[14]

The female urethra is composed of different regions along its length and can be understood by dividing the length of urethral lumen into fifths, each approximately 20% of the total length.[15] In the first quintile, the lumen of the urethra is surrounded by the vesical neck (0–20%). Next, the sphincter urethra and smooth muscle encircle the lumen from 20% to 60% of the total urethral length. The arch-shaped compressor urethra and urethrovaginal sphincter are found from 60% to 80% of the total urethral length, whereas the distal component includes only fibrous tissue and no significant contractile elements (Fig. 1.1.3 A).

Striated Urogenital Sphincter

The outermost layer is composed of striated muscle that has the following three components: (1) sphincter urethra, (2) compressor urethra, and (3) urethrovaginal sphincter. These components are collectively known as the striated urogenital sphincter muscle. The sphincter urethra encircles the midurethral wall, whereas the compressor urethra and sphincter urethrovaginalis arch over the distal urethra. Distally, under the arch of the pubic bone, these fibers diverge to insert into the walls of the vagina and the perineal membrane (compressor urethra and urethrovaginal sphincter) (Fig. 1.1.3 A). This structure is often referred to as the external urethral sphincter. This muscle is responsible for increasing intraurethral pressure during times of need, and it also contributes at least approximately one third of the resting tone of the urethra. Its composition of primarily slow-twitch, fatigue-resistant muscle fibers belies its constant activity. The muscle cells are smaller than ordinary skeletal muscle cells, at approximately 20μm in diameter.[16]

With increasing age, striated muscle loss at the bladder neck and along the dorsal wall of the urethra (Fig. 1.1.3 B) has been found. This leads to a horseshoe-shaped aspect of the striated muscle layer in the midurethral cross section.[17]

During times of urgent need, the striated urogenital sphincter muscle increases closure pressure by shortening its circumferentially oriented muscle fibers and constricting the lumen. In addition, contraction of the pubococcygeal muscle results in urethra compression against adjacent tissue. This compression depends on the fascial attachment of the urethra to the levator ani muscle (Fig. 1.1.4). A blockade of striated muscle activity, on the other hand, decreases resting urethral pressure by approximately 50%.[18] Constantinou and Govan[19] demonstrated that during a cough a urethral pressure increase is measurable at the level of the urethral sphincter, but not more proximally. This, together with increased electromyographic activity during cough and hold in healthy women,[20] suggests that the striated muscle does contribute to urethral closure pressure. Muscle contraction is not always present in women with SUI, and some stress-incontinent women who are capable of contracting their pelvic floor on demand do not have a muscle contraction visible during a cough.[21]

Urethral Smooth Muscle

The striated muscle layer surrounds a two-layered smooth muscle component. The two layers of the urethral smooth muscle consist of an outer circular layer (circular smooth muscle [CSM]) and an inner longitudinal layer (longitudinal smooth muscle [LSM]) (Fig. 1.1.4). The circular fibers contribute to urethral constriction, and the smooth muscle blockade reduces resting urethral closure pressure by approximately one third. The function of the longitudinal muscle is not entirely understood. The longitudinal muscles are considerably vaster than circular muscles; the reasons for this have yet to be determined.

The smooth muscle of the female urethra is associated with relatively few noradrenergic nerves, but it receives an extensive presumptive cholinergic parasympathetic nerve supply, which is identical in appearance to that which supplies the detrusor.[22] The innervation and longitudinal orientation of the majority of muscle fibers suggest that the urethral smooth muscle in the female is active during micturition, serving to shorten and widen the urethral lumen.

Urethral Submucosal Vasculature

The submucosal vasculature is remarkably prominent and is far more extensive than one would expect for such a small organ. It is likely responsible, in part, for the hermetic seal that maintains mucosal closure. Occlusion of arterial flow into this area decreases resting urethral closure pressure; therefore, these vessels are felt to participate in closure function.[12]

Urethral Glands

A series of glands are found in the submucosa, primarily along the dorsal (vaginal) surface of the urethra.[23] The glands are mostly concentrated in the distal and middle thirds of the urethra, varying in number and extent from one woman to the next. The location of urethral diverticula, which are derived from cystic dilation of these glands, follows this distribution, being most common distally and usually originating along the dorsal surface of the urethra.

Innervation of the Urethra

The innervation of the urethral sphincter complex is controversial. Nerve stimulation studies showed early evidence that the external urethral sphincter is innervated by the somatic fibers of the pudendal nerve. Several others have concluded that the external sphincter receives its innervation from the S2–S3 spinal roots via branches of the pudendal nerve.[24] Others have reported that the sphincter complex also receives autonomic innervation from the inferior hypogastric plexus via intrapelvic fibers. A recent study found that stimulation of the intrapelvic portion of the cavernous nerves results in contractions of the urethral sphincter even after the bladder and urethra have been surgically divided.[25] Although the importance of tonic sympathetic innervation to the male urethra has been well established, the role of sympathetic innervation in women is only now being explored.

The striated urethral sphincter is innervated by axons that originate from motor neurons in the sacral spinal cord and are carried into the pudendal nerve. The sphincter motor neurons are located in a circumscribed region of the sacral anterior horn called Onuf's nucleus. The potential for pharmacologically increasing muscle tone in the urethra is suggested by recent studies of duloxetine, which is a selective serotonin and norepinephrine reuptake inhibitor.[26,27]

Support of the Urethra and Pelvic Organs

Support of the urethra and pelvic organs rely on their attachments to the pubic bones, muscles, and connective tissue of the pelvis. The female pelvis and its supportive structures are not only important for micturition and defecation but must also accommodate cohabitation, as well as vaginal birth. These demands on the female pelvic floor may lead to a series of problems and disorders. The pelvic floor consists of several components lying between the peritoneum and the perineum (Fig. 1.1.1). The support for all of these structures comes from their connection to the bony pelvis and its attached muscles by a unique network of connective tissue.

The female pelvis can naturally be divided into anterior and posterior compartments because dense supportive tissues attach the pelvic organs to the lateral pelvic walls. The levator ani muscles form the bottom of the pelvis and are U-shaped (Fig. 1.1.5). The vagina and the pelvic

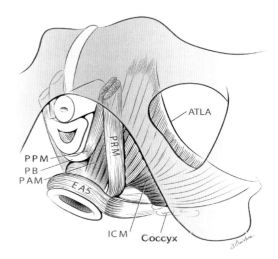

FIGURE 1.1.5. Schematic view of the U-shaped levator ani muscle from below, after the vulvar structures and perineal membrane have been removed, showing the arcus tendineus levator ani (ATLA); external anal sphincter (EAS); puboanal muscle (PAM); perineal body (PB) uniting the 2 ends of the puboperineal muscle (PPM); iliococcygeal muscle (ICM); puborectal muscle (PRM). Note that the urethra and vagina have been transected just above the hymenal ring. (*Source*: Copyright © DeLancey, 2003.)

1.1. Functional Anatomy of the Pelvic Floor and Lower Urinary Tract

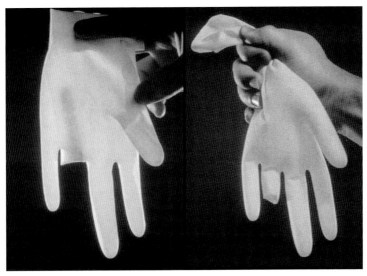

FIGURE 1.1.6. Bonney's analogy of vaginal prolapse. The eversion of an intussuscepted surgical glove finger by increasing pressure within the glove resulting is analogous to prolapse of the vagina. (*Source*: DeLancey, 2002, with permission.)

organs are attached to the levator ani muscles by connective tissue when they pass through the urogenital hiatus and are supported by these connections. The vagina has a similar relationship to the abdominal cavity as that of the inverted finger of a surgical glove (Fig. 1.1.6).[28] If the pressure in the glove is increased, it forces the finger to protrude downwards in the same way that increases in abdominal pressure force the vagina to prolapse. Vaginal support is a combination of constriction, suspension, and structural geometry. Figure 1.1.7 demonstrates this phenomenon

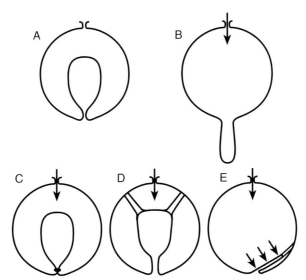

FIGURE 1.1.7. Diagrammatic display of vaginal support. (**A**) Invaginated area in a surrounding compartment. (**B**) The prolapse opens when the pressure (arrow) is increased. (**C**) The lower end of the vagina is held closed by the pelvic floor muscles, preventing prolapse by constriction. Pelvic floor muscle exercise is, therefore, commonly recommended for prolapse with less severe symptoms and to prevent pelvic organ prolapse by keeping the urogenital hiatus closed. (**D**) Ligament suspension. The vagina is suspended from the pelvic walls, and pelvic floor muscle exercise does not directly affect the suspension. (**E**) Flap valve closure where suspending fibers holds the vagina in a position against the wall, allowing increases in pressure (arrows) to pin it in place. (*Source*: DeLancey, 2002, with permission.)

and the strategies that the body uses to prevent prolapse.

Pelvic floor muscle exercise is commonly recommended for prolapse with less severe symptoms and to prevent pelvic organ prolapse by keeping the urogenital hiatus closed (Fig. 1.1.7 C).

The Levator Ani Muscle

Macroscopic Anatomy

The major structural component of the pelvic floor is the levator ani muscle group because they form the effective contractile support structure of the region. Four muscles make up the levator ani: pubococcygeus, puborectalis, iliococcygeus, and coccygeus. In practical terms, the pelvic floor is synonymous with the levator ani. The opening within the levator ani muscle through which the urethra and vagina pass is called the urogenital hiatus of the levator ani (Fig. 1.1.8). It is through this opening that genital prolapse occurs. Constant adjustments in muscular baseline activity of the levator ani muscle[29] keeps the urogenital hiatus closed, by compressing the vagina, urethra, and rectum against the pubic bone, the pelvic

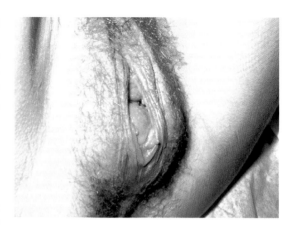

FIGURE 1.1.9. Descending perineum syndrome. Damage to the levator ani muscle can result in visible descent of the perineum and enlarged hiatus. (*Source*: DeLancey 2002, with permission.)

floor, and related organs in a cephalic direction (Figs. 1.1.5 and 1.1.8 A). Damage to the levator ani muscle can lead to muscle weakness and to relaxation of the pelvic floor, resulting in a visible descent of the perineum (descending perineum syndrome) (Fig. 1.1.9). In women with normal support and without previous surgery, the urogenital hiatus area is minimal. Increasing pelvic

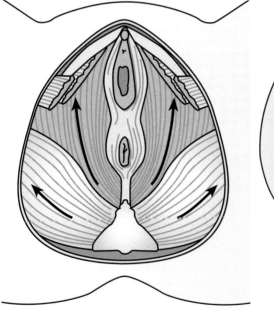
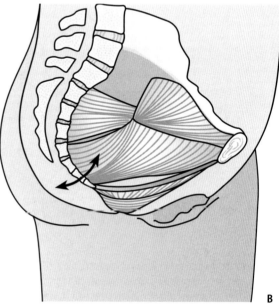

FIGURE 1.1.8. (**A**) All muscles of the pelvic floor insert into the coccyx. (**B**) The position of the coccyx varies because of pelvic floor muscle contraction or relaxation.

organ prolapse is associated with perineal descent and increasing urogenital hiatus size. The hiatus is larger after several failed operations than after successful surgery or single failure.[30]

The bony coccyx is also influenced by the activity of the PFMs. All muscles of the pelvic floor insert into the coccyx, and magnetic resonance imaging (MRI) studies have shown that PFM contractions lead to movement of the coccyx in a ventral cranial direction, thereby contributing to a closed urogenital hiatus (Fig. 1.1.8 B). During straining, the coccyx was pressed in a caudal, dorsal direction, thus, facilitating the opening of the urogenital hiatus.[31]

At the onset of micturition, the levator ani muscle relaxes, the urogenital hiatus opens, and the vesical neck rotates downward to the limit of the elasticity of the fascial attachments. At the end of micturition, the levator ani muscle resumes normal position and the urogenital hiatus is closed.

The connective tissue covering the levator ani muscles on the superior and inferior surfaces are called the superior and inferior fascia of the levator ani. When these muscles and their associated fascia are considered together, the combined structures form the pelvic diaphragm.

Two prominent lateral connective tissue structures are important in supporting the levator ani muscle. The ATLA and the ATFP (Figs. 1.1.10 and 1.1.11) are condensations of obturator internus

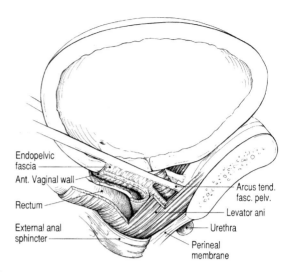

FIGURE 1.1.11. Lateral view of the pelvic floor structures related to urethral support seen from the side in the standing position, cut just lateral to the midline. Note that windows have been cut in the levator ani muscles, vagina, and endopelvic fascia so that the urethra and anterior vaginal walls can be seen. (*Source*: DeLancey, 2002, with permission; redrawn after DeLancey, 1994.)

and levator ani fascia and consist of dense aggregations of connective tissue, predominantly collagen, that provide lateral passive pelvic support. The ATLA inserts at the anterior pubic rami bilaterally and at the posterior region at or near the ischial spine. The ATLA overlies the obturator internus muscle. The ATFP lies medial to the ATLA and inserts at the lower sixth of the pubic bone, 1 cm from the midline, and the posterior region inserts into the ischium, just above the spine.

The levator ani muscle is composed of two portions. The iliococcygeal muscle is a thin layer of muscle that spans the potential gap from one pelvic sidewall to the other; it lies laterally and is relatively flat horizontal in the standing position. The iliococcygeal muscle originates at the ATLA, with a few fibers arising from the pubis. These fibers insert into the midline to form the anococcygeal raphe midline between the anus and coccyx. This region has also been called "the levator plate."

The more medial portions of the levator ani (pubococcygeal and puborectal) form a sling (Fig. 1.1.5) that arises ventrally from the pubic bones and encircles the pelvic organs. It arises

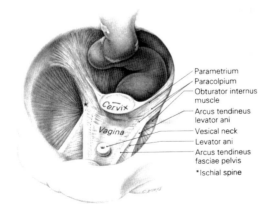

FIGURE 1.1.10. Attachments of the cervix and vagina to the pelvic walls demonstrating different regions of support with the uterus in situ. Note that the uterine corpus and the bladder have been removed. (*Source*: DeLancey, 2002, with permission.)

bilaterally from the pubis and wraps around the midline structures of the bladder, urethra, vagina, and rectum, sending fibers to insert into the perineal body, vagina, and anal sphincters, respectively. Various muscle subdivisions have been assigned to the medial portion of the pubococcygeus to describe different visceral attachments of the muscle to the vagina (pubovaginalis), perineal body (puboperineus), and anus (puboanalis). The puborectalis lies laterally and is not considered a part of the pubococcygeal group (puborectalis).[32]

The obturator internus and piriformis are the major muscles of the pelvic sidewalls (Fig. 1.1.10). The obturator internus is a large, fan-shaped muscle that arises from the bony margins of the obturator foramen, the pelvic surface of the obturator membrane, and the rami of the ischium and pubis. This muscle forms the lateral wall of the pelvis and can be palpated transvaginally. The piriformis muscles form the posterior wall of the pelvis. These muscles originate from the anterior and lateral aspect of the sacrum in its middle to upper portion, coursing through the greater sciatic foramen and inserting on the greater trochanter of the femur.

Muscle Fiber Type and Muscle Physiology

The levator ani muscle is a striated muscle. The constant activity of this muscle is analogous to the postural muscles of the spine. Their continuous contraction is similar to the continuous activity of the external anal sphincter muscle, and it closes the lumen of the vagina in a manner similar to how the anal sphincter closes the anus. This constant action eliminates any opening within the pelvic floor and prevents prolapse.

The muscle fibers of the levator ani include both, type I (slow twitch) and II (fast twitch) fibers. Fast-twitch fibers are metabolically suited for more rapid, forceful contraction, and slow-twitch fibers are suited for providing sustained muscular tone. Gilpin et al.[33] found that in women with no symptoms of urinary incontinence the anterior pubococcygeus muscle had a 33% population of type II fibers and the posterior pubococcygeus had a 24% population of type II fibers. A decrease in the percentage of type II fibers, along with an increase in the diameter of type I fibers, is a known adaptive response to inactivity, innervation damage, and ageing.[34] During whole-muscle contraction, motor units are recruited in order of increasing size.[35,36] A graded contraction that proceeds from low to high intensity (weak to strong) begins with the recruitment of the type I motor units, followed by the recruitment type II units. In a muscle that is 30% type II fibers, such as in the pubococcygeus, 70% of the muscle must be contracted before the type II units are recruited. To maintain the exercise training effect, exercise physiology suggests that once initial muscle strengthening has occurred, a reduced program of exercise can be adequate for maintaining strength.[37] The possible impact of age and parity on the pelvic floor has been studied by several authors.[33,38,39] They found that ageing and vaginal birth lead to histomorphologically visible changes that were consistent with changes of myogenic origin. These changes were more pronounced in the ventral part of the pelvic floor, leading to the assumption that the ventral part is the most vulnerable part of the muscle.

Innervation of the Levator Ani Muscle

The levator ani muscle is innervated by somatic nerve fibers that run in the pudendal nerve and emanate primarily from sacral root S3 and, to a lesser extent, from S2 and S4. The pudendal nerve carries both, motor and sensory fibers. Initially, the pudendal nerve lies superior to the sacrospinous ligament that is lateral to the coccyx. The nerve leaves the pelvis, crossing the ischial spine to reach the ischiorectal fossa via the lesser sciatic foramen. It extends forward in a fibrous tunnel called Alcock's canal on the medial side of the obturator internus muscle and distally gives rise to branches, which supply the levator ani and the membranous urethra. Some variation occurs in the pudendal nerve peripheral anatomy.

Levator Ani Function and Continence Mechanism

The levator ani muscles play a critical role in maintaining continence during the increase of abdominal pressure.[10,40,41] During normal abdominal pressure, it is postulated that the resting urethral closure pressure is maintained by the tone of the smooth and striated sphincters, intraurethral blood pressure, and the tendency of the urethral epithelium to close through coaptation.[3,9,12,42] In the event of a sudden or prolonged increase in intraabdominal pressure (such as during a cough or laughter), bladder pressure may exceed resting urethral pressure and result in leakage, unless compensated for by additional urethral closure pressure. This additional closure pressure is thought to be primarily developed by the striated levator ani because of the time constraint required for the smooth muscle contraction to take place within the approximate half-second pressure rise associated with a cough. During stress, the connective tissue, via its attachment to the striated levator ani muscle, helps create a firm "floor" of support underlying the vagina and urethra, onto which the urethra is compressed by intraabdominal pressure.[10] This extrinsic continence mechanism provides the additional closure pressure necessary to augment urethral closure during increased intraabdominal stress. The relative contributions to urethral closure made by the striated urethral sphincter muscles and the levator muscles have not yet been fully elucidated. Miller et al.[43] showed that a precisely timed volitional contraction of the PFM before and during stressful activities has considerable potential for helping to prevent urine loss in mild to moderate SUI. Pelvic floor muscle contraction in preparation for, and throughout, a cough can augment proximal urethra support during stress, thus, reducing the amount of dorsocaudal displacement.[44]

Miller et al.[43] observed that some women with SUI reported immediate benefits from Kegel exercises within 1–2 days after learning them. However, this time span was too short for any true strengthening to have taken place. The improvement was hypothesized to have come from the well-timed volitional use of the PFMs during an activity known to precipitate urine loss (such as a cough) and not from the actual strengthening of the muscles. The skill of a well-timed volitional PFM contraction was termed "The Knack."[43] Miller et al. showed that selected women with mild to moderate SUI could learn to significantly reduce urine loss within one week by intentionally contracting the PFMs before and during a cough. Contraction of the pubococcygeal muscle is hypothesized to result in urethral compression against adjacent tissues. Magnetic resonance imaging has shown that 11% of continent primiparous women demonstrated a major loss of the pubococcygeal portion of the levator ani muscle, with certain women showing evidence of complete muscle loss.[40] This observation offered the opportunity to compare increases in urethral closure pressure in women with and without intact pubococcygeal muscle. In consecutive studies, it was found that the likelihood of increasing urethral pressure with PFM contraction was decreased by 50% in the women with loss of the pubococcygeus.[45]

Connective Tissue of the Pelvic Floor: Endopelvic Fascia

On each side of the pelvis, connective tissue attaches the cervix and vagina to the pelvic wall (Figs. 1.1.10 and 1.1.11). This fascia forms a continuous sheet-like mesentery extending from the uterine artery at its cephalic margin to the point at which the vagina fuses with the levator ani muscles below. The body of connective tissue that attaches the pelvic organs to the sidewall is called endopelvic fascia, which is a heterogeneous group of tissues including collagen, elastin, smooth muscle, blood vessels, and nerves. The composition of the endopelvic fascia reflects its combined functions of neurovascular conduit and supportive structure. The part of the fascia that attaches to the uterus is called the parametrium, and the part that attaches to the vagina is called the paracolpium (Fig. 1.1.10).[46]

DeLancey[46] has introduced the concept of dividing the connective tissue support in the anterior part of the pelvis into three levels (Fig. 1.1.12), with levels I, II, and III representing apical, midvaginal, and distal vaginal support, respectively. Although assigning these levels artificially divides what is actually a continuum of connective tissue in the pelvis, dividing this support into regions (levels) proves useful tool to understand how the loss of support at different levels can correlate with different physical signs and symptoms that accompany a cystocele, enterocele, or rectocele. Damage to the upper suspensory fibers of the paracolpium (level I) results in uterine or vaginal vault prolapse

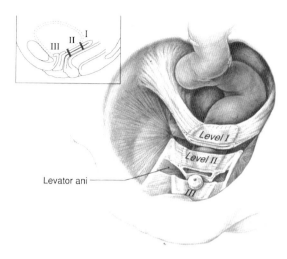

FIGURE 1.1.12. Levels of vaginal support after hysterectomy. Level I (suspension) and II (attachment). In level I, the paracolpium suspends the vagina from the lateral pelvic walls. Fibers of level I extend both vertically and posteriorly toward the sacrum. In level II, the vagina is attached to the arcus tendineus fasciae pelvis and the superior fascia of levator ani. In level III the vagina is directly attached without intervening paracopium. (*Source*: DeLancey, 1992, with permission.)

responsible for the diversity of clinically encountered problems.

The interaction between the PFMs and the supportive ligaments is critical to pelvic organ support. With proper function of the levator ani muscles, the pelvic floor remains closed and the ligaments and fascial structures are under minimal tension. The fasciae simply act to stabilize the organs in their position above the levator ani muscles. When the PFMs relax or are damaged, the pelvic floor opens and the vagina lies between the high abdominal pressure and low atmospheric pressure. As a result, the vagina must be held in place by the ligaments. Although the ligaments can sustain these loads for short periods, if the PFMs do not close the pelvic floor then the connective tissue will eventually fail, resulting in pelvic organ prolapse.

Anterior Wall Support and the Urethra

The anterior vaginal wall and urethra are intimately connected (Figs. 1.1.4, 1.1.5, and 1.1.11). Both PFMs and the pelvic fasciae determine the support and fixation of the urethra, and the activity of the muscles has significant influence on the urethral support.[43]

Disruption of this supportive system will result in downward descent of the anterior vaginal wall. The layer of tissue that provides urethral support

(Fig. 1.1.13 B). Damage to the level II and III portions of the vagina results in anterior prolapse (cystocele; Fig. 1.1.13 A) and posterior prolapse (rectocele; Fig. 1.1.13 C). These defects occur in varying combinations and these variations are

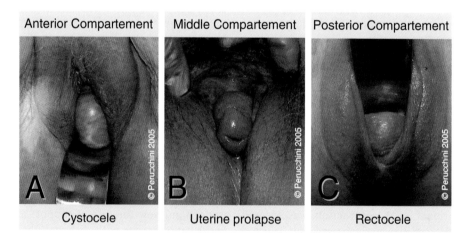

FIGURE 1.1.13. Illustration of different prolapse findings. Three types of movement occur in patients with pelvic organ prolapse: (A) The anterior vaginal wall can protrude through the introitus. This is called "cystocele." (B) The cervix (or vaginal apex) can move downward between the anterior and posterior supports. (C) The posterior wall can protrude through the introitus. This is called "rectocele."

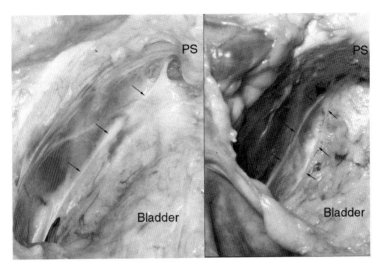

FIGURE 1.1.14. (Left) The attachment of the arcus tendineus pelvis to the pubic bone, the arcus tendineus pelvis (arrows). (Right) A paravaginal defect where the cervical fascia has separated from the arcus tendineus (arrows point to the sides of the split). PS = pubic symphysis. (*Source*: DeLancey, 2002, with permission.)

has two lateral attachments; a fascial attachment and a muscular attachment.[10] The fascial attachment of the urethral supports connects the periurethral tissues and anterior vaginal wall to the arcus tendineus fascia of the pelvis and has been called the paravaginal fascial attachments by Richardson et al.,[47] who observed that a lateral detachment (lateral defect) of the connections of this paravaginal fascia from the pelvic wall was associated with stress incontinence and anterior prolapse (Figs. 1.1.14 and 1.1.15). The muscular attachment connects these same periurethral

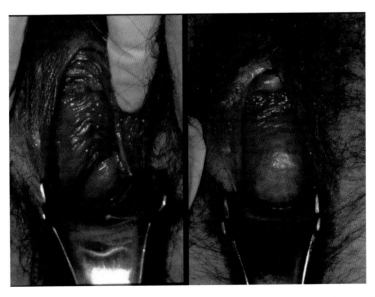

FIGURE 1.1.15. (Left) Displacement "cystocele" where the intact anterior vaginal wall has prolapsed downward because of paravaginal defect. Note that the right side of the patient's vagina and cervix has descended more than the left because of a larger defect on this side. (Right) Distension "cystocele" where the anterior vaginal wall fascia has failed and the bladder is distending the mucosa flattening out the vaginal tissues. (*Source*: DeLancey 2002, with permission.

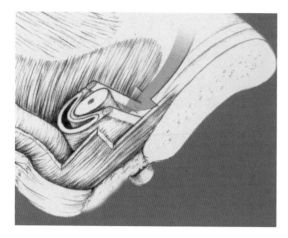

FIGURE 1.1.16. Lateral view of pelvic floor with the urethra and fascial tissues transected at the level of the proximal urethra. The arrow indicates compression of the urethra by a downward force against the supportive tissues and illustrates the influence of abdominal pressure on the urethra. (*Source*: DeLancey, 1994, with permission.)

tissues to the medial border of the levator ani muscle (Fig. 1.1.4). These attachments allow the levator ani muscle's normal resting tone to maintain the position of the vesical neck, which is supported by the fascial attachments. When the muscle relaxes at the onset of micturition, it allows the vesical neck to rotate downward to the limit of the elasticity of the fascial attachments, and then contraction at the end of micturition allows it to resume its normal position (Fig. 1.1.5). Damage to the connective tissue integrity can lead to prolapse.[48,49]

The vaginal wall, in turn, is supported by connections to the levator ani muscles laterally and to the arcus tendineus fascia pelvis. Simulated increases in abdominal pressure reveal that the urethra can be compressed against the vaginal wall, acting as a supporting hammock (Fig. 1.1.16).[10] In fact, it is the relative elasticity of this supporting apparatus, rather than the height of the urethra, that results in stress incontinence. In an individual with a firm supportive layer, the urethra would be compressed between abdominal pressure and pelvic fascia. If, however, the layer under the urethra becomes unstable and does not provide a firm backstop for abdominal pressure to compress the urethra against, the opposing force that causes closure is lost and the occlusive action diminished.

Uterovaginal Support

The cardinal and uterosacral ligaments attach the cervix and uterus to the pelvic walls.[50,51] Together these tissues are referred to as the parametrium (Fig. 1.1.10). The parametrium continues downward over the upper vagina to attach it to the pelvic walls. At this location, it is called the paracolpium.[46] The paracolpium provides support for the vaginal apex after a hysterectomy, and it has two portions. The uppermost portion of the paracolpium consists of a relatively long sheet of tissue that suspends the superior aspect of the vagina (Fig. 1.1.12, Level I) by attaching it to the pelvic wall. This is true whether or not the cervix is present. In the midportion of the vagina (Fig. 1.1.12, Level II), the paracolpium attaches the vagina laterally and more directly to the pelvic walls. This attachment stretches the vagina transversely between the bladder and rectum and has functional significance. The structural layer that supports the bladder ("pubocervical fascia") does not exist as a separate layer from the vagina, but rather, is composed of the anterior vaginal wall and its attachment through the endopelvic fascia to the pelvic wall (Figs. 1.1.4 and 1.1.16). Similarly, the posterior vaginal wall and endopelvic fascia (rectovaginal fascia) form the restraining layer that prevents the rectum from protruding forward, thereby blocking formation of a posterior prolapse. In the distal vagina (Fig. 1.1.12, level III), the vaginal wall is directly attached to surrounding structures without any intervening paracolpium. Anteriorly, it fuses with the urethra, and posteriorly it fuses with the perineal body, and laterally with the levator ani muscles (Fig. 1.1.12).

Prolapse of the uterus or the vagina after a hysterectomy has been performed is common. The nature of uterine support can be understood when the uterine cervix is pulled downward with a tenaculum in an anesthetized pelvic surgery patient. After a certain amount of descent, the parametria become tight and arrest further cervical descent. Similarly, descent of the vaginal apex after hysterectomy is resisted by the paracolpia. The inability of these ligaments to determine the resting position of the uterine cervix in normal healthy women is supported by the observation that the cervix may be drawn down to the level of the hymen with little difficulty.[52]

Perineal Membrane (Urogenital Diaphragm) and External Genital Muscles

In the anterior pelvis, below the levator ani muscles, there is a dense, triangularly shaped membrane called the perineal membrane. The term perineal membrane replaces the old term "urogenital diaphragm," reflecting the fact that this layer is not a single layer with a "diaphragm," but rather a set of connective tissues that surround the urethra.[53] The perineal membrane lies at the level of the hymenal ring and attaches the urethra, vagina, and perineal body to the ischiopubic rami (Fig. 1.1.17). The compressor urethra and urethrovaginal sphincter muscles are associated with the upper surface of the perineal membrane (Figs. 1.1.1 and 1.1.3 A). Previous concepts of the urogenital diaphragm show two fascial layers with a transversely oriented muscle, the "deep transverse perineal muscle," in between them. Observations based on serial histology and anatomic dissection, however, reveal a single connective tissue membrane, with the compressor urethra and urethrovaginal sphincter lying immediately above. These striated muscles have the largest bulk of the striated urogenital sphincter, and this fact explains why pressures during a cough are greatest in the distal urethra,[14,19] where they can compress the lumen closed in anticipation of a cough.[9,15]

Posterior Support

The posterior vaginal wall is supported by connections between the vagina, the bony pelvis, and

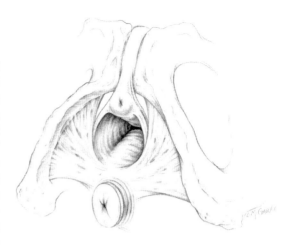

FIGURE 1.1.17. The perineal membrane spans the arch between the ischiopubic rami, with each side attached to the other through their connection in the perineal body. Note that separation of the fibers in this area leaves the rectum unsupported and results in a low posterior prolapse. (*Source*: DeLancey, 1999, with permission.)

the levator ani muscles.[54] The lower third of the vagina is fused with the perineal body (Fig. 1.1.17). This structure is the attachment between the perineal membranes on either side of it, and this connection prevents downward descent of the rectum in this region. If the fibers that connect one side with the other rupture (Fig. 1.1.18 A) then the bowel may protrude downward, resulting in a posterior vaginal wall prolapse (Fig. 1.1.18 B). The midposterior vaginal wall (Fig. 1.1.12, Level II) is connected to the inside of the levator ani muscles

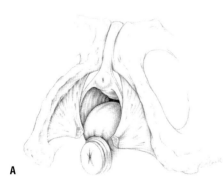
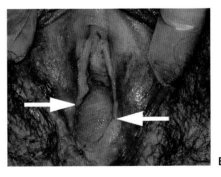

FIGURE 1.1.18. **(A and B)** Posterior prolapse caused by separation of the perineal body **(A)**. (DeLancey, 2002, with permission). Note the end of the hymenal ring that lies laterally on the side of the vagina, no longer united with its companion on the other side (arrows) **(B)**.

by sheets of endopelvic fascia. These connections prevent the ventral movement of the vagina during increases in abdominal pressure.

The attachment of the levator ani muscles into the perineal body is important. Damage to this part of the levator ani muscle during delivery is one of the irreparable injuries to pelvic floor (Fig. 1.1.18 B). Recent MRI has vividly depicted these defects, showing that up to 20% of nulliparous women have a visible defect in the levator ani muscle on MRI.[40] This muscular damage is likely an important factor associated with the recurrence of pelvic organ prolapse after initial surgical repair. Moreover, these defects were found to occur more frequently in those individuals complaining of stress incontinence. An individual with malfunctioning muscles has a problem that is not surgically correctable. A more complete understanding of the pelvic floor biomechanics is needed to understand the structural effects of these lesions better.

Anatomy of the Anal Sphincter Complex

Fecal incontinence is a devastating condition, which is often associated with childbirth. The anal sphincter complex and the puborectalis muscle provide the majority of control of anal continence. The anal sphincter complex is composed of the internal and external anal sphincter muscle and contains both smooth and striated muscle (Fig. 1.1.19). The internal anal sphincter muscle is a continuation of the circular smooth muscle layer of the rectum. The external anal sphincter surrounds the internal sphincter in its lower 2 cm by a muscular component that is tethered to the coccyx through the anococcygeal raphe.[55] Immediately cephalic and anterior to the external sphincter is the puborectalis muscle. The striated external anal sphincter muscle provides voluntary squeeze tone to the sphincter complex. The external anal sphincter is classically described as having three portions: deep, superficial, and subcutaneous.

The internal anal sphincter contributes approximately 75% of the maximum anal resting pressure; 25% comes from the external anal sphincter. If there is sudden distention, the external anal sphincter can contribute up to 60% of the anal canal pressure for a short time, but it cannot maintain sustained tone. Resting and squeezing anal pressures decline with ageing.[56]

Between the internal and external anal sphincter is the intersphincteric groove. This space receives the downward extension of the conjoined fibers of the levator ani muscles. These fibers suspend and elevate the anorectum, preventing its downward prolapse.

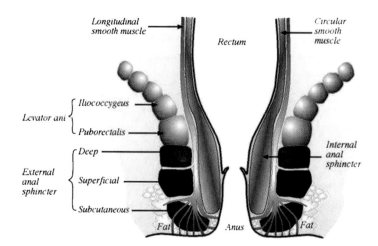

FIGURE 1.1.19. Diagrammatic representation of the anal sphincter mechanism. (Thakar and Sultan, 2003.)

The external anal sphincter is innervated by S2–S4 fibers that travel via the inferior hemorrhoidal portion of the pudendal nerve. The puborectalis muscle, as previously described, forms a U-shaped loop that begins from the pubic bones and passes behind the rectum. The muscle has constant muscular activity and is relaxed only at the time of defecation. It also acts by causing a kink in the rectum, so that there is a 90° angle between the anal and rectal canals. The contraction of this muscle can be assessed by the degree to which the anus is elevated ("levator ani") and pulled inward when the patient contracts her pelvic muscles.

Conclusion

A series of factors are important to urinary and fecal continence and to the normal support of the female pelvic organs. Many advances are yet to be made concerning the pelvic floor, the continence mechanism, and prolapse. Current and future researchers should aim to uncover the faults that are certainly present in the current paradigms so that prevention and treatment of pelvic floor disorders can pass from its current phase of clinical empiricism to scientific certainty.

References

1. Olsen AL, Smith VJ, Bergstrom JO. Epidemiology of surgically managed pelvic organ prolapse and urinary incontinence. Obstet Gynecol. 1997;89(4):501–506.
2. Mant J, Painter R, Vessey M. Epidemiology of genital prolapse: observations from the Oxford family planning association study. Br J Obstet Gynaecol. 1997;104(5):579–585.
3. Perucchini D, DeLancey JO, Ashton Miller J, et al. Age effects on urethral striated muscle I. Changes in number and diameter of striated muscle fibers in the ventral urethra. Am J Obstet Gynecol. 2002;186:351–355.
4. Nygaard IE, Thompson FL, Svengalis SL, Albright JP. Urinary incontinence in elite nulliparous athletes. Obstet Gynecol. 1994;84:183–187.
5. Burton TJ, Elneil S, Nelson CP, et al. Activation of epithelial Na(+) channel activity in the rabbit urinary bladder by cAMP. Eur J Pharmacol. 2000;404:273–280.
6. Deng FM, Ding M, Lavker RM, et al. Urothelial function reconsidered: a role in urinary protein secretion. Proc Natl Acad Sci USA. 2001;98:154–150.
7. Gil Vernet, S. Morphology and function of the vesico-prostato-urethral musculature. Edizioni Canova, Treviso, 1968.
8. Mcguire EJ. The innervation and function of the lower urinary tract. J Neurosurg. 1986;65(3):278–285.
9. DeLancey JO. Structural aspects of the extrinsic continence mechanism. Obstet Gynecol. 1988;72;296–301.
10. DeLancey JO. Structural support of the urethra as it relates to stress urinary incontinence: the hammock hypothesis. Am J Obstet Gynecol. 1994;170:1713–1720.
11. Wilson PD, Dixon JS, Brown AD, et al. Posterior pubo-urethral ligaments in normal and genuine stress incontinent women. J Urol. 1983;130(4):802–805.
12. Huisman AB. Aspects on the anatomy of the female urethra with special relation to urinary continence. Contrib Gynecol Obstet. 1983;10:1–31.
13. Rud T. Urethral pressure profile in continent women from childhood to old age. Acta Obstet Gynecol Scand. 1980;59:331–335.
14. Hilton P, Stanton, SL. Urethral pressure measurement by microtransducer: the results in symptom-free women and in those with genuine stress incontinence. Br J Obstet Gynaecol. 1983;90(10):919–933.
15. DeLancey JO. Correlative study of paraurethral anatomy. Obstet Gynecol. 1986;68:91–97.
16. Gosling JA, Dixon JS, Critchley HO, et al. A comparative study of the human external sphincter and periurethral levator ani muscles. Br J Urol. 1981;53(1):35–41.
17. Perucchini D, DeLancey JO, Ashton Miller J, et al. Age effects on urethral striated muscle II. Anatomic location of muscle loss. Am J Obstet Gynecol. 2002;186:356–360.
18. Thind P, Bagi P, Mieszczak C, et al. Influence of pudendal nerve blockade on stress relaxation in the female urethra. Neurourol Urodyn. 1996;15:31–36.
19. Constantinou CE, Govan DE. Spatial distribution and timing of transmitted and reflexly generated urethral pressures in healthy women. J Urol. 1982;127(5):964–969.
20. Constantinou, CE. Resting and stress urethral pressures as a clinical guide to the mechanism of

continence in the female patient. Urol Clin N Am. 1985;12(2):247–258.
21. Bump RC, Hurt WG, Fantl JA, et al. Assessment of Kegel pelvic muscle exercise performance after brief verbal instruction. Am J Obstet Gynecol. 1991;165(2):322–327.
22. Ek A, Alm P, Andersson K-E, et al. Adrenergic and cholinergic nerves of the human urethra and urinary bladder. A histochemical study. Acta Physiol Scand. 1997;99:345.
23. Huffman, J. Detailed anatomy of the paraurethral ducts in the adult human female. Am J Obstet Gynecol 1948;55:86.
24. Juenemann, KP, Lue TF, Schmidt RA, et al. Clinical significance of sacral and pudendal nerve anatomy. J Urol. 1988;139(1):74–80.
25. Nelson CP, Montie JE, Mcguire EJ, et al. Intraoperative nerve stimulation with measurement of urethral sphincter pressure changes during radical retropubic prostatectomy: a feasibility study. J Urol. 2003;169(6):2225–2228.
26. Thor KB. Serotonin and norepinephrin involvement in efferent pathways to the urethral rhabdosphincter: implication for treating stress urinary incontinence. Urology. 2003;62:3–9.
27. Millard RJ, Moore K, Rencken R, et al. Duloxetine UI study group. Duloxetine vs. placebo in the treatment of stress urinary incontinence: a four-continent randomized trial. BJU Int. 2004;93:311–318.
28. Bonney V. The principles that should underlie all operations for prolapse. Obstet Gynaecol Br Emp. 1934;41:669.
29. Taverner D. An electromyographic study of the normal function of the external anal sphincter and pelvic diaphragm. Dis Colon Rectum. 1959;2:153–158.
30. DeLancey JO, Hurd WW. Size of the urogenital hiatus in the levator ani muscles in normal women and women with pelvic organ prolapse. Obstet Gynecol. 1998;91(3):364–368.
31. Bo K, Lilleas F, Talseth T, et al. Dynamic MRI of the pelvic floor muscles in an upright and sitting position. Neurourol Urodyn. 2001;20:167–174.
32. Kearney R, Shawney R, DeLancey JO. Levator ani muscle anatomy evaluated by origin-insertion pairs. Obstet Gynecol. 2004;104(1):168–173.
33. Gilpin SA, Gosling JA, Smith AR, et al. The pathogenesis of genitourinary prolapse and stress incontinence of urine. A histologic and histochemical study. Br J Obstet Gynaecol. 1989;64(4):385–390.
34. Koelbl H, Strassegger H, Riss PA, et al. Morphologic and functional aspects of pelvic floor muscles in patients with pelvic relaxation and genuine stress incontinence. Obstet Gynecol. 1989;74(5):789–795.
35. Henneman E, Somjen C, Carpenter D. Functional significance of cell size in spinal motor neurons. J Neurophysiol. 1965;28:561–580.
36. Henneman E, Somjen G, Carpenter D. Excitability and inhibitibility of motoneurons of different sizes. J Neurophysiol. 1965;28:599–620.
37. Graves JE, Pollock ML, Leggett SH, et al. Effect of reducing training frequency on muscular strength. Inter J Sports Med. 1988;9:316–319.
38. Jundt K, Kiening M, Fischer P, et al. Is the histomorphological concept of the female pelvic floor and its changes due to age and vaginal delivery correct? Neurourol Urodyn. 2005;24:44–50.
39. Dimpfl T, Jaeger C, Mueller-Felber W, et al. Myogenic changes of the levator ani muscle in premenopausal women: the impact of vaginal delivery and age. Neurourol Urodyn. 1998;17(3):197–205.
40. DeLancey JOL, Kearney R, Chou Q, et al. The appearance of levator ani muscle abnormalities in magnetic resonance images after vaginal delivery. Obstet Gynecol. 2003;101:46–53.
41. Petros PEP, Ulmsten UL. An integral theory of female urinary incontinence. Experimental and clinical considerations. Acta Obstet Gynecol Scand. 1990;153:7–31.
42. Pandit M, DeLancey JO, Ashton-Miller J, et al. Quantification of intramuscular nerves within the female striated urogenital sphincter muscle. Obstet Gynecol. 2000;95:797–800.
43. Miller JM, Ashton-Miller JA, DeLancey JO. A pelvic muscle precontraction can reduce cough-related urine loss in selected women with mild SUI. J Am Geriatr Soc. 1998;46:870–874.
44. Miller JM, Perucchini D, Carchidi LT, et al. Pelvic floor muscle contraction during a cough and decreased vesical neck mobility. Obstet Gynecol. 2001;97:255–260.
45. Miller JM, Umek WH, DeLancey JO, et al. Can women without visible pubococcygeal muscle in MR images still increase urethral closure pressures? Am J Obstet Gynecol. 2004;191:171–175.
46. DeLancey JO. Anatomic aspects of vaginal eversion after hysterectomy. Am J Obstet Gynecol. 1992;166:1717–1728.
47. Richardson AC, Edmonds PB, Williams NL. Treatment of stress urinary incontinence due to paravaginal fascial defect. Obstet Gynecol. 1981;57:357–362.
48. DeLancey JOL. Fascial and muscular abnormalities in women with urethral hypermobility and anterior vaginal prolapse. Am J Obstet Gynecol. 2002;18:93–98.

49. Halban J, Tandler I. Anatomie und Aetiologie der Genitalprolapse beim Weibe. Vienna, Austria: Wilhelm Braumuller, 1907.
50. Campbell RM. The anatomy and histology of the sacrouterine ligaments. Am J Obstet Gynecol. 1950;59:1.
51. Range RL, Woodburne RT. The gross and microscopic anatomy of the transverse cervical ligaments. Am J Obstet Gynecol. 1964;90:460–462.
52. Bartscht KD, DeLancey JO. A technique to study the passive supports of the uterus. Obstet Gynecol. 1988;72(6):940–943.
53. Oelrich TM. The striated urogenital sphincter muscle in the female. Anat Rec. 1983;205:223–232.
54. DeLancey JO. Structural anatomy of the posterior pelvic compartment as it relates to rectocele. Am J Obstet Gynecol. 1999;180:815–823.
55. DeLancey JOL, Toglia MR, Perucchini D. Internal and external anal sphincter anatomy as it relates to midline obstetrical lacerations. Obstet Gynecol. 1997;90:924–927.
56. Haadem K, Dahlström JA, Ling L. Anal sphincter competence in healthy women. Clinical implication of age and other factors. Obstet Gynecol. 1991;78:823–827.

1.2
Neural Control of Pelvic Floor Muscles

David B. Vodušek

Key Message

The coordinated function of pelvic floor muscles (PFM) and related organs is an important prerequisite that enables women to be continent of urine and feces, to micturate and defecate, and to experience orgasm. It is dependent on a complex neurocontrol consisting of voluntary muscle actions and reflexes, genetically determined activation patterns allowing for a meticulous interplay of skeletal and autonomous muscle structures, and related organs. This scientifically complicated material is broken down in this chapter to a very understandable and practical level, which is a basic necessity for sophisticated pelvic floor physiotherapy.

Introduction

Neural control of pelvic organs is affected by a unique coordination of the somatic and autonomic motor nervous systems. Pelvic floor muscles are intimately involved in the function of the LUT and the anorectum, as well as in sexual function. The neural control of the participation of striated muscle in "visceral activity" transcends the simple somatic innervation necessary for control of striated muscles functioning alone. Sensory information and feedback to pelvic floor organs is supplied by both visceral and somatic sensory fibers, which also influence pelvic floor muscle excitability thresholds through central integrative mechanisms.

A clear understanding of anatomy and the neural control of the PFM and related visceral organs, as well as an insight into neuromuscular damage and repair, is needed for planning successful PFM treatment in the individual patient for any dysfunctional pelvic floor disorder. Particularly in stress urinary incontinence, the activation and coordination disturbances of PFM are clinically relevant. Proper identification of such PFM disturbances should lead to better selection of patients for particular modes of treatment and to an improvement in the outcomes of conservative treatments.

Pelvic Floor Muscle Activity

Muscles do not really have much activity of their own. They are dependent on neural control. The "denervated" muscle atrophies and turns into fibrotic tissue. As every tissue muscle consists of cells (muscle fibers). However, the functional unit within a striated muscle is not a single muscle cell, but a motor unit. A motor unit consists of one alpha (or "lower") motor neuron (from the motor nuclei in the brainstem or spinal cord), and all the muscle cells that this motor neuron innervates. In other words, the motor unit is the basic functional unit of the somatic motor system, and control of a muscle means control of its motor units. Thus, in discussing neural control of muscle, we only need to consider the motor neurons in the spinal cord and all the influences they are exposed to.

1.2. Neural Control of Pelvic Floor Muscles

The function of pelvic floor and sphincter lower motor neurons is organized quite differently from other groups of motor neurons. In contrast to the reciprocal innervation that is common in limb muscles, the neurons innervating each side of the PFM have to work in harmony and synchronously. Thus, they achieve the "closure unit" of the excretory tracts, the "support unit" for pelvic viscera, and an "effector unit" in the sexual response. In general, the muscles involved in the above functions act in a strictly unified fashion as "one muscle," as has been demonstrated not only in sphincter muscles but also for both pubococcygeus muscles.[1]

However, as each muscle in the pelvis has its own unilateral peripheral innervation, dissociated activation patterns are also possible and have been reported between the two pubococcygeus[2] and between the levator ani and the urethral sphincter.[3]

The differences in evolutionary origin of the sphincter muscles and levator ani further imply that unilateral activation may be less of an impossibility for the PFMs than for sphincters. It can be postulated that the neural mechanisms controlling the different muscles involved in sphincter mechanisms and pelvic organ support may not be as uniform as has been assumed. How much variability there is in the normal activation patterns of the PFM is not yet clarified. It is clear, however, that the coordination between individual PFMs can definitively suffer because of disease or trauma.

Tonic and Phasic Pelvic Floor Muscle Activity

The normal striated sphincter muscles demonstrate some continuous motor unit activity at rest, as revealed by kinesiological EMG (Fig.

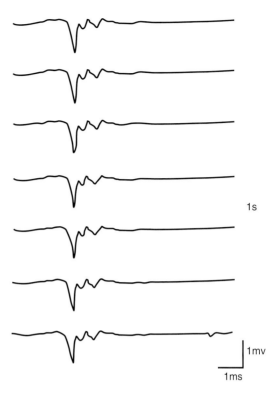

FIGURE 1.2.2. A falling leaf display of a continuously firing tonic motor unit. The vertical intervals between the beams indicate the interval between two consecutive activations of the motor unit. The regularity of the motor unit discharge should be noted. (From the urethral sphincter of a continent 46-year-old woman.) (*Source*: Schüssler et al., 1994.)

1.2.1). This differs between individuals and also continues after subjects fall asleep.[4] This physiological spontaneous activity may be called "tonic," and it depends on prolonged activation of certain tonic motor units (Figs. 1.2.2 and 1.2.3).[5] As a rule, it increases with bladder filling,

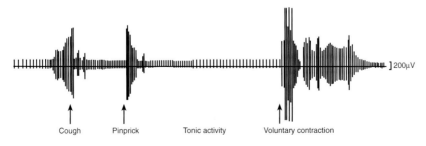

FIGURE 1.2.1. Kinesiological EMG recordings from urethral sphincter muscle. Concentric needle electrode recording in a continent 53-year-old female; recruitment of motor units on reflex maneuvers and on command to contract. (*Source*: Schüssler et al., 1994.)

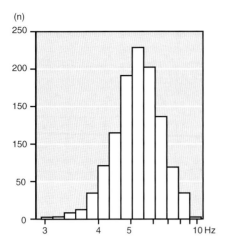

FIGURE 1.2.3. Frequency histogram of the discharge rate of a continuously firing tonic motor unit from the anal sphincter of a healthy adult female. The values on the y axis denote the number of discharges (n). The regularity of the motor unit discharge should be noted. (Schüssler et al., 1994.)

while depending on the rate of filling. Any reflex or voluntary activation is mirrored first in an increase of the firing frequency of these motor units. On the contrary, inhibition of firing is apparent on initiation of voiding (Fig. 1.2.4).

With any stronger activation or sudden increase in abdominal pressure (e.g., coughing), and only for a limited length of time, new motor units are recruited and may be called "phasic" motor units. As a rule, they have potentials of higher amplitudes and their discharge rates are higher and irregular. A small percentage of motor units with an "intermediate" activation pattern can also be encountered.[5] (It must be stressed that this typing of motor units is electrophysiological, and that no direct correlation to histochemical typing of muscle fibers has so far been achieved.)

In regard to tonic activity, sphincters differ from some other perineal muscles; tonic activity is encountered in many, but not all, detection sites for the levator ani muscle[1,5] and is practically never seen in the bulbocavernosus muscle.[5] In the pubococcygeus of the normal female there is some increase of activity during bladder filling, and reflex increases in activity during any activation maneuver performed by the subject, e.g., talking, deep breathing, and coughing.

On voiding, inhibition of the tonic activity of the external urethral sphincter, and the PFM, leads to relaxation. This can be detected as a disappearance of all EMG activity, which precedes detrusor contraction. Similarly, the striated anal sphincter relaxes with defecation and micturition.[6]

Reflex Activity of Pelvic Floor Muscles

As the human urethral and anal striated sphincters seem to have no muscle spindles, their reflex reactivity is, thus, intrinsically different from the levator ani muscle complex, in which muscle spindles and Golgi tendon organs have been demonstrated.[7] Thus, PFM have the intrinsic proprioceptive "servomechanism" for adjusting muscle length and tension, whereas the sphincter muscles depend on mucosa and afferents from skin. Both muscle groups are integrated in reflex activity that incorporates pelvic organ function.

The reflex activity of the PFM is clinically and electrophysiologically evaluated by eliciting the bulbocavernosus and anal reflexes. The bulbocavernosus reflex is evoked on nonpainful dorsal clitoral nerve stimulation. As recorded by EMG, it is a complex response; its first component is thought to be an oligosynaptic reflex and the later component a polysynaptic reflex.[8] The polysynaptic anal reflex is elicited by pinprick stimulation in the perianal region.

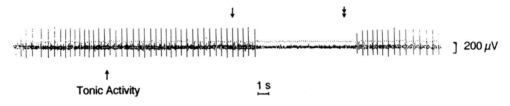

FIGURE 1.2.4. Inhibition of firing of tonic muscle activity on command (to fake micturition). The arrow indicates the point at which the command was given; the double arrowheads indicate when the command was terminated (from the urethral sphincter of a continent 53-year-old woman).

The constant tonic activity of sphincter muscles is thought to be the result of the characteristics of their "low threshold" motor neurons and the constant reflex inputs either of segmental or of suprasegmental origin. It is supported by cutaneous stimuli, pelvic organ distension, and intraabdominal pressure changes. Sudden increases in intraabdominal pressure, as a rule, lead to brisk reflex activity, which has been called the "guarding reflex"; it is organized at the spinal level.

To correspond to their functional (effector) roles as pelvic organ supporters (e.g., during coughing or sneezing), as sphincters for the lower urinary tract (LUT) and anorectum, and as an effector in the sexual arousal response, orgasm, and ejaculation, PFM also have to be involved in very complex involuntary reflex activity, which coordinates the behavior of pelvic organs (smooth muscle) and several different groups of striated muscles. This activity is to be understood as originating from so-called "pattern generators" within the central nervous system, particularly the brainstem. These pattern generators ("reflex centers") are genetically built-in and are responsible for complex reflex motor activity.

Muscle Awareness

The sense of position and movement of one's body is referred to as proprioception, and it is particularly important for sensing limb position (stationary proprioception) and movement (kinesthetic proprioception). Proprioception relies on special mechanoreceptors in muscle tendons and joint capsules. In muscles, there are specialized stretch receptors, called muscle spindles, and in tendons there are Golgi tendon organs, which sense the contractile force. In addition, stretch-sensitive receptors signaling postural information are also in the skin, and this cutaneous proprioception is particularly important for control of the movements of muscles without bony attachment (e.g., lips and anal sphincter).

Thus, the functional status of a striated muscle (or rather a certain movement) is represented in the brain. Indeed, muscle awareness reflects the amount of sensory input from various sites. Typically, feedback on limb muscle function (acting at joints) is derived not only from their input from muscle spindles, and receptors in tendons, but also from the skin, from visual input, etc. "Awareness," in fact, overlaps with the ability to voluntarily change the state of a muscle.

In contrast to limb muscles, PFMs and sphincters lack several of the aforementioned sensory input mechanisms and, hence, the brain is weakly informed about their status. Additionally, there is a gender difference, inasmuch as PFM awareness in females seems to be generally less developed compared with males. Although nearly all healthy males are able to voluntarily contract the pelvic floor, roughly every fourth healthy woman lacks this ability. The need for "squeezing out" the urethra at the end of micturition, as well as the close relationship of penile erection and ejaculation with pelvic floor function, seems to be the origin of this difference. This primarily weak or absent awareness of PFM in women seems to be further jeopardized by vaginal delivery.

To regain awareness (or initiate it if it had primarily been weak or absent) follows the same principles as any other motor rehabilitation. Motor learning has to aim for the activation of a "dormant" pattern generator.

Voluntary Activity of Pelvic Floor Muscles

Skilled movement of distal limb muscles requires individual motor units to be activated in a highly focused manner by the primary motor cortex. In contrast, the activation of axial muscles (necessary to maintain posture, etc.), while also under voluntary control, depends particularly on vestibular nuclei and reticular formation to create predetermined "motor patterns."

PFM are, in principle, under voluntary control, i.e., it is possible to voluntarily activate or inhibit the firing of their motor units (Figs. 1.2.1 and 1.2.4). There have been some claims that the sphincter muscles can be contracted voluntarily, but not relaxed at will. EMG studies have shown, however, that the activity of motor units in the urethral sphincter can be extinguished at both low and high bladder volumes, even without initiating micturition.[9,10] Nevertheless, the voluntary control of both the sphincter and other perineal muscles and PFMs is not as straightforward, unconditional, and reliable as it is for limb muscles. It is well known that many otherwise neurologically healthy women cannot contract their PFMs on command, which is also demonstrated by EMG recordings. Thus, the voluntary control of these muscles seems to be less straightforward, and maybe more fragile than is commonly the case with striated limb musculature.

In order to voluntarily activate a striated muscle we have to have the appropriate brain "conceptualization" of that particular movement, which acts as a rule within a particular complex "movement pattern." This evolves particularly through repeatedly executed commands and represents a certain "behavior."

Proprioceptive information is crucial for striated muscle motor control, both in the "learning" phase of a certain movement and for later execution of overlearned motor behaviors. Proprioceptive information is passed to the spinal cord by fast-conducting, large-diameter myelinated afferent fibers and is influenced not only by the current state of the muscle but also by the efferent discharge the muscle spindles receive from the nervous system via gamma efferents. In order to work out the state of the muscle, the brain must take into account these efferent discharges and make comparisons between the signals it sends out to the muscle spindles along the gamma efferents and the afferent signals it receives from the primary afferents. Essentially, the brain compares the signal from the muscle spindles with the copy of its motor command (the "corollary discharge" or "efferents copy"), which was sent to the muscle spindle intrafusal muscle fibers by the central nervous system via gamma efferents. The differences between the two signals are used in deciding on the state of the muscle. Experiments were done in limb muscles,[11] but it has been suggested[12] that similar principles rule in bladder neurocontrol; the brain would know which efferent discharges were caused by distension and which were caused by contractions because the latter would be initiated by the central nervous system. We propose that this is generally true, but that the mechanism, while present, is quite "weakly developed" for PFMs.

Innervation of Pelvic Floor Muscles

Somatic Motor Pathways

The motor neurons that innervate the striated muscle of the external urethral and anal sphincters and perineum originate from a localized column of cells in the sacral spinal cord called Onuf's nucleus.[13] In humans, it expands from the second to third sacral segment (S2–S3), and occasionally into S1.[14] Within Onuf's nucleus there is some spatial separation between motor neurons concerned with the control of the urethral and of the anal sphincter (Fig. 1.2.5).

Spinal motor neurons for the levator ani group of muscles seem to originate from S3–S5 segments and show some overlap.[15]

Sphincter motor neurons are uniform in size and smaller than the other alpha motor neurons. They also differ with respect to their high concentrations of amino acid–, neuropeptide–, norephinephrine–, serotonin–, and dopamine-containing terminals; these represent the

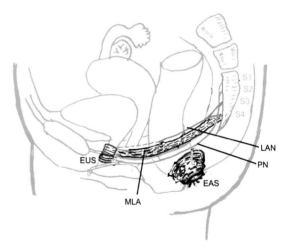

FIGURE 1.2.5. The pudendal nerve is derived from ventral rami of roots (S1, S2, S3, and S4). It continues through the greater sciatic foramen and enters in a lateral direction into the ischiorectal fossa. Its muscular branches innervate the external anal sphincter and the external urethral sphincter. There may be muscular branches for the levator ani, which is, as a rule, innervated by direct branches from the sacral plexus (from above), which is the levator ani nerve. EAS = external anal sphincter; EUS = external urethral sphincter, LAN = levator ani nerve, MLA = musculus levator ani, PN = pudendal nerve. (*Source*: Modified from Swash, 2002.)[41]

substrate for the distinctive neuropharmacologic responses of these neurons, which differ both from those of limb muscles, the bladder, and the PFM.

The somatic motor fibers leave the spinal cord at the ventral radices and fuse with the dorsal radices to constitute the spinal nerve. After passing through to the intravertebral foramen, the spinal nerve divides into a dorsal ramus and a ventral ramus.[16] Somatic fibers from the ventral rami (also called the sacral plexus) form the pudendal nerve. Traditionally, the pudendal nerve is described as being derived from the S2–S4 ventral rami, but there may be some contribution from the S1, and possibly little or no contribution from the S4.[17] The pudendal nerve continues through the greater sciatic foramen and enters in a lateral direction, through the lesser sciatic foramen, into the ischiorectal fossa (Alcock's canal). In the posterior part of the Alcock's canal, the pudendal nerve gives off the inferior rectal nerve; it then branches into the perineal nerve and the dorsal nerve of the penis/clitoris. Although still a controversial matter, it is generally accepted that the pudendal nerve also supplies the urinary and anal sphincter.

In contrast, it is mostly agreed that the main innervation for the PFM is through direct branches from the sacral plexus (from above) rather than predominantly by branches of the pudendal nerve (from below).

Significant variability of normal human neuroanatomy is probably the source of these controversies, originating from anatomical studies of the peripheral innervation of the pelvis, which, as a rule, are performed in only a small number of cases.

Descending inputs to PFM motor neurons are manifold, and mostly "indirect." More "direct" connections to Onuf's nucleus are from some nuclei in the brainstem (e.g., raphe or ambiguous) and from the paraventricular hypothalamus. Positron emission tomography (PET) studies revealed activation of the (right) ventral pontine tegmentum (in the brainstem) during the holding of urine in human subjects.[18] This finding is consistent with the location of the "L region" in cats, which is proposed to control PFM nuclei. These connections serve the coordinated inclusion of PFM into "sacral" (LUT, anorectal, and sexual) functions. It should be mentioned that PFM need to not only be neurally coordinated "within" a particular function (for instance, with bladder activity), but the single functions need to be neurally coordinated with each other (e.g., voiding and erection).

The aforementioned sacral function control system is proposed to be a part of the "emotional motor system." This is a system derived from brain or brainstem structures belonging to the limbic system. It consists of a medial and a lateral component.[19] The medial component represents diffuse pathways originating in the caudal brainstem and terminating on (almost all) spinal grey matter, using serotonin in particular as its neurotransmitter. This system is proposed to "set the threshold" for overall changes, such as, for instance, in muscle tone under different physiological conditions (sleeping, etc.). The lateral component of the emotional motor system consists of discrete areas in the hemispheres and the brainstem that are responsible for specific motor activities, such as micturition and mating. The pathways use spinal premotor interneurons to influence motor neurons in somatic and autonomic spinal nuclei, thus, allowing for confluent interactions of various inputs to modify the motor neuron activity.

Furthermore, PFM nuclei receive descending corticospinal input from the cerebral cortex. PET studies have revealed the activation of the superomedial precentral gyrus during voluntary PFM contraction, and of the right anterior cingulate gyrus during sustained PFM straining.[18] Not surprisingly, PFM contraction can be obtained by electrical or magnetic transcranial stimulation of the motor cortex in men (Fig. 1.2.6).[20, 21]

Afferent Pathways

Because PFM function is intimately connected to pelvic organ function, it is proposed that all sensory information from the pelvic region is relevant for PFM neural control.

The sensory neurons are bipolar. Their cell bodies are in spinal ganglia. They send a long process to the periphery and a central process into the spinal cord, where it terminates segmentally or, after branching for reflex connections, ascends, in some cases as far as the brainstem.[16]

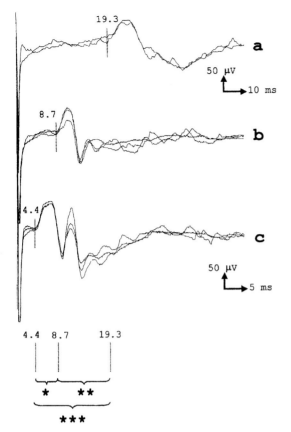

FIGURE 1.2.6. MEPs recorded by concentric needle in the external urethral sphincter of a 51-year-old woman. Cortical (**a**), thoracic (**b**), and sacral (**c**) stimulation. Central motor conduction time (CMCT) is calculated as cortical–lumbar latency (** = 10.6 ms). Cauda equina motor conduction time is calculated as lumbar–sacral latency (* = 4.3 ms). (*Source*: Brostrom et al, 2003.)

The afferent pathways from the anogenital region and pelvic region are commonly divided into somatic and visceral. Somatic afferents derive from touch, pain, and thermal receptors in skin and mucosa and from proprioceptors in muscles and tendons. (Proprioceptive afferents especially arise from muscle spindles and Golgi tendon organs.)

The visceral afferents accompany both parasympathetic and sympathetic efferent fibers, the somatic accompany the pudendal nerves and direct somatic branches of the sacral plexus. The different groups of afferent fibers have different reflex connections and transmit, at least to some extent, different afferent information.

The terminals of pudendal nerve afferents in the dorsal horn of the spinal cord are found not only ipsilaterally but also bilaterally, with ipsilateral predominance.[22]

The proprioceptive afferents form synaptic contacts in the spinal cord and have collaterals ("primary afferent collaterals") that run ipsilaterally in the dorsal spinal columns to synapse in the gracillis (dorsal column) nuclei in the brainstem. This pathway transmits information about of innocuous sensations from the PFMs.

The lateral columns of the spinal cord transmit information concerning pain sensations from perineal skin, as well as sexual sensations. In humans this pathway is situated superficially, just ventral to the equator of the cord, and is probably the spinothalamic tract.[23]

The spinal pathways, which transmit sensory information from the visceral afferent terminations in the spinal cord to more rostral structures, can be found in the dorsal, lateral, and ventral spinal cord columns.

Neural Control of Sacral Functions

Neural Control of Micturition

The LUT is innervated by three sets of peripheral nerves: parasympathetic nerves (S2–S4), which excite the bladder and relax the urethra (pelvic nerves); sympathetic nerves (Th11-L2), which inhibit the bladder body and excite the bladder base and urethra (hypogastric and pelvic nerves); and somatic nerves (S2–S4), which control the external urethral sphincter and PFM (pudendal and levator ani nerves). These nerves contain afferent (sensory) axons and motor (efferent) nerve fibers (see Fig. 1.2.7).

Central nervous system integrative ("reflex") centers responsible for micturition and urinary continence are located in the rostral brainstem in all species studied, including man, their activity being modulated by higher centers in the hypothalamus and other brain areas, including the frontal cortex. Centers in the pons (brainstem) coordinate micturition as such, but centers rostral to the pons are responsible for the timing of the start of micturition. The pontine micturition center (PMC) coordinates the activity of motor

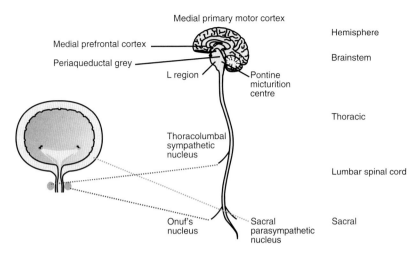

FIGURE 1.2.7. A schematic presentation of the main integrative areas of lower urinary tract neural control. The three sets of peripheral nerves are somatic (red), parasympathetic (green), and sympathetic (blue). The most important integrative center for coordinated storage and emptying of the bladder are in the brainstem (black).

neurons of the urinary bladder and the urethral sphincter (both nuclei are located in the sacral spinal cord), receiving afferent input via the periaqueductal grey. The central control of LUT function is organized as an on/off switching circuit (or, rather, a set of circuits) that maintains a reciprocal relationship between the urinary bladder and urethral outlet.

The PMC has been well studied in experimental animals, and it has also been demonstrated by PET in the right dorsomedial pontine tegmentum in human subjects. Apparently, it is more active on the right.[18] The PMC sends direct excitatory (glutaminergic) projections to the parasympathetic detrusor nucleus, and (possibly through the same pathway) to the GABA-ergic commissural nucleus at the S2–S3 spinal level.[19] The premotor interneurons from commissural nucleus inhibit the urethral sphincter.[24] The PMC receives descending input from brain areas, and afferent input from the LUT; the latter it receives indirectly via the periaqueductal grey (in the brainstem).[25] Without the PMC and its spinal connections coordinated bladder/sphincter activity is not possible; thus, patients with such lesions demonstrate bladder sphincter discoordination (dyssynergia). Patients with lesions above the pons do not show detrusor–sphincter dyssynergia; however, they suffer from urge incontinence (caused by detrusor overactivity) and an inability to delay voiding at inappropriate places and times.

Although voluntary micturition is a behavior pattern that starts with the relaxation of striated urethral sphincter and PFM, a voluntary PFM contraction during voiding can lead to a stop of micturition by the "reverse" activation of the micturition center (e.g., the "pattern generator of voiding"). Thus, PFM muscle contraction normally is necessarily "accompanied" by bladder inhibition. The voluntary "decision" is just a "push on the button," which activates a predetermined and integrated motor pattern.

Neural Control of Continence

At rest, continence is assured by a competent sphincter mechanism, including not only the striated and smooth muscle sphincter but also the PFM and an adequate bladder storage function. The kinesiological sphincter EMG recordings in normals show continuous activity of MUPs at rest, increasing with increasing bladder fullness. Reflexes mediating excitatory outflow to the sphincters are organized at the spinal level (the guarding reflex). The L region in the brain has also been called the "storage center".[18] This area was active in PET studies of those volunteers who could not void, but could contract their PFM.

The L region is thought to exert a continuous exciting effect on the Onuf's nucleus and, thus, on the striated urinary sphincter during the storage phase; in humans, it is probably part of a complex set of pattern generators for different coordinated motor activities, such as breathing, coughing, straining, etc.

During physical stress (e.g., coughing, and sneezing), the urethral and anal sphincters are not sufficient to passively withhold the pressures arising in the abdominal cavity and, hence, within the bladder and lower rectum. Activation of sphincter muscles, as well as the PFMs, is mandatory. (Individuals sensitive to their motor control, for example, are able to feel a tightening of the PFMs even during a minor stimulus like taking a deep breath.) This process occurs in two separate steps by two different activation processes:

1. Coughing, sneezing, and deep breathing are thought to be generated by individual pattern generators within the brainstem, and, thus, coactivation of the PFM is a preset coactivation, and not primarily a "reflex" reaction to increased intraabdominal pressure.
2. There is a reflex PFM response to increased abdominal pressure caused by distension of muscle spindles within the muscle. This reflex is an additional factor in sustaining reflex activation of the PFM.

In addition, and this is important for any treatment program, the PFM can be voluntarily activated concomitant with the very first awareness that sneezing or coughing will happen. Such voluntary contraction is part of normal behavior, e.g., the ability to control a maximally filled bladder or liquid stool–filled rectal ampulla during additional physical stress, such as coughing. For varying reasons, such coordinated, timed activity is not self-generated by patients, even if they still have the ability of voluntary pelvic floor contraction. But they may be able to learn the trick (the Knack procedure).[26]

Neural Control of Anorectal Function

The gastrointestinal tract has its own intrinsic innervation, the enteric nervous system, regulating enteric smooth muscle activity, as well as secretory and absorptive functions of the mucosa. It has its "own" afferents.

The colon is innervated by the parasympathetic system, the vagus, and the pelvic nerves. The latter originate from S2–S4 cord segments and innervate the descending and sigmoid colon and anorectum. Excitation promotes peristalsis, local blood flow, and intestinal secretion.

The medial prefrontal cortex and the anterior cingulate gyrus are thought to regulate the timing and initiation of defecation.

Rectal distension causes reflex relaxation of the smooth internal anal sphincter (the rectoanal inhibitory reflex), allowing the rectal contents to come into contact with the sensitive anal canal. Simultaneously, the external anal sphincter contracts to ensure continence. This physiologic process is called the "anal sampling reflex" and allows for identification of the rectal contents (gas, fluid, or solid). Indeed, it has been demonstrated to occur regularly (up to seven times per hour).

Feces stored in the colon is transported past the rectosigmoid "physiological sphincter" into the normally empty rectum, which can store up to 300 ml of contents. Rectal distension causes regular contractions of the rectal wall (effected by the intrinsic myenteric plexus) and prompts the desire to defecate.[27]

Stool entering the rectum is detected by stretch receptors in the rectal wall and PFM; their discharge leads to the urge to defecate. It starts as an intermittent sensation, which becomes more and more constant. Contraction of PFM may interrupt the process, probably by concomitant inhibitory influences to the defecatory neural "pattern generator," but also by "mechanical" insistence on sphincter contraction and the propelling of feces back to the sigmoid colon.[27]

PFM are intimately involved in anorectal function. Apart from the "sensory" role of PFM and the external anal sphincter function, the puborectalis muscle is thought to maintain the "anorectal" angle, which facilitates continence, and has to be relaxed to allow defecation.

Current concepts suggest that defecation requires increased rectal pressure coordinated with the relaxation of the anal sphincters and PFM. Pelvic floor relaxation allows opening of the anorectal angle and perineal descent, facilitating fecal expulsion. Puborectalis and external anal sphincter activity during evacuation is generally inhibited. However, observations by EMG and defecography suggest that the puborectalis may not always relax during defecation in healthy subjects. Puborectalis activity measured by EMG was unchanged in 9% and increased in 25% of healthy subjects.[28] Thus, although "paradoxical" puborectalis contraction during defecation is used to diagnose pelvic floor dyssynergia in patients with typical symptoms, this finding may be related to normal variations also.

Neural Control of the Sexual Response

Genitals have their own, but in principle similar, peripheral, somatic, parasympathetic, and sympathetic innervation from the same spinal cord segments by the same sets of nerves as the LUT.

Erection, seminal emission, ejaculation, and orgasm are part of the complex mating behavior integrated by the forebrain and hypothalamus. The nucleus paragigantocellularis has been implicated in the inhibition of climax. This nucleus, as well as the periaqueductal grey, receive afferent input from the genitals.

Sexual arousal in men has been associated with the activation of the insula and inferior frontal cortex in the right brain hemisphere.[29]

PFM are actively involved in the sexual response. Their activation has been mostly explored in males during ejaculation, where their repetitive activation during an interval of several seconds is responsible for the expulsion of semen from the urethra, particularly by the bulbospongiosus and the bulbocavernosus muscles.[29] Little is known about PFM activity patterns during other parts of the human sexual response cycle. It is assumed that apart from general changes in muscle tone set by the emotional motor system, it is the sacral reflex circuit that governs much of the PFM activity during the sexual response cycle.[30] The bulbocavernosus reflex behavior, as found in previous studies,[31] would allow for reflex activation of PFMs during genital stimulation. Tonic stimulation of the reflex is postulated to hinder venous outflow from the penis/clitoris, thus, helping erection. PFM reflex contraction should conceivably contribute to the achievement of the "orgasmic platform" (contraction of the levator ani and, in the female, the circumvaginal muscles). Climax in humans (in both sexes, and in experimental animals) elicits rhythmic contractions of the PFM/perineal muscles, which, in the male, drives the ejaculate from the urethra (assisted by a coordinated reflex bladder neck closure).

Pelvic Floor Muscles in Neurological (and Idiopathic) Lesions

Coordinated detrusor–sphincter activity is retained after suprapontine lesions (such as stroke, Parkinson's disease, etc.), indicating the importance of the PMC. However, normal behavioral patterns of sacral function may be altered and voluntary control of the PFM attenuated or lost.

Coordinated detrusor–sphincter activity is lost with lesions between the lower sacral segments and the upper pons (suprasacral infrapontine lesions), and sphincter inhibition is no longer seen preceding detrusor contractions. On the contrary, detrusor contractions are associated with increased sphincter EMG activity. This pattern of activity is called detrusor–sphincter dyssynergia.[32] Sphincter contraction (or failure of relaxation) during involuntary detrusor contractions have been reported in patients with Parkinson's disease.[33] Sphincter behavior in Parkinsonian patients has been also described as bradykinetic. The neurogenic uncoordinated sphincter behavior has to be differentiated from "voluntary" contractions, which may occur in poorly compliant patients; the sphincter contractions of the so-called nonneuropathic-voiding dyssynergia may be a learned abnormal behavior.[34]

Paradoxical external anal sphincter activation during defecation has been described in Parkinson's disease (anismus[35]). Nonrelaxation of the puborectalis muscle has been claimed as a major general cause of obstipation, but this is controversial. PFM spasm has been described in women as a functional problem in painful intercourse; the entity "vaginismus" is, however, controversial. No EMG changes have been found in one study[36] and increased basal activity has been reported by another.[37] Anismus, vaginismus, and pelvic pain derived from "PFM spasms" or PFM overactivity are attractive concepts derived from pathophysiological assumptions in various, as yet unexplained pathological pelvic conditions. The pathological PFM activity in such patients has been so far poorly documented, and the conditions remain controversial.

Peripheral lesions, either of the motor nuclei in spinal conus, the nerve roots within cauda equina or the peripheral nerves (the levator ani nerve and the pudendal nerve), lead to denervation injury of PFM.[38] This can be partial or total, and results in atrophy. Collateral reinnervation after partial lesions is the rule. Axonal reinnervation after total denervation may not occur for several physiological or anatomical reasons.

Neuromuscular Injury to the Pelvic Floor Caused by Vaginal Delivery

Many studies using different techniques have demonstrated both neurogenic and structural damage to PFM and sphincter muscles.[39] There is still a role for other lesion mechanisms, such as muscle ischemia. The implication of all these studies was that the PFM becomes weak; this can, indeed, be demonstrated.[40] Thus, the sphincter mechanisms and pelvic organ support become functionally impaired, with stress incontinence and prolapse being a logical consequence.

Although muscle weakness may well be a consequence of childbirth injury, there seem to be further possibilities for deficient PFM function because it is not only the strength of muscle contraction that counts. Normal neural control of muscle activity leads to coordinated and timely responses to ensure appropriate muscle function as required. These muscular behavioral patterns have been studied by kinesiological EMG recording.[1] Changes in muscular behavior may originate from minor repairable neuromuscular pelvic floor injury.[2]

In healthy nulliparous women, two types of behavioral patterns (tonic and phasic) can be found (Figs. 1.2.8 A and 1.2.8 B). The tonic pattern consists of a crescendo–decrescendo type of activity (probably derived from the grouping of slow-twitch motor units). The crescendo–decrescendo pattern may be the expression of constant (tonic) reflex input parallel to the breathing pattern.

The phasic pattern, which is probably related to fast-twitch motor unit activation, is motor unit activity seen only during voluntary (phasic) contraction (e.g., coughing) (Fig. 1.2.1).

In respect to these muscle activation patterns, parous women with stress urinary incontinence are subject to several possible changes of these activation patterns:[2]

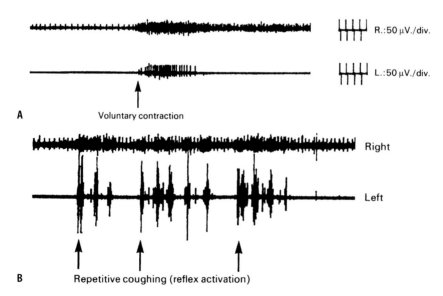

FIGURE 1.2.8. Patterns of activation of pubococcygeus muscles (right muscle, top trace; left muscle, bottom trace) in a normal continent nulliparous woman. **(A)** Voluntary contraction; **(B)** coughing; the tonic pattern from the detection site on the right, and the phasic pattern of muscle activation from the left. (*Source*: Reproduced from Deindl et al., 1993, with permission from Blackwell Publishing Ltd.)

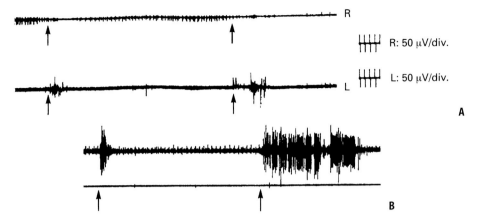

FIGURE 1.2.9. Patterns of activation of pubococcigeus muscles (right muscle, top trace; left muscle, bottom trace) in a parous stress-incontinent woman (Reproduced from Deindl et al., 1994, with permission from Blackwell Publishing Ltd.). **(A)** The expected recruitment of motor units during a cough occurs only on one side (although both pubococcigeus could be voluntarily activated). **(B)** A paradoxical inhibition of firing of motor units occurs during coughing.

1. Significant reduction of the duration of motor unit recruitment.
2. Unilateral recruitment of reflex response in the pubococcygeal muscle (Fig. 1.2.9 A).
3. Inhibition of continuous firing of motor units (Fig. 1.2.9 B).
4. Delay in PFM activation on coughing.

Although it has not been proven in studies, it is reasonable to assume that periods of pain and discomfort after childbirth (e.g., perineal tears and episiotomy), and particularly pain related to attempted PFM contraction, could lead to a temporary nonactivation of the PFM. This could be the origin of disturbances in behavioral patterns, which would need to be readjusted. In combination with a particularly vulnerable pelvic floor neural control, which only evolved in its complexity phylogenetically after the attainment of the upright stance, such a temporary disturbance of neural control after childbirth may persist, although the lesion(s) would have fully recovered.

Conclusion: Neurophysiologic Conceptualization of PFM Physiotherapy

PFMs are an axial muscle group under prominently reflexive and relatively weak voluntary control, which contributes few and poor sensory data to awareness. Furthermore, its neural control mechanism is fragile because of its relative phylogenetic youth, age, and exposure to trauma and disease because of its expansiveness of anatomy (from frontal cortex to the "tail"). Dysfunctional neural control induced by trauma, disease, or purely functional causes may manifest itself by over- or underactivity, and/or by discoordination of PFM. Often these disturbances are not "hardwired" into the nervous system, but are only a problem of neural control "software" (which can be "reprogrammed"). Therefore, physiotherapy should, in many patients, provide an appropriate, and perhaps the best, treatment.

References

1. Deindl FM, Vodusek DB, Hesse U, et al. Activity patterns of pubococcygeal muscles in nulliparous continent women. Br J Urol. 1993;72:46–51.
2. Deindl FM, Vodusek DB, Hesse U, et al. Pelvic floor activity patterns: comparison of nulliparous continent and parous urinary stress incontinent women. A kinesiological EMG study. Br J Urol. 1994;73:413–417.
3. Kenton K, Brubaker L. Relationship between levator ani contraction and motor unit activation in the urethral sphincter. Am J Obstet Gynecol. 2002;187:403–406.
4. Chantraine A. Examination of the anal and urethral sphincters. In: Desmedt JE, editor. New

developments in electromyography and clinical neurophysiology. Volume 2. Basel: Karger, 1973: 421–432.
5. Vodusek DB. Neurophysiological study of sacral reflexes in man (in Slovene). Institute of Clinical Neurophysiology. Ljubljana: University E. Kardelj in Ljubljana, 1982.
6. Read NW. Functional assessment of the anorectum in faecal incontinence. Neurobiology of incontinence (Ciba Foundation Symposium 151). Chichester, New York, Brisbane, Toronto, Singapore: John Wiley, 1990.
7. Borghi F, Di Molfetta L, Garavoglia M, et al. Questions about the uncertain presence of muscle spindles in the human external anal sphincter. Panminerva Med. 1991;33:170–172.
8. Vodusek DB, Janko M. The bulbocavernosus reflex. A single motor neuron study. Brain. 1990; 113(Pt 3):813–820.
9. Sundin T, Petersen I. Cystometry and simultaneous electromyography from the striated urethral and anal sphincters and from levator ani. Invest Urol. 1975;13:40–46.
10. Vodusek DB. Electrophysiology. In: Schussler B, Laycock J, Norton P, et al, editors. Pelvic floor re-education: principles and practice. London: Springer-Verlag, 1994.
11. McCloskey DI. Corollary changes: motor commands and perception. In: Brookhart JM, Mountcastle VB, editors. Handbook of physiology. Section I. The nervous system. Volume 2 (Part 2). Bethesda, MD: American Physiological Society, 1981.
12. Morrison JFB. Reflex control of the lower urinary tract. In: Torrens M, Morrison JF, editors. The physiology of the lower urinary tract. London: Springer-Verlag, 1987.
13. Mannen T, Iwata M, Toyokura Y, Nagashima K. The Onuf's nucleus and the external anal sphincter muscles in amyotrophic lateral sclerosis and Shy-Drager syndrome. Acta Neuropathol (Berl). 1982;58:255–260.
14. Schroder HD. Anatomical and pathoanatomical studies on the spinal efferent systems innervating pelvic structures. 1. Organization of spinal nuclei in animals. 2. The nucleus X-pelvic motor system in man. J Auton Nerv Syst. 1985;14:23–48.
15. Barber MD, Bremer RE, Thor KB et al. Innervation of the female levator ani muscles. Am J Obstet Gynecol. 2002;187:64–71.
16. Bannister LH, ed. Gray's anatomy. The anatomical basis of medicine and surgery. 38th edition. New York, London: Churchill Livingstone, 1995.
17. Marani E, Pijl ME, Kraan MC et al. Interconnections of the upper ventral rami of the human sacral plexus: a reappraisal for dorsal rhizotomy in neurostimulation operations. Neurourol Urodyn. 1993;12:585–598.
18. Blok BF, Sturms LM, Holstege G. A PET study on cortical and subcortical control of pelvic floor musculature in women. J Comp Neurol. 1997;389: 535–544.
19. Holstege G. The emotional motor system in relation to the supraspinal control of micturition and mating behavior. Behav Brain Res. 1998;92:103–109.
20. Vodusek DB. Evoked potential testing. Urol Clin North Am. 1996;23:427–446.
21. Brostrom S. Motor evoked potentials from the pelvic floor. Neurourol Urodyn. 2003;22: 620–637.
22. Ueyama T, Mizuno N, Nomura S, et al. Central distribution of afferent and efferent components of the pudendal nerve in cat. J Comp Neurol. 1984;222:38–46.
23. Torrens M, Morrison JF, editors. The physiology of the lower urinary tract. London: Springer-Verlag, 1987.
24. Blok BF, van Maarseveen JT, Holstege G. Electrical stimulation of the sacral dorsal gray commissure evokes relaxation of the external urethral sphincter in the cat. Neurosci Lett. 1998;249:68–70.
25. Blok BF, Holstege G. The central nervous system control of micturition in cats and humans. Behav Brain Res. 1998;92:119–125.
26. Miller JM, Ashton-Miller JA, DeLancey JO. A pelvic muscle precontraction can reduce cough-related urine loss in selected women with mild SUI. J Am Geriatr Soc. 1998;46:870–874.
27. Bartolo DC, Macdonald AD. Fecal continence and defecation. In: Pemberton JH, Swash M, Henry MM, editors. The pelvic floor. Its function and disorders. London: W. B. Saunders, 2002: 77–83.
28. Fucini C, Ronchi O, Elbetti C. Electromyography of the pelvic floor musculature in the assessment of obstructed defecation symptoms. Dis Colon Rectum. 2001;44:1168–1175.
29. Stoleru S, Gregoire MC, Gerard D, et al. Neuroanatomical correlates of visually evoked sexual arousal in human males. Arch Sex Behav. 1999; 28:1–21.
30. Petersen I, Franksson C, Danielson CO. Electromyographic study of the muscles of the pelvic floor and urethra in normal females. Acta Obstet Gynecol Scand. 1955;34:273–285.
31. Vodusek DB. Sacral reflexes. In: Pemberton JH, Swash M, Henry MM, eds. Pelvic floor. Its functions and disorders. London: Saunders, 2002.

32. Blaivas JG, Sinha HP, Zayed AA, Labib KB. Detrusor-external sphincter dyssynergia: a detailed electromyographic study. J Urol. 1981;125:545–548.
33. Pavlakis AJ, Siroky MB, Goldstein I, Krane RJ. Neurourologic findings in Parkinson's disease. J Urol. 1983;129:80–83.
34. Rudy DC, Woodside JR. Non-neurogenic neurogenic bladder: The relationship between intravesical pressure and the external sphincter electromyogram. Neurourol Urodyn. 1991;10:169.
35. Mathers SE, Kempster PA, Law PJ, et al. Anal sphincter dysfunction in Parkinson's disease. Arch Neurol. 1989;46:1061–1064.
36. van der Velde J, Laan E, Everaerd W. Vaginismus, a component of a general defensive reaction. An investigation of pelvic floor muscle activity during exposure to emotion-inducing film ex-cerpts in women with and without vaginismus. Int Urogynecol J Pelvic Floor Dysfunct. 2001;12:328–331.
37. Graziottin A, Bottanelli M, Bertolasi L. Vaginismus: a clinical and neurophysiological study. Urodinamica. 2004;14:117–121.
38. Podnar S, Oblak C, Vodusek DB. Sexual function in men with cauda equina lesions: a clinical and electromyographic study. J Neurol Neurosurg Psychiatry. 2002;73:715–720.
39. Vodusek DB. The role of electrophysiology in the evaluation of incontinence and prolapse. Curr Opin Obstet Gynecol. 2002;14:509–514.
40. Verelst M, Leivseth G. Are fatigue and disturbances in pre-programmed activity of pelvic floor muscles associated with female stress urinary incontinence? Neurourol Urodyn. 2004;23:143–147.

1.3
The Effects of Pregnancy and Childbirth on the Pelvic Floor

Kaven Baessler and Bernhard Schüssler

Key Messages

- Connective tissue remodeling before delivery and intrapartum trauma play major roles in pelvic floor dysfunction after childbirth.
- Frequency, stress incontinence, and, to a lesser extent, overactive bladder increase in the course of pregnancy and show a decline after delivery.
- Stress incontinence occurs in up to 85% of women during pregnancy, in 22% postpartum, and in 4–19% de novo after childbirth.
- Stage II pelvic organ prolapse, especially anterior vaginal wall prolapse, is not uncommon during pregnancy.
- After vaginal delivery, the position of the perineum and bladder neck are lower and the pelvic floor contraction strength is reduced.
- Reinnervation as a sign of denervation is present in 80% of women postpartum, but also progresses over time and with ageing.
- Obesity and age are major risk factors for the development of pelvic floor dysfunction.
- Risks factors for anal sphincter laceration include first delivery, occipitoposterior presentation, a prolonged second stage of labor, forceps delivery, higher birth weight, and maternal age greater than 35 years.
- Routine mediolateral and midline episiotomy do not prevent pelvic floor trauma or protect the baby, and they are obsolete.
- Cesarean section is only partially protective and might reduce, but not totally prevent, pelvic floor dysfunction.
- Antepartum pelvic floor exercises led to a decline of postpartum stress urinary incontinence.

Introduction

For many years, vaginal delivery has been considered a principle causal factor in the development of incontinence and pelvic organ prolapse.[1,2,3] Recent large, community-based epidemiological studies have revealed that pregnancy itself, with its associated hormonal, connective tissue, and physical changes, plays a major role, with vaginal and instrumental delivery only further increasing the risk of pelvic floor dysfunction.[4,5] Elective cesarean section is commonly believed to preserve pelvic floor function. Although it might significantly reduce the risk of pelvic organ prolapse and anal sphincter defects, it does not necessarily protect against urinary and anal incontinence in the long term.[5,6] Muscular, nerve, and connective tissue damage has been demonstrated in up to 80% of women after vaginal deliveries,[7] but these changes do not automatically result in pelvic floor dysfunction symptoms. Epidemiological studies have also emphasized the effect of promoting factors; inevitable influences like ageing and genetic predisposition[4,5,8] and often avoidable risk factors like obesity, constipation, and hormone replacement therapy. Recent research has questioned the value of previously routine interventions like midline and mediolateral episiotomy, but has also investigated the effects of forceps and vacuum, which are used for instrumental deliveries, on the pelvic floor.[9]

There are two main events before and during childbirth that concern the whole body, and the pelvic floor in particular; hormonal preparation for childbirth, which allows dense connective tissue to soften, and mechanical trauma during vaginal delivery. This chapter presents an update of our current knowledge on the effects of pregnancy and childbirth on the pelvic floor and pelvic floor symptoms. The influence of episiotomy, instrumental delivery, and positioning during vaginal birth are described and, subsequently, emerging possible strategies for pelvic floor protection are discussed. Many studies on this topic have underutilized instruments such as quality of life assessments, and the lack of consistency in terminology and definition of outcomes prevents comparison between studies.

Hormones: Changes and Impact

Hormonal alterations are physiological and essential to prepare the body and to adjust the musculature and connective tissue for vaginal birth. It seems reasonable to assume that some of these transformations might not have reversed in women with considerable postpartum anatomical and functional pelvic floor changes. This is particularly true in the case of evident damage after cesarean delivery, when mechanical trauma to the pelvic floor was avoided.

Diminishing pelvic floor function during pregnancy is at least partly related to the effects of the high progesterone levels during pregnancy. Progesterone is known to reduce the tonus in ureters, bladder, and urethra because of its smooth muscle–relaxing and estrogen-antagonizing effects.[10] Relaxin, which is a peptide hormone similar to insulin, increases markedly during pregnancy. It modifies the connective tissue and has a collagenolytic effect to allow for appropriate stretching during vaginal birth in guinea pigs.[11] This hormone also seems to be responsible for pelvic and symphyseal pain.[12] There might be many hormones involved in connective tissue remodeling and cervical ripening, e.g. increasing joint mobility in pregnancy is well known.[13] As a likely result of connective tissue remodeling in preparation for birth, Landon and colleagues[14] found that the connective tissue of the rectus sheath fascia and the obturator fascia could be stretched to greater lengths during pregnancy, but were also much weaker. In some women, these changes may be irreversible, or further stretching beyond physiological limits (e.g. during vaginal delivery) may result in permanent dysfunction.

Mechanical Trauma

There is little doubt that vaginal delivery constitutes a traumatic event to the pelvic floor, but we are not certain about the mechanisms that lead to complete restitution or result in anatomical or functional changes. Mechanical damage to the pelvic floor musculature, connective tissue, and nerve supply usually occurs during the second stage of labor when the fetal head distends and stretches the pelvic floor (Fig. 1.3.1; see also dynamic 3D image on DVD). Lien et al. have developed a pelvic model from MRI studies to examine behavior and interaction of the baby's head and the pelvic floor muscle. The most medial part of the pubococcygeus muscle is the part of the pelvic floor that undergoes the largest stretching, i.e. up to 3.26 times its original length.[15]

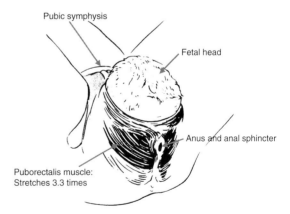

Figure 1.3.1. Pelvic floor distension during vaginal delivery. Note the stretching of the puborectalis and pubococcygeus muscle and the separation of the perineal membrane. According to Lien et al.,[15] the pubococcygeus muscle has the greatest rest/stretch ratio during vaginal delivery; it stretches to 3.26 times its original length. (*Source:* Modified from Schüssler B, Anthuber C, Warrell D. The pelvic floor before and after delivery. In: Schüssler B, Laycock J, Norton P, Stanton S, editors. Pelvic floor re-education. Principles and Practice. Springer-Verlag London Ltd; 1994:871994:105.)

Apart from the mechanical trauma – direct muscle rupture or, more commonly, stretching – there might also be biochemical damage to the soft tissue, especially during long second stages when the fetal head is compressing the pelvic floor. Currently, it is not possible to determine exactly whether pelvic floor symptoms are attributable to impairment caused by hormonal modification (and lack of reversal) or mechanical forces or a combination of both.

The Pelvic Floor During Pregnancy

Hormonal and mechanical changes, along with the growing uterus displacing and distorting the bladder, may contribute to pelvic floor dysfunction during pregnancy. The physiological weight gain during pregnancy might also play a role. An increased body mass index correlates with the intraabdominal pressure during urodynamics[16] and has repeatedly been shown to be an independent risk factor for urinary incontinence later in life.

Symptoms

Increased daytime and nighttime urinary frequency is a very common symptom in pregnant women.[17,18] Although the increase in urinary frequency might predominantly result from the increase in glomerular filtration rate, stress incontinence symptoms are clearly a sign of pelvic floor changes. Stress incontinence during pregnancy is reported by up to 85% of women,[17,18,19] with an increasing incidence from 10% at 12 weeks to 23% at 24 weeks and 26% at 36 weeks of gestation in one study.[20] Urgency symptoms are found in 23% of women, but urge incontinence is not as common in 8% (Fig. 1.3.2).[18] Voiding difficulties can also occur during pregnancy, with 43% of women in one study complaining of a poor stream and 37% of incomplete bladder emptying.[21]

Findings During Pregnancy: Pelvic Organ Prolapse and Urodynamics

Examination

During pregnancy, the blood flow to the pelvic organs and the pelvic floor increases, connective tissue becomes more elastic, and there is hypertrophy of smooth muscle and hyperplasia of mucous membranes.[22] Three recent studies have drawn attention to the occurrence of clinically significant pelvic organ prolapse during pregnancy in nulliparous women. All three studies employed the validated quantification of pelvic organ prolapse of the International Continence Society. O'Boyle et al. used a case-control study of 21 nulliparous women at 14–39 weeks of gestation and found stage 2 pelvic organ prolapse (leading edge of prolapse +/− 1 cm of the hymenal remnants) in 48%, whereas none of the 21 age- and race-matched nonpregnant controls demonstrated stage 2 prolapse.[23] Another study assessed

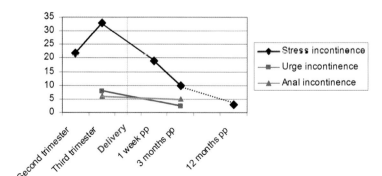

FIGURE 1.3.2. Stress and urge incontinence and anal incontinence during pregnancy and after delivery in primiparous women.[17,18,20]

pelvic organ prolapse at 36 weeks of gestation in 94 women.[24] It revealed stage 2 prolapse in 26%.[24] Cross-sectional comparisons of pelvic organ prolapse in the first, second, and third trimester also confirmed that there is more stage 2 pelvic organ prolapse in the third trimester (18/52; 35%).[25] The most frequent site of prolapse was the anterior vaginal wall in all of these studies. Interestingly, the latter study also demonstrated a clinically significant increase in perineal body length of almost 1 cm from the first to the third trimester. Genital hiatus and total vaginal length increased by approximately 5 mm.[25] The lengthening of the perineal body during pregnancy is an intriguing finding, given that a short perineum (<4 cm during the first stage of labor) was associated with higher episiotomy and vaginal tear rates, although the overall number of episiotomy in nulliparous women in that study was rather high at 76%.[26]

Imaging

Many studies have compared bladder neck mobility before and after delivery, but, unfortunately, there is a paucity of prospective data on changes that occur during pregnancy. However, comparative studies demonstrated that the urethrovesical angle at rest is increasing during pregnancy[27] and is significantly wider compared to nonpregnant controls,[28] and that the bladder neck shows an increased "backward displacement" antenatally compared with a nonpregnant control group.[29] This correlates well with the aforementioned increase in anterior vaginal wall prolapse during pregnancy.

Urodynamic Studies

Urodynamic studies during pregnancy have been rather inconsistent in their findings. The urethral closure pressure and the functional urethral length were shown to increase during the course of pregnancy according to one study,[30] whereas they remained unchanged in another.[31] Compared with nulliparous nongravid controls, the maximum urethral closure was lower during pregnancy.[29]

Childbirth and Pelvic Floor Function

Hormonal changes will affect all pregnant women; yet, the extent may vary widely. If a woman undergoes cesarean section, the extent might depend on the timing. Premature delivery or elective or emergency cesarean section after labor has already progressed might have different effects because of the different hormonal changes and trauma. This part of the chapter describes the immediate effects of childbirth on the pelvic floor anatomy and function and the impact of intrapartum interventions, and it attempts to place the findings into the long-term picture drawn by the new epidemiological studies.

Findings on Clinical Examination After Vaginal Delivery

The position of the perineum is lower after vaginal delivery (Fig. 1.3.3, A and B).[32] Employing the International Continence Society (ICS) pelvic organ prolapse standardization, Sze et al. prospectively studied nulliparous women and found that postpartum 52% had stage 2 prolapse, 37% had developed a new prolapse, and 15% revealed a more severe prolapse compared to antenatal examinations.[24] Pelvic floor muscle contraction strength is reduced after vaginal delivery, which has been demonstrated with several investigation techniques used: vaginal cones,[33] standardized physical examination assessment of pelvic floor muscle strength,[34] intravaginal squeeze pressure measurement,[7,35,36] and perineal ultrasound.[35] However, there seems to be some recovery of the lowered perineal position[37] and pelvic floor contraction strength[34,35] 6 weeks to 3 months postpartum. The exact mechanisms of recovery or deterioration remain unknown.

Neurophysiology

Intact innervation of the levator ani muscle, anal, and urethral sphincters is critical to normal pelvic function. Techniques to measure the competency of the nerve supply are often invasive (needle electromyography [EMG]) or not very

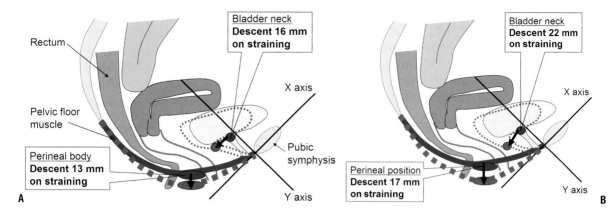

FIGURE 1.3.3. Position of the bladder neck and perineum at rest and on straining during pregnancy in nulliparas (A) and after vaginal delivery (B). While the positions at rest remain unchanged postpartum, the descent of the bladder neck and the perineum during straining increases significantly.[32,35]

specific and prone to disturbances (surface EMG) or the exact implications are not clear (pudendal nerve terminal motor latency [PNTML]).

Single-fiber EMG, as well as concentric-needle EMG, can identify partial denervation by signs of reinnervation, such as increased fiber density. Prospective EMG studies performed before and after childbirth substantiated the evidence of childbirth-induced pelvic floor denervation, detecting increased fiber density after vaginal delivery.[7,38] There is evidence of reinnervation in 80% of women after vaginal delivery.[7] Snooks et al.[37] investigated 14 multiparous women from their previous studies[38,39] after five years and demonstrated that pelvic floor denervation progressed, indicating that age is a contributory factor.[37] Similarly, progressive denervation with time up to 15 years postpartum was found in another prospective study, corroborating the ageing factor.[40a]

Prolongation of the PNTML is thought to be a result of pudendal nerve damage during vaginal delivery. Significantly prolonged mean PNTML's have been found in women two to three days after vaginal delivery, compared to a multiparous[39] and a nulliparous control group.[38] At follow-up five years later, prolongation of PNTML persisted.[37] Two prospective analyses demonstrated a prolongation of the PNTML antenatally to six to eight weeks after vaginal delivery, particularly after the first delivery.[32,40] But, again, many of these changes seem to be temporary, as two-thirds of the women with an abnormally prolonged PNTML after delivery had normal measurements six months later.[32]

Findings on Urodynamics

Prospective urodynamics studies performed during pregnancy and six to nine weeks postpartum revealed a notable decrease in urethral closure pressure[31] and urethral length[30,31,36] after vaginal delivery. Women with persisting urinary incontinence after childbirth were found to have a shorter functional urethral length and a lower urethral closure pressure compared with continent women.[41]

Findings on Perineal Ultrasound

Bladder neck mobility increases and the bladder neck position is lower after vaginal delivery, as imaged by perineal ultrasound (Fig. 1.3.3, A and B).[19,35,36] Women with postpartum stress incontinence were found to have greater bladder neck mobility during straining before delivery compared with continent women.[19] It remains unclear, however, whether the lower observed bladder neck position is caused by impaired connective tissue structures, by reduced levator ani muscle tone, or by both. Specific pelvic floor muscle trauma can be depicted with MRI and 3D ultrasound (see Imaging Chapter 2.4).

Urinary Incontinence After Childbirth

Up to 22% of women complain of urinary incontinence after delivery.[17–19,36] De novo stress urinary incontinence has been reported in 4–19% of women who gave birth vaginally.[17,18] Women who had prepregnancy urinary incontinence were more likely to complain of daily incontinence 6 weeks and 6 months postpartum.[42] Long-term persistence is high; 5 years postpartum, the risk of urinary incontinence was 92% in women who developed symptoms during pregnancy or puerperium (Fig. 1.3.2). In women who had remission of incontinence 3 months postpartum, 42% had recurrence 5 years later.[43] Compared with women who delivered by cesarean section, urinary incontinence is significantly more common after vaginal delivery.[17,42,44]

Obstetric and Maternal Risk Factors for Urinary Incontinence

The literature on the contribution of obstetrical factors to the development of urinary incontinence is inconsistent. In some studies, the duration of the second stage of labor[17] and birth weight[45] were associated with a higher incidence of stress incontinence. Other authors did not find significant correlation between stress incontinence, second stage of labor,[17,44] or birth weight. However, obstetric risk factors, which show a statistical significant correlation with the incontinence incidence immediately after delivery, might disappear three months later.[17] The large Norwegian Epidemiology of Incontinence in the County of Nord-Trøndelag (EPINCONT) study asked women more than 30-years-old about incontinence symptoms after their first vaginal delivery. The data were linked with a meticulous birth register. The study confirmed known risk factors like age, body mass index, and parity, but also demonstrated that birth weight >4,000 g, breech, and instrumental delivery and epidural are obstetric factors that seem to lose their influence on incontinence (if they had any in the first place), and that the ageing and weight gaining processes take over.[45] Forceps delivery has also been identified to increase the risk of postpartum urinary incontinence by 50% compared with vaginal birth.[42]

Wilson et al. examined maternal risk factors for postpartum urinary incontinence and found obesity to be associated with the development of incontinence three months after childbirth.[44] Three studies, including a randomized controlled trial, observed that antepartum pelvic floor exercises significantly reduced the prevalence of postpartum incontinence.[19,44,46] One study failed to predict postpartum urinary and fecal incontinence with markers of collagen weakness like striae, varicose veins, hemorrhoids, and joint hypermobility.[18]

Anal Sphincter Laceration and Anal Incontinence

Clinically visible anal sphincter tears occur in up to 7% of vaginal deliveries.[6,47–50] Studies using endoanal ultrasound revealed that there are also occult sphincter defects in 33% of primiparous and 4% of multiparous women after vaginal delivery.[47,51] The long-term impact of these occult sphincter defects is not known.

The contraction pressure of the anal sphincter decreases significantly after vaginal delivery[37,47,52] and does not return to antenatal strength,[47,51] even after 5 years.[37] Anal sphincter pressures at rest are reduced after vaginal delivery, but not after cesarean section.[51,53,54] Resting pressures tend to recover 5 weeks to 6 months postpartum,[53,54] whereas squeeze pressures increase, but might not reach antenatal values,[53] particularly if the anal sphincter was damaged.[54] Anal sphincter ruptures can cause decreased contraction pressures compared with spontaneous deliveries without perineal tears.[51] The decrement of the resting pressure was significantly greater in women with a sonographically demonstrated internal sphincter defect, and squeeze pressures were lower in women with an external sphincter defect.[51]

Anal canal sensation assessed by mucosal electrosensitivity might be impaired immediately after vaginal delivery but seems to recover six months postpartum.[54] In contrast, two studies with 232 women did not observe any changes in anal sensation after vaginal delivery.[52,53]

After vaginal delivery, 0.04% to 5% of women develop anal incontinence symptoms.[18,55,56] This prevalence rises to levels of 17% to 57% after anal sphincter rupture, despite primary repair.[48,51,54,57] Even nine to twelve months after vaginal delivery, incontinence of flatus was also more common in women with anal sphincter rupture.[58] Most of these anal symptoms are transient, but might persist long term.[59] Fecal incontinence symptoms correlated with decreased anal pressures and sphincter defects imaged by endoanal ultrasound.[51,57]

Obstetric and Maternal Risk Factors for Anal Incontinence

Several obstetric risk factors for the development of sphincter lacerations have been identified. It is the first vaginal delivery that causes most of the damage and results more frequently in anal sphincter tears.[48,50,60] The occipitoposterior presentation,[47] a higher birth weight,[47,48,50,60,61] prolonged second stage of labor,[48,57] and maternal age of more than 35 years[50] were also associated with an increased sphincter laceration rate. An unchangeable feature – a subpubic arch angle of less than 90 degrees – was associated with prolonged labor and the development of anal incontinence.[62]

Sphincter lacerations occur more frequently with forceps-assisted deliveries than with vacuum extraction.[47,57,60,61] Only sonographically visible occult sphincter defects were also more common after forceps delivery.[51] Fecal incontinence symptoms correlate with sphincter defects and are seen in 38% of women delivered by forceps, in 12% of women delivered by vacuum, and in 4% of spontaneous deliveries, with a significant difference between forceps-assisted and spontaneous delivery.[51] In general, the risk of developing fecal incontinence is increased after operative vaginal delivery[55] but more so after use of forceps.[63] A systematic review estimated that one anal sphincter tear is avoided for every 18 women who had a vacuum delivery instead of forceps.[9] In up to 91% of women who sustained a third-degree tear, endoanal sonography showed residual sphincter defects indicating an inadequate primary sphincter repair.[47,49,64]

The higher risk of pelvic floor damage when using forceps in contrast to vacuum extraction might be explained by the ability to pull the baby's head vigorously along an undesirable line. The vacuum would be lost if not guided in an optimal direction.

Subsequent deliveries increase the incidence of sphincter defects and anal incontinence.[65,66] Prolonged pudendal nerve terminal latency, instrumental delivery, and higher maternal age[56] were also associated with an increased risk of postpartum alteration of fecal continence.

Although the immediate anal incontinence rate is significantly higher after sphincter lacerations, it seems that ageing is probably another important factor in deteriorating pelvic floor function. In a retrospective-matched control study approximately 30 years postpartum, 29 women with anal sphincter disruption were compared with 89 women with episiotomy and 33 women delivered by cesarean section.[6] The prevalence of flatus and fecal incontinence did not differ significantly in the 3 groups (episiotomy group 31%, sphincter rupture group 42.7%, cesarean section group 36.4%). The prevalence of fecal incontinence also did not differ significantly between episiotomy group (18%) and sphincter rupture group (6.9%), but women who were delivered by cesarean section did not develop fecal incontinence.[6] Although this study did not include clinical or sonographical examination to correlate findings and symptoms, the results display that successful management of childbirth is not the ultimate end of care. Further direction might be necessary to maintain an initially good outcome.

Only one protective factor has been described; Klein et al. showed that a strong exercise profile (e.g. jogging or cycling 3 or more times per week) was associated with fewer third- and fourth-degree tears.[61]

Intrapartum Interventions: Impact on Pelvic Floor Anatomy and Function

Routine Mediolateral and Midline Episiotomy Are Obsolete

Woolley[67,68] reviewed the literature on the benefits and risks of episiotomies from 1980 to 1994. Woolley concluded that a mediolateral or midline episiotomy does not prevent damage to the anal sphincter and its sequelae. Episiotomy does not protect the newborn from intracranial hemorrhage or intrapartum asphyxia. Episiotomy does, however, prevent anterior lacerations. Midline episiotomy carries a significant risk of extension into the anal sphincter. Episiotomy increases maternal blood loss. In the first week after vaginal delivery, perineal pain is more common after episiotomy than after a spontaneous perineal laceration. However, these differences disappear by 1 month postpartum.[67,68] These conclusions are still valid; routine and selective midline episiotomies were associated with higher incidences of sphincter lacerations[50,61] and a reduction in the use of midline episiotomies resulted in an increase in the rate of intact perineum.[69] Reduced midline, as well as mediolateral episiotomy rates, resulted in an increase in anterior and vaginal lacerations.[61,69–71]

Most studies show that mediolateral episiotomies do not prevent sphincter ruptures,[69,72] and a systematic and a Cochrane review of randomized controlled trials comparing restrictive versus routine use of episiotomy concluded that the restrictive episiotomy policies are beneficial in terms of posterior vaginal trauma and pain.[73] Three retrospective case-control studies reported a protective effect of the mediolateral episiotomy for the occurrence of severe sphincter lacerations.[50,60,74] However, one of these studies calculated that among primiparas, 48 episiotomies would have to have been performed to prevent one severe tear. This number rises to 106 episiotomies among multiparas.[74] On review of the literature, it seems that the mediolateral episiotomy rate in primiparas can be lowered to 20% and 30% without an increase in anal sphincter damage (Fig. 1.3.4).[70,71,75]

Perineal pain, disturbed wound healing, and dyspareunia are more common in women delivered by mediolateral than by midline episiotomy or with spontaneous perineal tear.[76,77] Women belonging to the group with restricted use of episiotomy and women with an intact perineum started sexual intercourse earlier.[70]

Prospective randomized controlled trials and retrospective studies have failed to demonstrate a protective effect of the mediolateral or midline episiotomy on stress urinary incontinence after vaginal delivery.[17,70,71] There was no difference in stress urinary and anal incontinence rates or pelvic organ prolapse between women with or without mediolateral episiotomy.[77] Viktrup et al. found a higher rate of stress incontinence when a mediolateral episiotomy was performed.[17] The incidence of stress urinary incontinence did not differ after episiotomies compared with spontaneous perineal tears.

Episiotomy did not influence PNTML[38,39] or the urethral closure pressure.[31] It did not result in increased or decreased pelvic floor denervation.[7] The ability to hold vaginal cones was more reduced after vaginal delivery when an episiotomy was performed.[33]

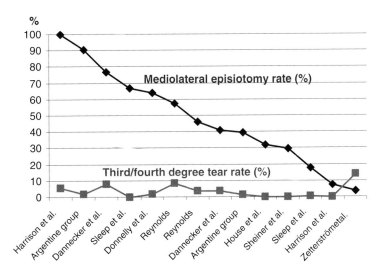

FIGURE 1.3.4. Different rates of mediolateral episiotomies in nulliparous women with corresponding third- and fourth-degree sphincter tears. It seems obvious that episiotomy rates can be restricted to at least 20–30% without an increase in severe sphincter lacerations.[58,72,87–91]

Upright or Lateral Position for Birth, Perineal Massage, and Continuity of Care

In several studies, supine or lithotomy position for childbirth had a negative effect on the perineum.[61,78] The adoption of upright positions might protect the perineum[56,79,80] and tend to reduce forceps-assisted deliveries.[78] Perineal massage during the weeks before delivery was shown to reduce the risk of sutured perineal trauma.[9] Perineal massage and gentle stretching during contractions in the second stage of labor resulted in fewer third-degree tears, but bladder and bowel function did not differ in a randomized controlled trial of 1,340 women.[81] Adoption of upright birth positions, warm compresses to the perineum, flexion, and counter pressure to the baby's head may be protective for both use of episiotomy and occurrence of sphincter lacerations.[80] Continuity of care by midwives during pregnancy, childbirth, and puerperium seems to be beneficial with regard to antenatal hospital admission, analgesia during childbirth, rates of episiotomies, and the woman's satisfaction.[82]

Cesarean Section: Only Partially Protective

Immediate postpartum, as well as long-term, studies have demonstrated that cesarean section might reduce, but not totally prevent, urinary and anal incontinence. Although cesarean section might prevent direct anal sphincter trauma, both elective and emergency cesarean section do not necessarily prevent the development of anal incontinence.[55,83] Pelvic organ prolapse seemed considerably lower after cesarean section. However, in a rare prospective study before and 6 weeks after childbirth, Sze et al. described stage II prolapse (ICS standardization) in 32% of 41 women who had spontaneous vaginal deliveries and in 35% of 26 women after cesarean section during active labor.[24]

Pregnancy itself caused no changes in PNTML,[84] and although no changes in the motor unit potentials were found after elective cesarean section, they were increased after emergency cesarean sections, when labor had already progressed.[7] Anal sphincter pressures at rest were reduced after vaginal delivery, but not after cesarean section.[51,53,54] The same applies to anal canal sensation.[54]

In summary, cesarean section is likely to avoid immediate direct trauma to the pelvic floor, as well as early functional impairment, but long-term protection is not necessarily given. The benefits of short-term prevention of pelvic floor dysfunction have to be weighed against the risk of cesarean section as an operation, as well as the psychosocial impacts.

The Woman's Contribution to Reduce Loss of Pelvic Floor Function

Antepartum pelvic floor exercises led to a decline of postpartum stress urinary incontinence[19,44] and an increase in vaginal pressures during voluntary pelvic floor contraction.[85] Obese women have a higher risk of urinary incontinence during pregnancy and postpartum[44,86] and several epidemiological studies have highlighted that obesity is associated with deteriorating pelvic floor function. Whether the physiological increase in weight during pregnancy contributes to incontinence, for instance, cannot be determined. The odds for anal sphincter trauma are decreased in women who regularly participate in sporting activities like jogging.[61]

Conclusions: Can Vaginal Delivery Be Made Safer for the Pelvic Floor?

It is evident that pregnancy and childbirth can result in pelvic floor dysfunction. In everyday life, many precautions against potential damage and injury are taken, but so far no predictive factor for immediate postpartum pelvic floor dysfunction has been established. Even cesarean section does not completely eliminate the risk of pelvic floor dysfunction. Why should women undergo cesarean section when it does not prevent pelvic floor symptoms in the long term? Women have much control over some risk factors. They should stay fit and healthy with a good diet and regular low-impact sport, avoid constipation and obesity, perform antenatal pelvic floor exercises, and consider having babies before the age of 35.

Can vaginal delivery be made safer if midwives and obstetricians utilize the available data to minimize trauma and subsequent pelvic floor dysfunction: no midline episiotomies, restrictive use of mediolateral episiotomy for fetal indications, and vacuum rather than forceps delivery, continuous care during pregnancy and childbirth, and lateral or upright birth position to reduce instrumental deliveries? Prospective longitudinal interventional studies are warranted to answer these questions.

References

1. Samuelsson E, Victor A, Svardsudd K. Determinants of urinary incontinence in a population of young and middle-aged women. Acta Obstet Gynecol Scand. 2000;79:208–215.
2. Uustal Fornell E, Wingren G, Kjolhede P. Factors associated with pelvic floor dysfunction with emphasis on urinary and fecal incontinence and genital prolapse: an epidemiological study. Acta Obstet Gynecol Scand. 2004;83:383–389.
3. Olsen AL, Smith VJ, Bergstrom JO, et al. Epidemiology of surgically managed pelvic organ prolapse and urinary incontinence. Obstet Gynecol. 1997;89:501–506.
4. MacLennan AH, Taylor AW, Wilson DH, et al. The prevalence of pelvic floor disorders and their relationship to gender, age, parity and mode of delivery. BJOG. 2000;107:1460–1470.
5. Rortveit G, Daltveit AK, Hannestad YS, et al. Urinary incontinence after vaginal delivery or cesarean section. N Engl J Med. 2003;348:900–907.
6. Nygaard IE, Rao SS, Dawson JD. Anal incontinence after anal sphincter disruption: a 30-year retrospective cohort study. Obstet Gynecol. 1997;89:896–901.
7. Allen RE, Hosker GL, Smith AR, et al. Pelvic floor damage and childbirth: a neurophysiological study. B J Obstet Gynaecol. 1990;97:770–779.
8. Grodstein F, Fretts R, Lifford K, et al. Association of age, race, and obstetric history with urinary symptoms among women in the Nurses' Health Study. Am J Obstet Gynecol. 2003;189:428–434.
9. Eason E, Labrecque M, Wells G, et al. Preventing perineal trauma during childbirth: a systematic review. Obstet Gynecol. 2000;95:464–471.
10. Miodrag A, Castleden CM, Vallance TR. Sex hormones and the female urinary tract. Drugs. 1988;36:491–504.
11. Wahl LM, Blandau RJ, Page RC. Effect of hormones on collagen metabolism and collagenase activity in the pubic symphysis ligament of the guinea pig. Endocrinol. 1977;100:571–579.
12. MacLennan AH, MacLennan SC. Symptom-giving pelvic girdle relaxation of pregnancy, postnatal pelvic joint syndrome and developmental dysplasia of the hip. The Norwegian Association for Women with Pelvic Girdle Relaxation (Landforeningen for Kvinner Med Bekkenlosningsplager). Acta Obstet Gynecol Scand. 1997;76:760–764.
13. Calguneri M, Bird HA, Wright V. Changes in joint laxity occurring during pregnancy. Ann Rheum Dis. 1982;41:126–128.
14. Landon CR, Crofts CE, Smith ARB, et al. Mechanical properties of fascia during pregnancy: a possible factor in the development of stress incontinence of urine. Contemp Rev Obstet Gynaecol. 1990;2:40–46.
15. Lien KC, Mooney B, DeLancey JO, et al. Levator ani muscle stretch induced by simulated vaginal birth. Obstet Gynecol. 2004;103:31–40.
16. Noblett KL, Jensen JK, Ostergard DR. The relationship of body mass index to intra-abdominal pressure as measured by multichannel cystometry. Int Urogynecol J Pelvic Floor Dysfunct. 1997;8:323–326.
17. Viktrup L, Lose G, Rolff M, et al. The symptom of stress incontinence caused by pregnancy or delivery in primiparas. Obstet Gynecol. 1992;79:945–949.
18. Chaliha C, Kalia V, Stanton SL, et al. Antenatal prediction of postpartum urinary and fecal incontinence. Obstet Gynecol. 1999;94:689–694.
19. King JK, Freeman RM. Is antenatal bladder neck mobility a risk factor for postpartum stress incontinence? Br J Obstet Gynaecol. 1998;105:1300–1307.
20. Kristiansson P, Samuelsson E, von Schoultz B, et al. Reproductive hormones and stress urinary incontinence in pregnancy. Acta Obstet Gynecol Scand. 2001;80:1125–1130.
21. Cutner A, Cardozo LD, Benness CJ. Assessment of urinary symptoms in early pregnancy. Br J Obstet Gynaecol. 1991;98:1283–1286.
22. van Rooyen AJ. Pregnancy and the lower urinary tract. I. A study of histological changes in the lower urinary tract during pregnancy. S Afr Med J. 1969;43:749–753.
23. O'Boyle AL, Woodman PJ, O'Boyle JD, et al. Pelvic organ support in nulliparous pregnant and nonpregnant women: a case control study. Am J Obstet Gynecol. 2002;187:99–102.
24. Sze EH, Sherard GB III, Dolezal JM. Pregnancy, labor, delivery, and pelvic organ prolapse. Obstet Gynecol. 2002;100:981–986.
25. O'Boyle AL, O'Boyle JD, Ricks RE, et al. The natural history of pelvic organ support in pregnancy. Int Urogynecol J Pelvic Floor Dysfunct. 2003;14:46–49.

26. Rizk DE, Thomas L. Relationship between the length of the perineum and position of the anus and vaginal delivery in primigravidae. Int Urogynecol J Pelvic Floor Dysfunct. 2000;11:79–83.
27. Wijma J, Weis Potters AE, de Wolf BT, et al. Anatomical and functional changes in the lower urinary tract during pregnancy. BJOG. 2001;108:726–732.
28. Wijma J, Potters AE, de Wolf BT, Tinga DJ, Aarnoudse JG. Anatomical and functional changes in the lower urinary tract following spontaneous vaginal delivery. BJOG. 2003;110:658–663.
29. Meyer S, Bachelard O, De Grandi P. Do bladder neck mobility and urethral sphincter function differ during pregnancy compared with during the non-pregnant state? Int Urogynecol J Pelvic Floor Dysfunct. 1998;9:397–404.
30. Iosif S, Ingemarsson I, Ulmsten U. Urodynamic studies in normal pregnancy and in puerperium. Am J Obstet Gynecol. 1980;137:696–700.
31. van Geelen JM, Lemmens WA, Eskes TK, et al. The urethral pressure profile in pregnancy and after delivery in healthy nulliparous women. Am J Obstet Gynecol. 1982;144:636–649.
32. Sultan AH, Kamm MA, Hudson CN. Pudendal nerve damage during labour: prospective study before and after childbirth. Br J Obstet Gynaecol. 1994;101:22–28.
33. Rockner G, Jonasson A, Olund A. The effect of mediolateral episiotomy at delivery on pelvic floor muscle strength evaluated with vaginal cones. Acta Obstet Gynecol Scand. 1991;70:51–54.
34. Peschers UM, Schaer GN, DeLancey JO, et al. Levator ani function before and after childbirth. Br J Obstet Gynaecol. 1997;104:1004–1008.
35. Peschers U, Schaer G, Anthuber C, et al. Changes in vesical neck mobility following vaginal delivery. Obstet Gynecol. 1996;88:1001–1006.
36. Meyer S, Schreyer A, De Grandi P, et al. The effects of birth on urinary continence mechanisms and other pelvic-floor characteristics. Obstet Gynecol. 1998;92:613–618.
37. Snooks SJ, Swash M, Mathers SE, et al. Effect of vaginal delivery on the pelvic floor: a 5-year follow-up. B J Surg. 1990;77:1358–60.
38. Snooks SJ, Swash M, Henry MM, et al. Risk factors in childbirth causing damage to the pelvic floor innervation. Int J Colorectal Dis. 1986;1:20–24.
39. Snooks SJ, Setchell M, Swash M, et al. Injury to innervation of pelvic floor sphincter musculature in childbirth. Lancet. 1984;2:546–550.
40. Tetzschner T, Sorensen M, Rasmussen OO, et al. Pudendal nerve damage increases the risk of fecal incontinence in women with anal sphincter rupture after childbirth. Acta Obstet Gynecol Scand. 1995;74:434–440.
40a. Dolan LM, Hosker GL, Mallett VT, Allen RE, Smith AR. Stress incontinence and pelvic floor neurophysiology 15 years after the first delivery. BJOG. 2003;110:1107–1114.
41. Iosif S, Ulmsten U. Comparative urodynamic studies of continent and stress incontinent women in pregnancy and in the puerperium. Am J Obstet Gynecol. 1981;140:645–650.
42. Farrell SA, Allen VM, Baskett TF. Parturition and urinary incontinence in primiparas. Obstet Gynecol. 2001;97:350–356.
43. Viktrup L, Lose G. The risk of stress incontinence 5 years after first delivery. Am J Obstet Gynecol 2001;185:82–87.
44. Wilson PD, Herbison RM, Herbison GP. Obstetric practice and the prevalence of urinary incontinence three months after delivery. Br J Obstet Gynaecol. 1996;103:154–161.
45. Rortveit G, Daltveit AK, Hannestad YS, et al. Vaginal delivery parameters and urinary incontinence: the Norwegian EPINCONT study. Am J Obstet Gynecol. 2003;189:1268–1274.
46. Reilly ET, Freeman RM, Waterfield MR, et al. Prevention of postpartum stress incontinence in primigravidae with increased bladder neck mobility: a randomised controlled trial of antenatal pelvic floor exercises. BJOG. 2002;109:68–76.
47. Sultan AH, Kamm MA, Hudson CN, et al. Third degree obstetric anal sphincter tears: risk factors and outcome of primary repair. BMJ. 1994;308:887–891.
48. Wood J, Amos L, Rieger N. Third degree anal sphincter tears: risk factors and outcome. Aust N Z J Obstet Gynaecol. 1998;38:414–417.
49. Fitzpatrick M, Fynes M, Cassidy M, et al. Prospective study of the influence of parity and operative technique on the outcome of primary anal sphincter repair following obstetrical injury. Eur J Obstet Gynecol Reprod Biol. 2000;89:159–163.
50. Jander C, Lyrenas S. Third and fourth degree perineal tears. Predictor factors in a referral hospital. Acta Obstet Gynecol Scand. 2001;80:229–234.
51. Sultan AH, Kamm MA, Hudson CN, et al. Anal-sphincter disruption during vaginal delivery. N Engl J Med. 1993;329:1905–1911.
52. Chaliha C, Sultan AH, Bland JM, et al. Anal function: effect of pregnancy and delivery. Am J Obstet Gynecol. 2001;185:427–432.
53. Small KA, Wynne JM. Evaluating the pelvic floor in obstetric patients. Aust N Z J Obstet Gynaecol. 1990;30:41–44.
54. Cornes H, Bartolo DC, Stirrat GM. Changes in anal canal sensation after childbirth. Br J Surg. 1991;78:74–77.

55. MacArthur C, Bick DE, Keighley MR. Faecal incontinence after childbirth. Br J Obstet Gynaecol. 1997;104:46–50.
56. Zetterstrom JP, Lopez A, Anzen B, et al. Anal incontinence after vaginal delivery: a prospective study in primiparous women. Br J Obstet Gynaecol. 1999;106:324–330.
57. Abramowitz L, Sobhani I, Ganansia R, et al. Are sphincter defects the cause of anal incontinence after vaginal delivery? Results of a prospective study. Dis Colon Rectum. 2000;43:590–596.
58. Zetterstrom J, Lopez A, Anzen B, et al. Anal sphincter tears at vaginal delivery: risk factors and clinical outcome of primary repair. Obstet Gynecol. 1999;94:21–28.
59. Haadem K, Gudmundsson S. Can women with intrapartum rupture of anal sphincter still suffer after-effects two decades later? Acta Obstet Gynecol Scand. 1997;76:601–603.
60. Poen AC, Felt Bersma RJ, Dekker GA, et al. Third degree obstetric perineal tears: risk factors and the preventive role of mediolateral episiotomy. Br J Obstet Gynaecol. 1997;104:563–566.
61. Klein MC, Janssen PA, MacWilliam L, et al. Determinants of vaginal-perineal integrity and pelvic floor functioning in childbirth. Am J Obstet Gynecol. 1997;176:403–410.
62. Frudinger A, Halligan S, Spencer JA, et al. Influence of the subpubic arch angle on anal sphincter trauma and anal incontinence following childbirth. BJOG. 2002;109:1207–1212.
63. Fitzpatrick M, Behan M, O'Connell PR, et al. Randomised clinical trial to assess anal sphincter function following forceps or vacuum assisted vaginal delivery. BJOG. 2003;110:424–429.
64. Poen AC, Felt Bersma RJ, Strijers RL, et al. Third-degree obstetric perineal tear: long-term clinical and functional results after primary repair. Br J Surg. 1998;85:1433–1438.
65. Faltin DL, Sangalli MR, Roche B, et al. Does a second delivery increase the risk of anal incontinence? BJOG. 2001;108:684–688.
66. Oberwalder M, Connor J, Wexner SD. Meta-analysis to determine the incidence of obstetric anal sphincter damage. Br J Surg. 2003;90:1333–1337.
67. Woolley RJ. Benefits and risks of episiotomy: a review of the English-language literature since 1980. Part II. Obstet Gynecol Surv. 1995;50:821–835.
68. Woolley RJ. Benefits and risks of episiotomy: a review of the English-language literature since 1980. Part I. Obstet Gynecol Surv 1995;50:806–820.
69. Bansal RK, Tan WM, Ecker JL, et al. Is there a benefit to episiotomy at spontaneous vaginal delivery? A natural experiment. Am J Obstet Gynecol 1996;175:897–901.
70. Sleep J, Grant A, Garcia J, et al. West Berkshire perineal management trial. Br Med J (Clin Res Ed) 1984;289:587–90.
71. Sleep J, Grant A. West Berkshire perineal management trial: three year follow up. Br Med J (Clin Res Ed). 1987;295:749–751.
72. Reynolds JL. Reducing the frequency of episiotomies through a continuous quality improvement program. CMAJ. 1995;153:275–282.
73. Carroli G, Belizan J. Episiotomy for vaginal birth. Cochrane Database Syst Rev. 2000;Cd000081.
74. Anthony S, Buitendijk SE, Zondervan KT, et al. Episiotomies and the occurrence of severe perineal lacerations. Br J Obstet Gynaecol. 1994;101:1064–1067.
75. Henriksen TB, Bek KM, Hedegaard M, et al. Methods and consequences of changes in use of episiotomy. BMJ. 1994;309:1255–1258.
76. Abraham S, Child A, Ferry J, et al. Recovery after childbirth: a preliminary prospective study. Med J Aust. 1990;152:9–12.
77. Sartore A, De Seta F, Maso G, et al. The effects of mediolateral episiotomy on pelvic floor function after vaginal delivery. Obstet Gynecol. 2004;103:669–673.
78. Gardosi J, Sylvester S, B Lynch C. Alternative positions in the second stage of labour: a randomized controlled trial. Br J Obstet Gynaecol. 1989;96:1290–1296.
79. Bofill JA, Rust OA, Schorr SJ, et al. A randomized prospective trial of the obstetric forceps versus the M-cup vacuum extractor. Am J Obstet Gynecol. 1996;175:1325–1330.
80. Gupta JK, Hofmeyr GJ. Position for women during second stage of labour. Cochrane Database Syst Rev. 2004;Cd002006.
81. Stamp GE, Kruzins GS. A survey of midwives who participated in a randomised trial of perineal massage in labour. Aust J Midwifery. 2001;14:15–21.
82. Hodnett ED. Continuity of caregivers for care during pregnancy and childbirth. Cochrane Database Syst Rev. 2000;Cd000062.
83. Lal M, H Mann C, Callender R, et al. Does cesarean delivery prevent anal incontinence? Obstet Gynecol. 2003;101:305–312.
84. Tetzschner T, Sorensen M, Jonsson L, et al. Delivery and pudendal nerve function. Acta Obstet Gynecol Scand. 1997;76:324–331.
85. Henderson JS. Effects of a prenatal teaching program on postpartum regeneration of the pubococcygeal muscle. JOGN Nurs. 1983;12:403–408.

86. Hojberg KE, Salvig JD, Winslow NA, et al. Urinary incontinence: prevalence and risk factors at 16 weeks of gestation. Br J Obstet Gynaecol. 1999;106: 842–850.
87. Anonymous. Routine vs selective episiotomy: a randomised controlled trial. Argentine Episiotomy Trial Collaborative Group. Lancet. 1993;342: 1517–1518.
88. Harrison RF, Brennan M, North PM, et al. Is routine episiotomy necessary? Br Med J (Clin Res Ed). 1984;288:1971–1975.
89. Dannecker C, Hillemanns P, Strauss A, et al. Episiotomy and perineal tears presumed to be imminent: the influence on the urethral pressure profile, analmanometric and other pelvic floor findings–follow-up study of a randomized controlled trial. Acta Obstet Gynecol Scand. 2005;84: 65–71.
90. Donnelly V, Fynes M, Campbell D, et al. Obstetric events leading to anal sphincter damage. Obstet Gynecol. 1998;92:955–961.
91. Sheiner E, Levy A, Walfisch A, et al. Third degree perineal tears in a university medical center where midline episiotomies are not performed. Arch Gynecol Obstet. 2005;271:307–310.

1.4
Muscle Function and Ageing

Brenda Russell and Linda Brubaker

Key Messages

Normal muscle physiology in relation to the pelvic floor is discussed to enable an understanding of dysfunction of the pelvic floor, particularly of the anal and urethral sphincters, as a consequence of denervation, vaginal delivery, and ageing. Clinical, electrophysiological, and magnetic resonance imaging (MRI) assessments are reviewed, and the mechanisms of repair and rehabilitation are outlined.

Introduction

Several important pelvic functions rely on healthy, functioning muscles. The urinary stress continence system includes the striated pelvic floor muscles and urethral sphincter muscles, with both striated and smooth elements. Support of pelvic organs is also believed to rely on healthy muscle functions. This chapter will review clinically relevant aspects of normal and abnormal muscle, mechanisms of injury and repair, techniques for muscle assessment, and opportunities to alter muscle for prevention or treatment of pelvic floor disorders.

Overview of Normal Skeletal Muscle

Muscles are richly innervated so that contraction can be controlled physiologically to shorten and provide tension when necessary. Muscle tissue is the major consumer of energy in humans because it comprises over a third of the body mass and uses adenosine triphosphate (ATP) for its contractile functions. Nearly 85% of body heat is produced as the byproduct of muscle activity, making muscle a key player in thermal homeostasis. Extensive vascularization perfuses the tissue to provide oxygen and nutrients and remove waste. Muscle is also a major endocrine organ with an important role in metabolic regulation. Disuse of muscle in quadriplegics, for example, leads to type II diabetes.

Gross Anatomical Aspects

The gross organization of muscle tissue has the familiar bundles of muscle fibers wrapped in connective tissue layers.[1] These fascial layers are slippery so that the muscles slide over each other, enabling independence in function. Fibers terminate firmly on bone or anchor on adjacent fascia so that the force developed by the muscle can be transmitted to the surrounding structures. In limb muscles, the ends of the fibers move toward each other when force is produced, resulting in shortening. In situations where both ends are fixed, no relative movement can occur. This results in isometric contraction for production of tension without a change in length, thus, providing support.

An individual skeletal muscle fiber is a long, thin cell with many nuclei. This structure is formed by the fusion of many myoblasts as a fiber develops to make the multinucleated syncytia. There is no limit to the length of a fiber, and a single fiber can span from tendon to tendon in a limb muscle, being over a meter long in a tall person. The diameter of a muscle fiber ranges from 10 µm in the finger to over 1,000 µm in a body builder's biceps.

Structure and Function of Normal Skeletal Muscle

Muscle is highly anisotropic, yielding a very different appearance in the longitudinal and transverse planes of section. In the longitudinal orientation, with light microscopy we see that the bulk of the muscle fiber is striated (Fig. 1.4.1). The dark bands are called the A-bands, and the light bands are called the I-bands. In the middle of the I-band it is also possible to see the thin, dark Z-line. The unit of contraction is the sarcomere, which is defined between two adjacent Z-lines. In cross-section, the fibers are circular or polygonal (Fig. 1.4.2). Ultrastructure seen with electron

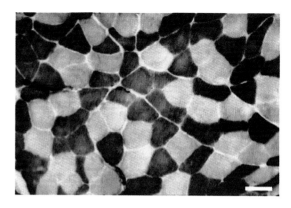

FIGURE 1.4.2. Human skeletal muscle cut in cross section and stained with a myofibrillar ATPase to show slow type I fibers as dark and fast type II fibers as light. Note the polygonal shape of the fibers and the mosaic mixture between the types. Scale bar, 50 µm. (*Source:* Brenda Russell, University of Illinois at Chicago.)

microscopy[2] reveals details about the striations and other organelles (Fig. 1.4.3). The striations are composed of interdigitating thick and thin filaments bundled into myofibrils. The thick myofilaments in the A-band are composed mainly of myosin molecules and the thin myofilaments in the I-band are polymers of actin. The thin myofilaments are anchored at the Z-line and extend into the A-band. The H-zone is the paler central region of the A-band into which the thin filaments have not penetrated. There are numerous additional molecules in the sarcomere, whose functions are to modulate contraction, transmit force effectively, sense mechanical changes,

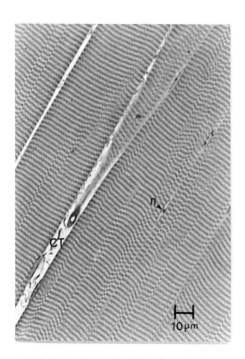

FIGURE 1.4.1. Light micrograph of skeletal muscle in longitudinal section. White vastus muscle of the guinea pig. Light micrograph of plastic-embedded muscle cut in a 0.5-cm-thick longitudinal section. Fibers are striated with A-bands (dark) and I-bands (light). Note peripherally located nuclei (n) and connective tissue (CT). Scale bar 10 µm. (*Source:* Reproduced with permission from Eisenberg, 1983.)

FIGURE 1.4.3. Low-power electron micrograph of rabbit skeletal muscle in longitudinal section. (*Source:* Reprinted from Williams PL. Grays Anatomy. 38th edition. 742. Copyright 1996 Churchill Livingston, with permission from Elsevier.)

1.4. Muscle Function and Ageing

signal the need for adaptation, and regulate assembly and disassembly in training or disuse.[1-4]

Contraction occurs when the thick and thin filaments slide past each other, making a shorter sarcomere with the I-band becoming smaller. Note that the filaments do not change in length. The molecular motors driving this contraction are the myosin cross-bridges, which cyclically bind, rotate, and detach from the actin filaments using energy provided by ATP hydrolysis. The force generated throughout the fiber at the billions of cross-bridges is transmitted by the intracellular and extracellular cytoskeleton. Forces go longitudinally through the ends of the fiber and laterally to the membrane at the Z-disc.

Muscle fibers have a very extensive membrane system that function to link the excitation of the outer sarcolemal membrane with the release of calcium from the inner membrane reticulum to initiate contraction (Fig. 1.4.4). A series of small transverse tubules (T-tubules) open to the extracellular space at the sarcolema and envelop the myofibrils twice, each sarcomere at the ends of the A-band. The T-tubules carry the action potential from the muscle surface to the interior of the fiber. The endoplasmic reticulum is a separate inner membrane system, called the sarcoplasmic reticulum, which surrounds the myofibrils over their entire length and pumps, stores, and releases calcium ions. The excitation spreads inwards through the T-tubule triggers and releases calcium via the junction between the T-tubules and sarcoplasmic reticulum. The pumps that return the calcium to the reticulum are located throughout the sarcoplasmic reticulum. Fat

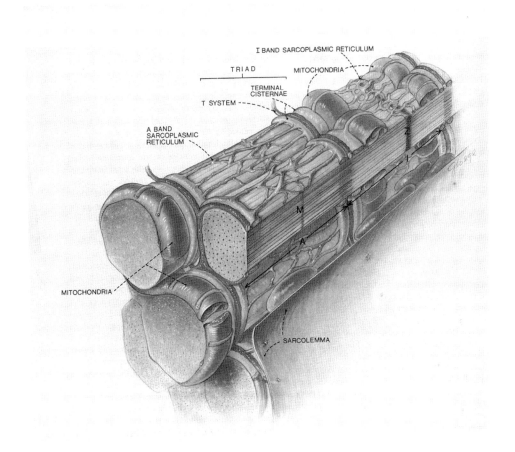

FIGURE 1.4.4. Schematic three-dimensional diagram showing the subcellular structure of skeletal muscle. (*Source:* Reproduced from The Journal of Cell Biology 1974;60:732–754, by copyright permission of the Rockefeller University Press.)

droplets and glycogen store energy, and rows of mitochondria lie in close proximity to the myofibrils, ready to supply ATP for the contractile process.

Normal Skeletal Muscle Adaptation

Normal muscle is highly adaptive so that the molecular composition and size of the fibers can be changed to match the needs of the person over time. Thus, training can easily increase the fiber strength by enlarging the fiber diameter, and improve endurance by increasing blood flow, mitochondrial content, and oxidative capacity. Conversely, disuse causes fiber atrophy and limits ability. In extreme situations, the speed of contraction can be altered by exchange of the isoforms of the contractile proteins.[2,5]

Individual fibers can be evaluated by techniques such as histochemistry, biochemistry, molecular biology, morphology, or physiology. Quantification by these methods has yielded enormous heterogeneity of function in the human muscles. Nonetheless, it has been useful to lump these functions and classify broad fiber types. The relative proportion of fiber types within a muscle varies from person to person and is partly determined by genetics. However, this ratio can be modified with exercise and training.

Muscle fibers that contract at a relatively slow speed and can sustain activity over long periods without loss of tension are called type I, slow-twitch, slow-oxidative, or slow fatigue-resistant fibers. They generate energy via aerobic metabolism from both fat and carbohydrates. Slow fibers have a high concentration of mitochondria, increased intracellular myoglobin to store and transport oxygen, and a rich capillary bed. Myoglobin, mitochondria, and blood all contain red pigments, so these slow fatigue-resistant muscles are red.

A twitch with a rapid contractile rise time defines the second major group as fast fibers, which are also known as type II fibers. They have myosin ATPase enzymes with a rate of shortening 3 to 5 times that of slow muscle fibers. Fast type II fibers are further subdivided into types IIA, IIB, and IIX, according to their metabolism and fatigue-resistant properties. Small lab animals and humans differ in these fast subtypes. In human, the fast-twitch muscle fibers are predominantly type IIX and IIA, but type IIB fibers with very low oxidative capacity are seldom found. Human fast fiber characteristics include moderate to high concentrations of mitochondria, fast myosin ATPase, and variable glycogen content.

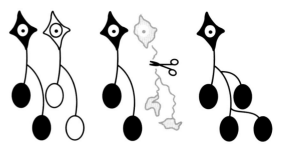

FIGURE 1.4.5. Diagram showing innervation of a slow, type I (dark) and fast, type II (light) fibers. When the nerve is cut (indicated by scissors in middle panel) the slow fibers atrophy. The denervated fibers are later reinnervated by sprigs from a surviving motor neuron and convert the fibers from their original slow to the new fast (dark) fibers resulting in a group of fibers with the same fiber type and loss of the normal mosaic pattern. (*Source:* Adapted from figure provided by Brenda Russell, University of Illinois at Chicago.)

Skeletal muscles perform complex motor functions. Those muscles lying deep or close to the bones provide posture and sustain tension over long periods of time and are predominantly slow-twitch. The most superficial muscles often have higher leverage over the joints and are recruited to perform rapid voluntary movements, such as throwing. These fibers in human are predominantly fast-fatigable type IIX. However, most muscles perform mixed functions and normally have different fiber types intermixed in a mosaic pattern (Fig. 1.4.5).

Within the Pelvic Floor

The pelvic floor and the sphincter muscles contract constantly, except during voiding or defecation.[6] This requires strength and fatigue resistance, but not speed, so as expected, these fibers have a high percentage of slow type 1 fibers. The pelvic floor muscles act as a functional group to keep the vagina, urethra, and anal canal closed while at rest. A pelvic muscle contraction assists with stabilization of the urethra and raises urethral pressure.[7,8] Changes in muscle function occur as a result of ageing and vaginal delivery.

In normal women, the urethral sphincter also contracts continuously, except during voiding, again, being best served by slow fatigue-resistant type I fibers.[9,10] With cough, the intraurethral pressure increases more and before the bladder pressure increases.[11] It is believed that disruptions in this urethral responsiveness may play a role in stress urinary incontinence.[12] In limb muscles, repetitive muscle contractions can cause fatigue.[13] Verelst et al. found no differences in the time to fatigue between continent and stress incontinent women.[14]

Mechanisms of Skeletal Muscle Weakness

Anything that affects production and transmission for useful mechanical work also limits muscle function. Normal skeletal muscle function obviously requires the motor machinery to contract, but also depends on innervation to provide the correct patterns of motor unit recruitment, vascular flow to provide the oxygen, and an intact connective tissue and skeleton for useful transmission of force.

Mechanical and Myopathic Damage

Extreme exercise, overstretch, or trauma can damage muscle tissue.[15] In the pelvic floor region, vaginal delivery with ripping of the muscle at its insertion or in the center is a common traumatic injury. Also, the birth process, with its extreme demands on endurance and partial nerve or vascular damage, might be expected to reduce muscle performance that would be further exacerbated by underlying genetic differences.

There have been great advances in genetic understanding in muscle diseases in the past decades. Many single nuclear polymorphisms (SNPSs) have been identified in myopathies and reproduced in transgenic mouse models.[16] Although over half of these genetic changes have been found in contractile proteins, they also affect force transmission, such as dystophin in muscular dystrophy. Given the importance of calcium handling to trigger contraction, and of metabolism to sustain it, changes in those genes can also compromise cardiac and skeletal muscle function. These SNPSs are more common than one might imagine, but are often masked until a second event, such as trauma or extreme exercise, or a pathological incident further stresses the muscle.

Neural

A common reason for lack of muscle function is pathology of innervation. This can be from any of the major neurological diseases, such as multiple sclerosis, but it can also arise from trauma to the spinal cord or peripheral nerves. Recovery depends on regeneration and reprogramming of the motor recruitment pathways.

Cardiovascular

It is often forgotten that continuous oxygenation is essential for endurance because of the constant demand for fuel by muscle fibers. Thus, patients with heart failure, pulmonary problems, or anemia will have secondary locomotor difficulties with continuous exercise, although short bursts are feasible. Regional blood flow can be compromised by many factors, including trauma or diabetes, which secondarily affect peripheral nerves.

Muscle Repair and Rehabilitation

Cellular Mechanisms of Repair

Muscle has a great ability to repair itself, and complete recovery can be expected as long as the underlying nerve and vascular systems are functional. Muscle fibers have a population of adult stem cells called satellite cells.[17-19] One in about every 20 muscle nuclei are actually these dormant satellite cells locally situated under the basement membrane, but not actually fused into the syncytium of the muscle fiber.[20] Local injury elicits an inflammatory response that is maximal at 3–5 days and which triggers these satellite cells to divide and to fuse with an adult fiber in the immediate vicinity.[21] Major damage results in the migration of satellite cells from surrounding healthy fibers. Note that research on adult stem cells is currently very active and the proliferation, mobilization, targeting, and differentiation into skeletal muscle is being rapidly characterized.[22] There is much hope that deeper understanding of these processes will permit improvement of repair in the future.

If a muscle tissue is strained to a breaking point, satellite cells will fuse and bridge the gap to reconnect healthy portions on the distal and proximal ends of the injury. The greatest amount of muscle repair occurs at the ends of the fibers,[23] although there is also repair in the central regions.[15,24]

Denervation leads to loss of muscle control and muscle atrophy. However, the denervated motor endplate regions are readily innervated by ingrowing dendrites from nearby healthy nerves. There is also potential for the formation of entirely new motor endplates by incoming axons. The reinnervated muscle fiber will receive the pattern of activity from the newly connected nerve and the fiber will change its type to meet this new pattern. This sprouting of nerves can result in groups of neighboring fibers sharing identical innervation and, hence, leading to a group of one kind of fiber rather than the usual mosaic seen in cross-section (Fig. 1.4.6).

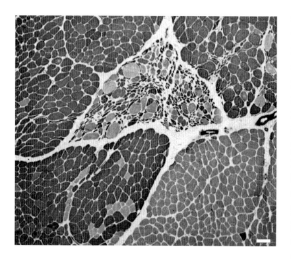

FIGURE 1.4.6. Human skeletal muscle of a patient with a long-standing denervation disorder. The muscle is cut in cross-section and stained with a myofibrillar ATPase to show slow type I fibers as dark and fast type II fibers as light. Note that the normal mosaic pattern has been replaced by denervation and reinnervation. Scale bar 100 μm. (*Source:* Brenda Russell, University of Illinois at Chicago.)

Clinical Aspects of Rehabilitation

The pelvic floor muscles may be a primary or secondary target of muscle rehabilitation. For most pelvic floor disorders, these muscles are a primary target, and are trained with progressive strengthening programs that eventually incorporate real-life strategies for using muscles to reduce urinary incontinent episodes. There is no consensus for the optimal technique, although a wide variety of programs appear to be adequate clinically. Long-term compliance remains a challenge.[25] Pelvic muscle training may also be a secondary target as part of trunk stabilization programs.[26,27] Interested readers are referred to these primary sources for further information.

Research in this area is often challenging because pelvic muscle training is typically only a part of the overall nonsurgical intervention. Commonly, a myriad of low-risk, but potentially high-impact, behavioral techniques are recommended simultaneously with muscle training. This is most logical from a clinical care standpoint. However, to date, the most important aspect of this treatment "package" remains unclear.

Clinical Assessment

Clinical assessment of muscle function is important for planning treatment, as well as for pelvic floor research. There is no consensus on the optimal method of muscle measurement for clinical or research purposes. There are several methods for assessing pelvic muscle function, including vaginal palpation, vaginal pressure measurement, electromyography, and transanal palpation. Although imaging plays a role, it remains a research tool at this time and will not be reviewed in detail here. Digital assessment is often performed in a casual clinical manner and may be sufficient for gross assessment of muscle function and symmetry. Several scoring systems are available for further refinement of the digital assessment. All of these hands-on measures have undergone some degree of testing, but none of them qualify or are accepted as the gold standard for assessing pelvic muscle function.

Vaginal palpation is the most commonly used method of assessing pelvic muscle function because it is simple to perform, requires no equipment, and costs little. Two methods for palpation have been described in the literature; the Oxford grading system[28] and the Brink scale.[29]

The Oxford grading system applies a 6-point categorical scale (0 = nil, 1 = flicker, 2 = weak, 3 = medium, 4 = strong, and 5 = very strong) for rating pelvic muscle strength. Determination of pelvic muscle strength using the Brink scale is based on three muscle contraction variables; intensity of the "squeeze" generated by the muscle contraction, vertical displacement of the examiners fingers as the muscles lateral to the vagina contract, and duration of the muscle contraction.

Bo and Finckenhagen reported the interrater reliability of the Oxford scale and compared Oxford ratings with vaginal pressure measures obtained from a vaginal balloon device using a small sample of young females with a mean age of 25 years, mostly nulliparous, and the majority without symptoms of pelvic floor dysfunction.[30] The Oxford scale had only fair intertester agreement. Vaginal squeeze pressure, measured by the balloon device, did not differ between women categorized as having weak, moderate, good, or strong muscle contractions; however, the study was limited by the small sample size.

Brink et al. reported the interrater reliability, test–retest reliability, and validity of the Brink scale from a sample of 208 women attending pretreatment clinic visits. Test–retest had moderate correlations ($r = 0.51–0.65$) for pressure, displacement, duration, and total score. Interrater reliability in 36 women was also moderate, ranging from 0.52 for duration to 0.74 for pressure.

A combined approach using visual and digital assessment was tested for interobserver reliability in two groups of women, continent and incontinent.[31] These investigators reported high reliability in both groups. One commonly cited limitation of digital grading systems is the lack of discrimination between clinically important groups.[32-35] Comparisons of the digital methods have been compared with intravaginal pressure measurements with conflicting findings.[36-38]

Electromyographic (EMG) measures of pelvic muscle strength typically use intravaginal surface probes that may receive signals from other pelvic muscles, such as gluteus or adductors. Digital assessments are only modestly correlated with EMG using surface patches.[29,39] The Oxford system and pressure perineometry in continent women has not shown good correlation in the single small study reporting these results.[38]

Vaginal dynamometers and perineometers have been introduced in an effort to provide a more standardized assessment of pelvic floor muscle strength. Morin et al. reported their experience in 89 premenopausal women who underwent a standardized assessment using the modified Oxford grading system and a dynamometric speculum.[40] These investigators confirm prior reports that digital grading often results in clinical overlap.[38] They reported that the correlation between the modified Oxford grading system and the dynamometric assessment was moderate to good.

Magnetic Resonance Imaging has been used for pelvic muscle assessment by several investigators.[41,42] DeLancey has reported MRI evidence of levator abnormalities in nearly 30% of women following vaginal delivery.[43] In a subset of such women, Miller et al. reported that women without visible pubococcygeus muscle in MRIs are less likely to be able to voluntarily increase their urethral closure pressure greater than 5 cm H20 on standardized urodynamic sphincter assessment.[44] Although the clinical significance of such findings requires further analysis, clearly the pubococcygeus is an important pelvic muscle. Hoyte et al.[45] has highlighted the differences in levator ani thickness in asymptomatic and symptomatic women using color thickness mapping technique (Fig. 1.4.7, A and B). However, given the expense of these investigations, MRI is not currently recommended as a clinical tool.

Bo et al. have reported a small series of female physical therapists who were assessed by traditional palpation techniques, as well as ultrasound.[46] In addition, they compared two sets of instructions – one aimed at the transverses abdominis and one aimed at pelvic floor muscles. They suggest that in such healthy women, ultrasound is more valid than palpation and that instructions aimed at pelvic floor muscles are more useful than instructions aimed at transverses abdominis muscles. As with much research in pelvic floor muscle assessment, this finding awaits reproduction in a clinical population. Thompson and O'Sullivan used ultrasound in

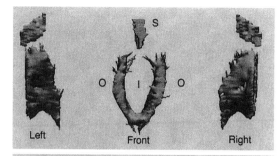

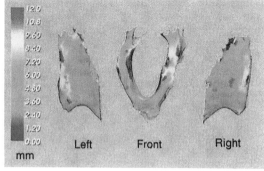

FIGURE 1.4.7. **A.** A reconstructed levator is shown in brown, with the symphysis (*S*). Views of the front, left, and right sides are shown. In the front view, the inner layer faces the label *I*, and the outer layer faces the label *O*. Additional images and video can be found online at www.us.elsevierhealth.com/ajog. **B.** Three views of a color-mapped levator are shown. The color bar on the left shows the thickness (in millimeters) corresponding to the colors seen. Additional images and video can be found online at www.us.elsevierhealth.com/ajog. (*Source:* Reprinted from Hoyte et al. Copyright 2004, with permission from Elsevier.)

104 women with incontinence and/or prolapse to assess levator plate movement.[42] The relationship between the instructions to elevate the levator plate and the ultrasound appearance was variable and inconsistent. Relationships to specific clinical conditions, such as prolapse, require further analysis. However, the investigators raise awareness that muscle dysfunction may play a role in the genesis, propagation, or treatment outcomes of common pelvic floor disorders such as prolapse and incontinence. Bernstein reported that ultrasound is able to detect increasing muscle thickness in women who are compliant with pelvic muscle training.[47] He also confirmed that levator muscles are thicker in asymptomatic women than those with stress urinary incontinence. This finding is consistent with the later work using MRI techniques.

Clinical Relevance

Vaginal parity affects the appearance and function of the pelvic floor muscles.[48-50] Multiple authors have reported statistically significant decreases in the pelvic floor muscle strength of incontinent women compared with continent controls.[51-53] Bo compared the pelvic floor muscle strength of physical education students in Norway and was unable to detect significant differences in strength in her small sample.[54] In several clinical case series, investigators have reported that women may not correctly perform a voluntary contraction of their pelvic floor muscles.[7,55-58]

Opportunities for Prevention or Clinical Treatment

Several investigators have assessed the role of pelvic muscle training in the prevention of symptoms, especially during pregnancy.[59-61] In addition, one novel randomized trial assessed the effect of pelvic muscle training on the duration of the second stage of labor, finding that although training did not reduce the duration of the second stage overall, women in the training group had a lower rate of prolonged second stage labor (training group 34% vs. no training 21%).[62]

Kegel is credited with the first observation that a structured training program may reduce clinical symptoms of incontinence.[63] He described a strengthening program that is now known to increase muscle tone and may elevate or improve support of the pelvic organs in selected women.

A recent Cochrane review concludes that pelvic floor muscle exercise is consistently superior to no treatment (or placebo).[64] Miller et al. report that performing a voluntary contraction of the pelvic floor muscles before and during increases in abdominal pressure reduced cough leakage by 73–98%, depending on the strength of the cough.[65] A separate Cochrane review regarding weighted vaginal cones for urinary incontinence suggests that there may be a role for these devices in conservative treatment.[66] These reviewers recommend that larger, well-conducted studies should be performed in order to detect differences in several techniques of muscle training.

The optimal techniques for muscle training in the pelvis are not established. However, experts agree that individualized programs are essential. Such programs must take into account the current skills and strength of the individual woman, and further strength training must be specific for each individual. Sapford et al. report that contraction is associated with an increase in pelvic floor muscle activity in health volunteers.[67] Sapsford et al. further reported that abdominal contraction is a normal response to pelvic floor exercises in health subjects.[68] Neumann and Gil also reported this finding.[69] Thus, the often taught technique of abdominal relaxation during pelvic muscle contraction may be incorrect. In Neumann and Gil's small series of four nulliparous lean women, testing was performed twice, one week apart; they noted that strong pelvic floor muscular contractions invariably activated the abdominal wall muscles. To date, these findings have not been reproduced in a clinical population. Sapsford emphasizes the importance of trunk stabilization in training[27] although this belief has not undergone scientific testing. Beliefs and biases strongly impact descriptions of pelvic floor muscle training, similar to the impact of expert opinion and bias in many other areas of pelvic floor health. This area is ready for multiple high quality research trials to determine the optimal techniques for pelvic muscle training.

Normal Ageing Changes

Systemic changes in the nerve and cardiovascular systems result in concomitant decline in muscle function.[70] However, the muscle fibers themselves become atrophic because some growth factors and hormones decline with age.[71] At present, there are intense studies on growth hormones to enhance muscle growth with age and in disease. The hope for a boost in the muscle growth rate that can compensate for the deficits in muscular dystrophies is a major instigator for these hypertrophy studies. The insulin growth factor has muscle-specific isoforms that are clearly important for increasing muscle mass.[72] Furthermore, in males, testosterone jumps as puberty begins, but reverses in the latter decades of life, which can be compensated for by anabolic and natural steroids. Nonetheless, the major application for muscle mass increases is in the sports industry where steroids and other drugs are widely abused.

The regenerative properties of most tissues decline with age because of changes in the stem cells and to environmental cues from the surrounding extracellular matrix. The finding that youthfulness can be returned to old satellite cells has long been known to depend on the age of the tissue matrix, which is partially attributable to the finding that remodeling of the extracellular hydrolytic and proteolytic systems differs in muscles of old animals compared to young animals.[73] In muscle, satellite cell activation and cell fate determination are controlled by the Notch signaling pathway.[74] Interestingly, the reduced activation and proliferative capacity of satellite cells in the elderly human population means that the same bout of exercise leads to a smaller increase in muscle mass and damage is much less readily repaired.[75] These ageing processes may be tightly coupled to the muscle-specific growth factors, which are important for the division of the satellite cells.[76]

Clinical Impact of Muscle Dysfunction

Striated urethral sphincteric contraction is an important part of maintaining normal continence. It may also play a role in compensatory techniques for maintaining continence. This is particularly true during moments of increased abdominal pressure, such as cough. This has led investigators to describe the "cough anal reflex."[6,77]

The role of muscle dysfunction in pelvic floor disorders is often discussed as an important etiological contributor. Stress urinary incontinence is increasingly recognized as a neuromuscular disorder with concomitant anatomical alterations. Regardless of "which came first," measurable neuromuscular dysfunction is documented by multiple modes of evaluation.

Amarenco et al. describe the relationship between intravesical pressure and pelvic floor muscle electromyographic activity using both urodynamic assessment and electrophysiologic techniques. In a small series, 16 women who were referred for frequency and/or urgency without urge or stress incontinence underwent urodynamics with patch electrodes recording from the three and nine o'clock perianal positions.[78] At five different volumes (0, 100, 200, 300, and 400) a series of successively increasing coughs occurred. These investigators report a close relationship between the intravesical pressure (used to compare the intensity of the cough effort) and the integrated perianal EMG signal. They reported that bladder filling did not alter the recruitment curve, but, instead, the perianal EMG signal increased only with increases in intravesical pressure.

Aukee et al. reported their series of 31 women with urodynamic stress incontinence and a comparable group of 35 women without urinary symptoms.[79] Using vaginal EMG measures, they recorded three rapid contractions in both supine and standing positions. They reported a significant difference across groups, with incontinent women demonstrating a lower mean decrease in EMG activity with the standing position (incontinent, 12.9 [5.0–33.0] vs. continent, 18.2 [8.0–43.5]). Because of the small number of subjects and the use of a relatively indirect assessment with the vaginal EMG probe, they could not substantiate further claims about muscle dysfunction. However, their data is suggestive of differences between groups by continence status.

Future Directions

Multiple investigators are creating three-dimensional models of the pelvic muscles to simulate the clinical situations of birth and surgery with the hope of further understanding the pathophysiology of pelvic floor disorders.[45,80,81] Our understanding of pelvic muscle morphology and function is critical to the accurate creation of appropriate models. Janda et al. describe their efforts to obtain a complete data set to describe the morphological parameters of the pelvic floor muscles structures using a single embalmed cadaver from a 72-year-old female. Through a carefully described series of sequential techniques, the investigators attempted to measure and summarize muscle parameters for all muscles and all muscle elements of the levator ani complex.[80] These investigators have demonstrated the technical feasibility of transferring cadaveric data to an MRI data set for ultimate use in modeling. However, DeLancey et al. have previously pointed out the limitations of certain embalming techniques and the significant artifacts that may arise from measurements and observations using cadaveric specimens that have been embalmed in the supine, non-suspended positions.[82]

Conclusion

The muscles of the pelvic floor are an intriguing area of physiology. In health, these muscles assist with pelvic functions, including urinary and bowel control, as well as sexual function. In disease, these muscles may suffer from primary muscular injury or secondary dysfunction when neuropathy is present. Given their central role in pelvic floor health, continued emphasis on understanding the anatomy, physiology, and pathophysiology is important. Research regarding the optimal methods of muscle assessment will be an essential prerequisite to meaningful prevention and clinical treatment. Our understanding of optimal clinical treatment requires high-quality clinical trials in well-characterized populations.

References

1. Salmons S. Striated muscle. In: Gray's Anatomy. 38th Edition. Bannister LH, Berry MM, Collins P, et al., editors. London: Churchill Livingstone; 1995:739.
2. Eisenberg BR. Section 10. Quantitative ultrastructure of mammalian skeletal muscle. In: Handbook of Physiology. Peachey LD and Adrian RH, editors. Bethesda, MD: American Physiological Society; 1983:73112.
3. Russell B, Motlagh D, Ashley WW. Form follows function: how muscle shape is regulated by work. J Appl Physiol. 2000;88:1127 1132.
4. Tidball JG. Mechanical signal transduction in skeletal muscle growth and adaptation. J Appl Physiol. 2005;98:1900–1908.
5. Eisenberg BR, Salmons S. The reorganization of subcellular structure in muscle undergoing fast to slow type transformation: a stereological study. Cell Tissue Res. 1981;220:449–471.
6. Deindl FM, Vodusek DB, Heese U, et al. Activity patterns of pubococcygeal muscles in nulliparous continent women. Br J Urol. 1993;72:46–51.

7. Bump RC, Hurt WG, Fantl JA, et al. Assessment of Kegel exercise performance after brief verbal instruction. Am J Obstet Gynecol. 1991;165:322–329.
8. Bo K, Talseth T. Change in urethral pressure during voluntary pelvic floor muscle contraction and vaginal electrical stimulation. Int Urogynecol J Pelvic Floor Dysfunct. 1997;8:3–7.
9. Hale DS, Benson T, Brubaker L, et al. Histologic analysis of needle biopsy of urethral sphincter from women with normal and stress incontinence with comparison of electomyographic findings. Am J Obstet Gynecol. 1999;180:342–348.
10. Heidkamp MC, Leong FC, Brubaker L, et al. Pudendal denervation affects the structure and function of the striated, urethral sphincter in female rats. Int Urogynec J. 1998;9:88–93.
11. Constantinou CE, Govan DE. Spatial distribution and timing of transmitted and reflexly generated urethral pressures in healthy women. J Urol. 1982;127:964–969.
12. Petros PE, Ulmsten U. Role of the pelvic floor in bladder neck opening and closure I: muscle forces. Int Urogynecol J Pelvic Floor Dysfunct. 1997;8:74–80.
13. Lieber RL. Physiologic basis of fatigue. In: Butler JP, editor. Skeletal muscle. Structure and function. Baltimore:Williams and Wilkins; 1992:100–105.
14. Verelst M, Leivseth G. Are fatigue and disturbances in pre-programmed activity of pelvic floor muscles associated with female stress urinary incontinence? Neurourol Urodyn. 2004;23:143–147.
15. Lieber RL, Friden J. Mechanisms of muscle injury gleaned from animal models. Am J Phys Med Rehabil. 2002;81:S70–S79.
16. Chien KR, Karsenty G. Longevity and lineages: toward the integrative biology of degenerative diseases in heart, muscle, and bone. Cell. 2005;120:533–544.
17. Partridge T. Reenthronement of the muscle satellite cell. Cell. 2004;119:447–448.
18. Sherwood RI, Christensen JL, Conboy IM, et al. Isolation of adult mouse myogenic progenitors: functional heterogeneity of cells within and engrafting skeletal muscle. Cell. 2004;119:543–544.
19. Hawke TJ, Garry DJ. Myogenic satellite cells: physiology to molecular biology. J Appl Physiol. 2001;91:534–551.
20. Kennedy JM, Eisenberg BR, Reid SK, et al. Nascent muscle fibre appearance in overloaded chicken slow tonic muscle. Am J Anat. 1988;181:201–215.
21. Pizza FX, Peterson JM, Baas JH, et al. Neutrophils contribute to muscle injury and impair its resolution after lengthening contractions in mice. J Physiol. 2005;562:899–913.
22. Dreyfus PA, Chretien F, Chazaud B, et al. Adult bone marrow-derived stem cells in muscle connective tissue and satellite cell niches. Am J Pathol. 2004;164:773–779.
23. Dix DJ, Eisenberg BR. Myosin mRNA accumulation and myofibrillogenesis at the myotendinous junction of stretched muscle fibers. J Cell Biol. 1990;111:1884–1894.
24. Yu JG, Carlsson L, Thornell LE. Evidence for myofibril remodeling as opposed to myofibril damage in human muscles with DOMS: an ultrastructural and immunoelectron microscopic study. Histochem Cell Biol. 2004;121:219–227.
25. Chiarelli P, Murphy B, Cockburn J. Women's knowledge, practises and intentions regarding correct pelvic floor exercises. Neurourol Urodyn. 2003;22:246–249.
26. Graves JE, Webb DC, Pollock ML, et al. Pelvic stabilization during resistance training: its effect on the development of lumbar extension strength. Arch Phys Med Rehabil. 1994;75:210–215.
27. Sapsford R. Rehabilitation of pelvic floor muscles utilizing trunk stabilization. Manual Therapy. 2004;9:3–12.
28. Laycock J. Assessment and treatment of pelvic floor dysfunction [master's thesis]. West Yorkshire, UK: University of Bradford; 1992.
29. Brink CA, Wells TJ, Sampselle CM, et al. A digital test for pelvic muscle strength in women with urinary incontinence. Nurs Res. 1994;43:352–356.
30. Bo K, Finckenhagen HB. Vaginal palpation of pelvic floor muscle strength: inter-test reproducibility and comparison between palpation and vaginal squeeze pressure. Acta Obstet Gynecol Scand. 2001;80:883–887.
31. Devreese A, Staes F, De Weerdt W, et al. Clinical evaluation of pelvic floor muscle function in continent and incontinent women. Neurourol Urodyn. 2004;23:190–197.
32. Jundt K, Reckemeyer I, Namdar N, et al. Changes in pelvic floor function after childbirth. Neurourol Urodyn. 2002;21:364–365.
33. Blowman D, Pickles C, Emery S, et al. Prospective double blind controlled trial of intensive physiotherapy with and without stimulation of pelvic floor in treatment of genuine stress incontinence. Physiotherapy. 1991;77:661–664.
34. Schwartz S, Cohen ME, Herbison GJ, et al. Relationship between two measures of upper extremity strength: manual muscle test compared to hand-held myometry. Arch Phys Med Rehabil. 1992;73:1063–1068.

35. Herbison GJ, Cohen ME, et al. Strength post-spinal cord injury: myometer vs. manual muscle test. Spinal Cord. 1996;34:543–548.
36. Hahn I, Milsom I, Ohlsson BL, et al. Comparative assessment of pelvic floor function using vaginal cones, vaginal digital assessment and vaginal pressure measurements. Gynecol Obstetric Invest. 1996;41:269–274.
37. Kerschan-Schindl K, Uher E, Wiesinger G, et al. Reliability of pelvic floor muscle strength measurement in elderly incontinent women. Neurourol Urodyn. 2002;21:42–47.
38. Bo K, Fickenhagen HB. Vaginal assessment of pelvic floor muscle strength: Inter-test reproducibility and comparison between assessment and vaginal squeeze pressure. Acta Obstet Gynecol Scand. 2001;80:883–887.
39. Romanzi L, Polaneczky M, Glazer HI. Simple test of pelvic muscle contraction during pelvic examination: Correlation to surface electromyography. Neurourol Urodyn. 1999;18:603–612.
40. Dumoulin D, Bourbonnais D, Lemiuex M-C. Development of a dynameter for measuring the isometric force of the pelvic floor musculature. Neurourol Urodyn. 2003a;22:648–653.
41. Thompson DL, Smith DA. Continence nursing: a whole person approach. Holist Nurs Pract. 2002;16:14–31.
42. Thompson JA, O'Sullivan PB. Levator plate movement during voluntary pelvic floor muscle contraction in subjects with incontinence and prolapse: a cross-sectional study and review. Int Urogynecol J. 2003;14:84–88.
43. DeLancey JO, Kearney R, Chou Q, et al. The appearance of levator ani muscle abnormalities in magnetic resonance images after vaginal delivery. Obstet Gynecol. 2003;101:46–53.
44. Miller JM, Umek WH, DeLancey JOL, et al. Can women without visible pubococcygeal muscle in MR images still increase urethral closure pressures? Am J Obstet Gynecol. 2004:171–175.
45. Hoyte L, Jakab M, Warfield SK, et al. Levator ani thickness variations in symptomatic and asymptomatic women using magnetic resonance-based 3-dimensional color mapping. Am J Obstet and Gynecol. 2004;191:856–861.
46. Bo K, Sherburn M, Allen T. Transabdominal ultrasound measurement of pelvic floor muscle activity when activated directly or via a transversus abdominis muscle contraction. Neurourol Urodyn. 2003; 22:582–588.
47. Bernstein IT. The pelvic floor muscles: muscle thickness in healthy and urinary-incontinent women measured by perineal ultrasonography with reference to the effect of pelvic floor training. Estrogen receptor studies. Neurourol Urodyn. 1997;16:237–275.
48. Miller JM, Perucchini D, Carchidi LT, et al. Pelvic floor muscle contraction during a cough and decreased vesical neck mobility. Obstet Gynecol. 2001;97:255–260.
49. Peschers U, Schaer G, Anthuber C, et al. Changes in vesical neck mobility following vaginal delivery. Obstet Gynecol. 2001;88:1001–1006.
50. Haderer JM, Pannu HK, Genadry R, et al. Controversies in female urethral anatomy and their significance for understanding urinary continence: observations and literature review. Int Urogynecol J. 2002;13:236–252.
51. Gunnarsson M. Pelvic floor dysfunction: a vaginal surface EMG study in healthy and incontinent women. Lund, Sweden: Lund University; 2002.
52. Hahn I, Milsom I, Ohlson BL, et al. Comparative assessment of pelvic floor function using vaginal cones, vaginal digital palpation and vaginal pressure measurement. Gynecol Obstetric Invest. 1996; 41:269–274.
53. Morkved S, Salvesen K, Bo K, et al. Pelvic floor muscle strength and thickness in continent and incontinent nulliparous women. Neurourol Urodyn. 2002;21:358–359.
54. Bo K, Stien R, Kulseng-Hanssen S, et al. Clinical and urodynamic assessment of nulliparous young women with and without stress incontinence symptoms: a case-control study. Obstet Gynecol. 1994;84:1028–1032.
55. Benvenuti F, Caputo GM, Bandinelli S, et al. Reeducative treatment of female genuine stress incontinence. Am J of Phys Med. 1987;66:155–168.
56. Bo K, Larsen S, Oseid S, et al. Knowledge about the ability to correct pelvic floor muscle exercises in women with urinary stress incontinence. Neurourol Urodyn. 1988;7:261–262.
57. Hesse U, Schussler B, Frimberger J, et al. Effectiveness of a three step pelvic floor reeducation in the treatment of stress urinary incontinence: a clinical assessment. Neurourol Urodyn. 1990;9:397–398.
58. Kegel AH. Stress incontinence and genital relaxation. Ciba Clin Sympos. 1952;2:35–51.
59. Morkved S, Bo K, Schei B, et al. Pelvic floor muscle training during pregnancy to prevent urinary incontinence: a single-blind randomized controlled trial. Obstet Gynecol. 2003;101:313–319.
60. Reilly ET, Freeman RM, Waterfield MR, et al. Prevention of postpartum stress incontinence in primigravidae with increased bladder neck mobility: a randomised controlled trial of antenatal pelvic floor exercises. BJOG. 2002;109:68–76.

61. Sampselle CM, DeLancey JO. The urine stream interruption test and pelvic muscle function. Nurs Res. 1992;41:73–77.
62. Salvesen KA, Morkved S. Randomised controlled trial of pelvic floor muscle training during pregnancy. BMJ. 2004;329:378–380.
63. Kegel AH. Progressive resistance exercise in the functional restoration of the perineal muscles. Am J Obstet Gynecol. 1948;56:238–249.
64. Hay-Smith E, Bo K, Berghmans L, et al. Pelvic floor muscle training for urinary incontinence in women. Available in the Cochrane Library [database on disk and CD ROM]. Updated quarterly. The Cochrane Collaboration, 2001.
65. Miller JM, Ashton-Miller JA, DeLancey JO. A pelvic muscle precontraction can reduce cough-related urine loss in selected women with mild SUI. J Am Geriatr Soc. 1998;46:870–874.
66. Herbison P, Plevnik S, Mantle J. Weighted vaginal cones for urinary incontinence. [update of Cochrane Database Syst Rev. 2000;(2):CD002114; PMID: 10796862]. Cochrane Database of Systematic Reviews, 2002;1.
67. Sapsford RR, Hodges PW. Contraction of the pelvic floor muscles during abdominal maneuvers. Archs Phys Med Rehab. 2001;82:1081–1088.
68. Sapsford RR, Hodges PW, Richardson CA, et al. Co-activation of the abdominal and pelvic floor muscles during voluntary exercises. Neurourol Urodyn. 2001;20:31–42.
69. Neumann P, Gill V. Pelvic floor and abdominal muscle interaction: EMG activity and intra-abdominal pressure. Int Urogynecol J. 2002;13:125–132.
70. Brooks SV, Faulkner JA. Skeletal muscle weakness in old age: underlying mechanisms. Med Sci Sports Exerc. 1994;26:432–439.
71. Goldspink G, Harridge SD. Growth factors and muscle ageing. Exp Gerontol. 2004;39:1433–1438.
72. Goldspink G. Age-related loss of skeletal muscle function; impairment of gene expression. J Musculoskelet Neuronal Interact. 2004;4:143–147.
73. Bar-Shai M, Carmeli E, Coleman R, et al. The effect of hindlimb immobilization on acid phosphatase, metalloproteinases and nuclear factor-kappaB in muscles of young and old rats. Mech Ageing Dev. 2005;126:289–297.
74. Conboy IM, Rando TA. Ageing, stem cells and tissue regeneration: lessons from muscle. Cell Cycle. 2005;4:407–410.
75. Goldspink G, Harridge SD. Growth factors and muscle ageing. Exp Gerontol. 2004;39(10):1433–1438.
76. Machida S, Booth FW. Insulin-like growth factor 1 and muscle growth: implication for satellite cell proliferation. Proc Nutr Soc. 2004;63:337–340.
77. Bo K, Stien R. Needle EMG registration of striated urethral wall and pelvic floor muscle activity patterns during cough, Valsalva, abdominal, hip adductor, and gluteal muscle contractions in nulliparous healthy females. Neurourol Urodyn. 1994;13:35–41.
78. Amarenco G, Ismael SS, Lagauche D, et al. Cough anal reflex: strict relationship between intravesical pressure and pelvic floor muscle electromyographic activity during cough. Urodynamic and electrophysiological study. J Urol. 2005;173:149–152.
79. Aukee P, Penttinen J, Airaksinen O. The effect of ageing on the electromyographic activity of pelvic floor muscles. A comparative study among stress incontinent patients and asymptomatic women. Maturitas. 2003;44:253–257.
80. Janda S, van der Helm FC, de Blok SB. Measuring morphological parameters of the pelvic floor for finite element modelling purposes. J Biomech. 2003;36:749–757.
81. Parikh M, Rasmussen M, Brubaker L, et al. Three dimensional virtual reality model of the normal female pelvic floor. Ann Biomed Eng. 2004;32: 292–296.
82. DeLancey JO, Sampselle CM, Punch MR. Kegel dyspareunia: levator ani myalgia caused by overexertion. Obstet Gynecol. 1993;82:658–659.

1.5
Urinary Incontinence and Voiding Dysfunction

Annette Kuhn and Bernhard Schüssler

Key Messages

In this chapter, normal bladder function and its neurological control, as well as the types of incontinence, are classified and explained. Voiding dysfunction is similarly classified, and its symptoms, etiology, and outline of treatment are presented.

Introduction

Physiological bladder function consists of two phases; the storage phase, defined by the International Continence Society (ICS) as "the phase in which the bladder volume increases without a significant rise in pressure (accommodation)," and the voiding phase as when "a bladder contraction is initiated and bladder neck, urethra, and pelvic floor are relaxed in synchrony."[1]

These definitions are what they should be: short, precise, and easy to comprehend. In contrast, the physiological mechanism of bladder function is complicated and not yet completely elucidated.

It consists of two separate reflex circuits for storage and voiding, which the cortical brain is able to control and modify. This voluntary interference is distinct from most other visceral organs, such as the heart, intestine, etc., which are regulated exclusively by involuntary (autonomic) reflexes. As with the anal sphincter, cortical control highlights the fact that voluntary sphincter control of these two organs is a mandatory prerequisite for adequate social behavior.

The hardware driving these reflexes consists of different centers within the central nervous system, as well as of a network of autonomous and somatic nerves; as a consequence, the system involves smooth and skeletal muscles.

Based on careful perception of an individual's bladder cycle during filling, storage, and voiding, a few examples may help for a better understanding of the close entanglement of all of these structures:

- No sensation at the very beginning of bladder filling indicates a central threshold of bladder perception.
- First sensation of the fluid in the bladder and its voluntary suppression is carried out by the cortical control mechanism.
- Earlier and more intensive urgency to void in situations of psychic agitation apparently indicates connections to emotional areas within the brain. This mechanism is also active in situations when the bladder is adequately filled and voiding is urgently intended, but the individual could not voluntarily initiate the process because of social embarrassment (e.g. passing urine during urodynamic assessment).
- Voluntary suppression of an urgent need to void by contracting the pelvic floor muscle for a short period of time, relaxation of the pelvic floor as a mandatory prerequisite for the initiation of voiding, and the ability to interrupt voiding voluntarily by a contraction of the

pelvic floor muscles all indicate a connection between the autonomic system and the skeletal muscle system and brain, the latter one further highlighted by recent PET studies that revealed that the most medial portion of the motor cortex is activated during pelvic floor contraction.[2]

Physiology of Storage and Voiding Phase

As shown in Figure 1.5.1, A and B, storage and voiding of urine is controlled by 2 reflex arcs and involve descending nerves, the thoracolumbar sympathetic and pelvic parasympathetic nerves, the pudendal nerves, and the afferent nerve fibers. These circuits act as "On/Off" switches that are controlled by the presence or absence of glutamate at the level of the sacral micturition center. The central control of these reflexes is located in the brain stem at the pontine micturition center and the pontine voiding center.

The pelvic parasympathetic nerves, which arise at the sacral level of the spinal cord, excite the bladder. The thoracolumbar sympathetic nerves, which arise at the thoracolumbar level of the spinal cord, inhibit the bladder body and excite the bladder base and urethra. The pudendal nerves, which arise at S3–S5 (Onuf's nucleus), excite the urethral sphincter. Direct somatic nerve branches, which stem from the same area, excite the pelvic floor muscles. All of these nerves contain sensory (afferent) fibers.

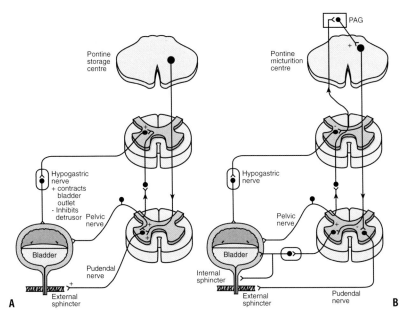

FIGURE 1.5.1. **A.** Storage reflex. During the storage phase, distension of the bladder produces low-level vesical afferent firing, which stimulates the sympathetic outflow of the bladder outlet and pudendal outflow to the external urethral sphincter. These responses occur by spinal reflex pathways and represent "guarding reflexes," which promote continence. (*Source*: Adapted with permission from Abrams PH, Cardozo L, Khoury S, Wein A, editors. Incontinence. Plymouth, UK: Health Publication Ltd.; 2005:370.)

B. Voiding reflex. During elimination of urine, intense bladder afferent firing activates spino–bulbo–spinal pathways passing through the pontine micturition center, which stimulates the parasympathetic outflow to the bladder and inhibits the sympathetic and pudendal outflow to the urethral outlet. (*Source*: Adapted with permission from Abrams PH, Cardozo L, Khoury S, Wein A, editors. Incontinence. Plymouth, UK: Health Publication Ltd.; 2005:370.)

During storage of urine while glutamate is present at Onuf's nucleus, distension of the bladder wall produces low-level vesical afferent firing, which in turn stimulates sympathetic outflow to the bladder outlet (bladder neck and urethra) and outflow of the pudendal nerves to urethral sphincter and direct branches to the pelvic floor respectively, increasing the tone of smooth and skeletal muscle at these sites. These responses occur by spinal reflex pathways and represent the "guarding reflexes," which promote continence. It is further enhanced before a cough and other events challenging the continence mechanism, thus, preparing the sphincter mechanism adequately. Sympathetic firing also inhibits the detrusor muscle and modulates transmission in bladder ganglia. A region in the pontine storage center increases urethral sphincter activity (Fig. 1.5.1 A).

During elimination of urine while glutamate is absent, intense bladder afferent firing activates spino–bulbo–spinal pathways passing through the pontine micturition center, which stimulates the parasympathetic outflow to the bladder muscle and internal sphincter smooth muscle, thus, inhibiting the sympathetic and pudendal outflow to the urethral outlet (Fig. 1.5.1 B).

Relaxation of the urethral smooth muscle is also mediated by activation of a parasympathetic pathway to the urethra that triggers the release of nitric oxide, which is an inhibitory transmitter, and by the removal of adrenergic and somatic cholinergic excitatory inputs to the urethra.[3]

Studying these reflexes by constantly measuring intravesical pressure during filling and voiding parallel to a sphincter/pelvic floor electromyograph (EMG), as well as uroflow, illustrates how these reflex circuits interact with each other in harmony (Fig. 1.5.2 A).

As already mentioned, centers above the pontine micturition center control and voluntarily interfere with these two reflex circuits. Although lesions in this area are not able to abolish the reflex activity, they do, however, disconnect cortical interference. This is why patients with cortical brain damage (e.g., after a stroke) may suffer from detrusor overactivity and/or may lose the ability to delay voiding at an appropriate place and time. Lesions below the pontine micturition center and above the level of the sacral reflex arc, as they occur in paraplegic patients, lead to reflex incontinence combined with a detrusor sphincter dyssynergia (Fig. 1.5.2 B).

Recently, interest has been focused on the layer as a further independent mode of bladder function. To date, there is good evidence, at least from animal data, that there are local interactions between the smooth muscle, urothelium, afferent

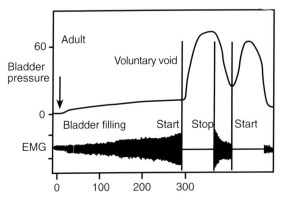 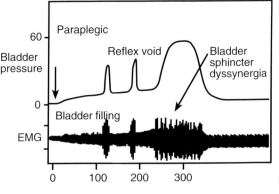

FIGURE 1.5.2. A. Voluntary voiding response in an adult. On the left side of the trace the arrows indicate the start of the bladder filling. Note increased pelvic floor activity (EMG) with increasing bladder filling. No pelvic floor activity during voiding is physiological. (Source: Reproduced with permission from Abrams PH, Cardozo L, Khoury S, Wein A, editors. Incontinence. Plymouth, UK: Health Publication Ltd.; 2005:371.) B. Voiding in a paraplegic patient. The reciprocal relationship between bladder and sphincter is abolished and detrusor sphincter dyssynergia occurs. During bladder filling, transient uninhibited bladder contractions occur in association with sphincter activity. Further filling leads to more prolonged and simultaneous contractions of the bladder and sphincter. (Source: Reproduced with permission from Abrams PH, Cardozo L, Khoury S, Wein A, editors. Incontinence. Plymouth, UK: Health Publication Ltd.; 2005:371.)

nerve terminals, and various neurotransmitters. The result seems to be an information system for the central nervous system and for phasic activity of the bladder wall during filling. The latter might be able to better explain the phenomena of bladder wall compliance, as it is not yet well understood how the bladder is able to keep normal bladder pressure increments as low as 30–40 cm H_2O during bladder filling of 400 ml and more.

Incontinence

Incontinence is a pathological condition of the storage phase and could be divided into functional and anatomical types. Various epidemiological studies have shown that stress urinary incontinence is the predominant type. Its prevalence has been shown to range between 27–50%.[4,5] The overactive bladder syndrome (OAB) ranges between 14%[5] and 26%,[6] and patients suffering from mixed incontinence, e.g. a combination of stress urinary incontinence and the OAB, contribute between 22%[4] and 55%.[6]

Overflow incontinence is a rare form of urinary incontinence, especially in females, mostly a result of a longstanding overdistension of the bladder either by subvesical obstruction, e.g. prolapse or obstructive incontinence surgery, or caused by bladder denervation following radical pelvic surgery.

Fistulae, as well as remnants of the embryonic kidney system, may also lead to incontinence, but are not functional disorders.

Stress Urinary Incontinence

Stress urinary incontinence is defined by the ICS as the complaint of involuntary leakage on effort or exertion, or on sneezing and coughing, at the absence of any bladder contraction.[1] This implies that weakness of the urethral closure mechanism is the underlying pathology of this condition.

At rest, involuntary urinary leakage is prevented by the maximum urethral closure pressure, which ranges between 50 and 80 cm H_2O and decreases with age.[7] As the bladder pressure – as a result of bladder wall compliance – is kept very low, even when the bladder is full, very low urethral closure pressures are sufficient for continence. Voluntary pelvic floor contraction could increase this pressure, as could contraction of the striated sphincter muscle. The situation changes when physical stress (e.g., sneezing and coughing) increases the intraabdominal pressure dramatically and, depending on its extent, easily exceeds any urethral closure pressure. The intactness of the following two mechanisms is critical: the "guarding reflex," which actively increases muscle tone, and the simultaneous passive transmission of the abdominal pressure (e.g. during a cough) to the urethra. It is obvious that the greater the loss of pressure transmission to the urethra the greater the likelihood of stress incontinence. It is also clear that a very low resting closure pressure needs excellent pressure transmission to maintain continence. Pressure transmission is dependent not only on muscle tone but also on the elasticity of the connective fibers both within and adjacent to the muscle. The laxer and more mobile the pelvic floor and urethra are, the less likely it is that continence is maintained in a stress situation.

When stress urinary incontinence occurs it is the result of three main causes:

1. Pregnancy and delivery (see Chapter 1.3)
2. Ageing (see Chapter 1.4)
3. The individual continence threshold (see Chapter 1.5, Fig. 1.5.2).

It is obvious that the lower the individual threshold, the more an individual is prone to develop stress urinary incontinence later in life.

There is good evidence that the closure threshold varies individually. This is supported by data from young nulliparous female athletes who present evidence for sphincter insufficiency in high-impact situations. Urine leakage occurred in 50% of them while playing tennis, 67% during gymnastics, and 29% during athletics.[8]

There are other risk factors besides high-impact sports, such as pregnancy,[9] parity, and an increased body mass index,[10,11] as well as influences such as cigarette smoking, hysterectomy, and menopause, but evidence in support of these factors is mixed.[12]

As stress continence is the result of adequate urethral closure pressure and a stabilized urethra at the very moment when stress occurs, decreased closure pressure (intrinsic urethral insufficiency) and hypermobility, both of which can occur with a funneled bladder neck, are the

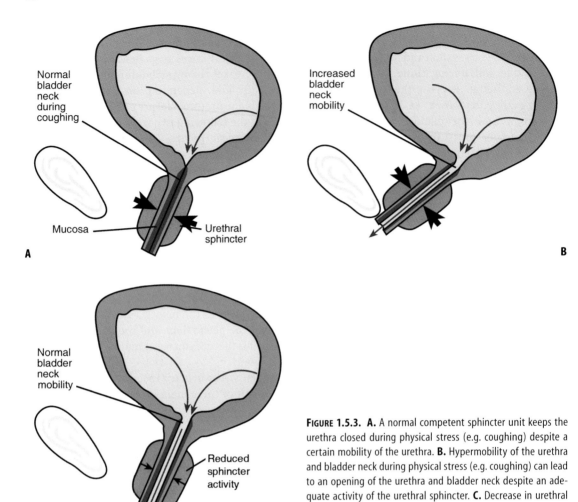

FIGURE 1.5.3. **A.** A normal competent sphincter unit keeps the urethra closed during physical stress (e.g. coughing) despite a certain mobility of the urethra. **B.** Hypermobility of the urethra and bladder neck during physical stress (e.g. coughing) can lead to an opening of the urethra and bladder neck despite an adequate activity of the urethral sphincter. **C.** Decrease in urethral sphincter activity can lead to an opening of the urethra and bladder neck without increase of normal mobility of the urethra.

critical challenges to continence. When stress urinary incontinence occurs, either of these issues may be the underlying pathology alone, but it is usually a combination of both (Fig. 1.5.3, A–C).

It is obvious that in patients with stress urinary incontinence caused by a low-pressure urethra (intrinsic sphincter insufficiency) alone, physiotherapy is less likely to be successful. The same is true for another entity of stress urinary incontinence that is often the final result of recurrent incontinence surgery or pelvic irradiation. In these cases, the urethra is scarred and completely immobile, mostly fixed to the back surface of the pubic bone. Consequently, this form of stress urinary incontinence is often combined with voiding disorder.

Overactive Bladder Syndrome

Overactive bladder syndrome is defined by the ICS as any involuntary detrusor contraction during the filling phase, which may be spontaneous or provoked, and that the patient cannot completely suppress.[1]

The OAB occurs either with (OAB/wet) or without (OAB/dry) urinary leakage. Overactive

TABLE 1.5.1. Common causes for the development of an overactive bladder

Gynecological	Prolapse
	Pregnancy
	Pelvic surgery
	Radiation
	Atrophy
	Pelvic mass
Urological	Intravesical tumor, Carcinoma in situ
	Interstitial cystitis
	Chronic urinary retention
	Tuberculosis
	Bladder calculus or other foreign bodies
Infections	Urinary tract infection
Endocrine	Diabetes mellitus
	Diabetes insipidus
Medical	Drugs
	Congestive heart failure
	Impaired renal function
	Neurological disorders
Psychological	Anxiety
	Drinking habit

bladder syndrome/wet refers to motor and sensory urge incontinence, whereas OAB/dry refers to frequency and urgency syndrome with the unsuppressible need to void as a predominant symptom. The maximal capacity of the bladder is almost always reduced. Consequently, these patients suffer from a bothersome increased daytime and nighttime frequency. Frequency and urgency caused by high fluid intake (which is fashionable today and sometimes exceeds more than three liters per day) is not a pathological condition, as long as bladder capacity at micturition is normal.

The etiology of the OAB is not completely understood today. Although there are diseases that are accompanied by OAB, in the majority of patients it is an idiopathic condition (Table 1.5.1).

Urodynamically, two different forms ob OAB/wet could be distinguished: Motor urge incontinence and, very rarely, sensory urge incontinence. In motor urge incontinence, an involuntary detrusor contraction leads to a rise in bladder pressure and, hence, to an immediate incontinence episode. High-pressure motor urge episodes are more likely to be affiliated with neurological diseases.

Sensory urge incontinence totally mimics the voiding phase apart from the fact that it occurs much earlier and without the ability of the individual to suppress the need to void. The best example for sensory urge incontinence is an acute urinary tract infection. OAB/dry may or may not be accompanied by autonomous detrusor contractions.

Mixed Incontinence

Mixed incontinence is a combination of stress urinary incontinence and overactive bladder. It has been defined by the ICS as the complaint of involuntary leakage of urine associated with urgency and also with exertion and effort, like sneezing or walking.[1] The diagnosis is usually made after careful review of the patient's history. In patients with a predominant motor urge incontinence component presenting with a low bladder capacity, care has to be taken to diagnose mixed incontinence. In these cases, neither the patient's symptoms nor clinical examination or urodynamics are able to categorize the stress component correctly.

Between individuals with mixed incontinence, both components vary in regard to the extent of the incontinence and the bother to the patient, which has implications for the treatment.

Voiding Dysfunction

Voiding in the female occurs as a result of a synchronized autonomic contraction of the detrusor muscle and relaxation of urethral smooth and skeletal muscles. Although relaxation of the pelvic floor muscle is necessary for an unimpeded micturition, voluntary gradual contraction is able not only to steer urine outflow but also to influence the direction of the urine stream. Increasing the abdominal pressure by a Valsalva maneuver also contributes to an increase in urine flow.

Retention occurs when the basic mechanism fails, e.g., when the detrusor is unable to contract, the urethra fails to relax, additional infravesical resistance is present, or when there is a failure in the synchronization of detrusor contraction and urethral relaxation.

Voiding normally leads to an empty bladder without residual urine. However, constant residual urine whatever the origin and its extent is, is a medical concern only if leading to symptoms that can mimic a variety of other lower urinary tract pathology (Table 1.5.2).

Although not as prevalent as in men, voiding difficulties in women are not uncommon, with prevalence rates of 12–17%.[14,15]

Chronic retention is the condition of a nonpainful bladder that remains palpable or percussable after the patient has passed urine. It commonly presents with residual urine of at least 50% of normal capacity.

An acute retention is the condition of a palpable or percussable bladder when the patient is unable to pass any urine and requires catheterization. Whether it is painful or painless depends on the neurological status of the patient.

The etiology of voiding dysfunction is representing defects at any conceivable level of the normal micturition reflex, expanding from mechanical infravesical obstruction as the most common cause via muscular and neurogenic failure up to psychogenic causes (Table 1.5.3).

Voiding dysfunction, even if it is caused by infravesical obstruction, does not exclude symptoms of any kind of urinary incontinence, and the patient may end up with incontinence once the pressure of the overdistended bladder wall exceeds urethral pressure (overflow incontinence). Based on the underlying pathology, physiotherapy may play an important role in reestablishing micturition or improving voiding dysfunction. Pelvic floor relaxation techniques may not only improve detrusor/pelvic floor dyssynergia conditions but also emptying of a hypocontractile bladder.

Prevention and Treatment

Prevention or early recognition of retention may avoid long term voiding difficulties. Not only

TABLE 1.5.2. Classification of voiding difficulties and retention

Condition	Symptom
Asymptomatic voiding difficulty	Frequency
	Urgency due to urinary infection or no symptoms
Symptomatic voiding difficulty	Poor stream
	Incomplete emptying
	Straining
	Frequency
Acute retention	Painful or painless
	Sudden onset
Chronic retention	Reduced sensation
	Hesitancy
	Straining to void
	Frequency, nocturia
	Urgency, incontinence
	Urinary tract infection
Acute chronic retention	Painful or painless
	Sudden onset
	Incontinence

Source: Modified from Shag and Dasgupta, 2002.[13]

TABLE 1.5.3. Etiology of voiding dysfunction

Etiology	Mechanism
Postsurgical causes	Any continence procedure or surgery close to the bladder neck can provoke retention by physical obstruction and postoperative edema.
Prolapse	Because of urethral kinking, cystourethrocele may cause inability to void.[16]
Acute inflammation	Patients with acute vulvitis, cystitis, urethritis may retain urine because of painful micturition.
Pharmacological	Tricyclic antidepressants, alpha-adrenergic substances, anticholinergics and spinal or epidural anesthetic.
Age	Widespread degenerative changes in all components of the detrusor, as smooth muscle cells and axons are the morphological equivalent for impaired contractility.
Neurological	Upper or lower motor neurone disease
Fowler's syndrome	A primary defect within the striated urethral sphincter which is hypertrophied and fails to relax.[17]
Detrusor/pelvic floor dyssynergia	This is a learned or functional voiding dysfunction, which sometimes is also called nonneurogenic neurogenic bladder. The patient contracts her pelvic floor while there is a detrusor contraction, which leads to an intermittent stream and high micturition pressure.
Psychogenic	Psychogenic causes for impaired voiding are well known in hysteria.[18,19]
Infravesical obstruction	Urethral stenosis is uncommon in female patients, but possible after traumatic catheterization. Distal urethral stenosis may be caused by urogenital atrophy in postmenopausal women. Extrinsic causes for obstruction may be fibroids.[16]
Endocrine	Diabetic neuropathy

long-time but also acute overdistension of the detrusor may result in chronic voiding disorders because of detrusor muscle damage.[20]

Therefore, surgery which may lead to voiding difficulties, e.g. incontinence operations or extensive reconstructive or radical pelvic surgery, will require prophylactic postoperative bladder drainage. Suprapubic catheterization is superior to a transurethral catheter, as it allows easy trials of postoperative voiding without traumatic transurethral recatheterization if voiding attempts are unsuccessful. The catheter can be clamped and unclamped according to the patient's progress.

If adequate voiding could not be reestablished, clean intermittent self-catheterization is the primary treatment for long-term voiding difficulties. Patients who are properly counseled by professional staff will easily learn the technique, which allows them to lead an independent, normal life.

Sacromodulation involves stimulation of the S3 root using a nerve stimulator. Encouraging early results have been reported.

References

1. Abrams P, Blaivas JC, Stanton SL. Standardization of terminology of lower urinary tract function. Neurourol Urodyn. 2002;21:167–178.
2. Blok BF, Willemsen AT, Holstege G. A PET study on brain control of micturition in humans. Brain. 1997;120:111.
3. Bennett BC, Kruse MN, Rappolo JR, et al. Neural control of urethral outlet activity in vivo: the role of nitric oxide. J Urol. 1995;53:2004–2009.
4. Diokno AC, Brock BM, Brown MB, et al. Prevalence of urinary incontinence and other urologic symptoms in the noninstitutionalized elderly. J Urol. 1986;136:1022–1025.
5. Hannestad YS, Rortveid G, Sandvik H, et al. A community base epidemiological survey of female urinary incontinence: the Norvegian EPICONT study. J Clin Epidemiol. 2000;53:1150–1157.
6. Wells TJ, Brink CA, Diokno AC. Urinary incontinence in the elderly women: clinical findings. J Am Geriatr Soc. 1987;35:933–939.
7. Sorensen S, Waechter PB, Constantinou CE, Kirkeby HJ, Djuruus JC. Urethral pressure and pressure variations in healthy fertile and postmenopausal women with reference to the female sex hormones. J Urol. 1991;146(5):1434–1440.
8. Nygaard IE, Thompson FL, Svengalis SL, et al. Urinary incontinence in elite nulliparous athletes. Obstet Gynecol. 1994;84(3):342.
9. Hunskaar S, Arnold EP, Burgio K, et al. Epidemiology and natural history of urinary incontinence. Int Urogyn J Pelvic Floor Dysfunct. 200;11:301–319.
10. Rortveit G, Daltveit AK, Hannestad YS, et al. Urinary incontinence after vaginal delivery or caesarean section. New Engl J Med. 2003;348:900–907.
11. Sampselle CM, Harlow SD, Skurnick J, et al. Urinary incontinence predictors and life impact in ethnically diverse perimenopausal women. Obstet Gynecol. 2002;100(6):1230–1238.

12. Sampselle CM, Hunskaar S, Burgio K, et al. Epidemiology and natural history of urinary incontinence. In: Incontinence. Second edition. Abrams P, Cardozo L, Khoury S, et al., editors. Plymouth, UK: Plymbridge Distributors Ltd.; 2002.
13. Shag PJR, Dasgupta P. Voiding difficulties and retention in Clinical Urogynaecology. Stanton SL, Monga A, editors. London: Churchill Livingstone; 2000.
14. Stanton SL, Ozsoy C, Hilton P. Voiding difficulties in the female: prevalence. Clinical and urodynamic review. Obstet Gynecol. 1983;61:144–147.
15. Brieger GM, Yip SK, Lin LY, et al. The prevalence of urinary dysfunction in Hong Kong Chinese women. Obstet Gynecol. 1996;88:1041–1044.
16. Monga AK, Woodhouse C, Stanton SL. A case of simultaneous urethral and ureteric obstruction. Br J Urol. 1996;77:606–607.
17. Fowler CJ, Kirby RS. Abnormal electromyographic activity in the striated urethral sphincter in 5 women with urinary retention. Br J Urol. 1985;57:67–70.
18. Barrett DM. Evaluation of psychogenic urinary retention. J Urol. 1978;120:191–192.
19. Krane R, Siroky M. Psychogenic voiding dysfunction In: Clinical neurourology. Krane R, Siroky M, editors. Boston: Little, Brown & Co; 1978:275–284.
20. Dean AM, Worth PHL. Female chronic urinary retention. Br J Urol. 1985;57:24–26.

1.6
Pelvic Organ Prolapse

Peggy A. Norton

Key Messages

In this chapter, the prevalence of prolapse and the causative factors are reviewed. Important features of the clinical examination are described, and the objective Pelvic Organ Prolapse Quantification (POP-Q) and Schüssler Quantitative systems of clinical examination are illustrated and explained.

Introduction

Pelvic organ prolapse is a common clinical entity that may affect 1 in 3 women, where the pelvic structures lose their support and descend to and beyond the vaginal introitus. In the United States, it accounts for 3 times as many surgical procedures as stress urinary incontinence, despite the fact that much of the condition is managed without surgery.[1] The number of women who seek treatment for pelvic floor disorders is expected to increase by approximately 50% in the near future.[2]

Clinical Findings

Pelvic organ prolapse includes defects of the anterior vaginal wall (cystocele), vaginal apex (uterine prolapse, vaginal vault prolapse, and enterocele), and posterior vaginal wall (rectocele and enterocele.) The major symptoms associated with POP include a sensation of fullness, pressure, bulging, or a lump in the vagina or at the introitus. Patients report difficulties with voiding or defecation, which is sometimes relieved by splinting with a finger or thumb to effect emptying, lower back pain, which improves after lying down, and dyspareunia. Unlike urinary and fecal incontinence, POP is difficult to assess in study populations by symptom questionnaire; this means that for epidemiological or outcome research into POP an examination is the only way to assess whether a woman has the condition or not. Two quality of life instruments specific to POP have been described;[3] although these instruments correlate with physical findings in women with POP presenting for evaluation, they have not been validated as outcome measures or as population screening tools.

Etiology and Risk Factors

Similar to other pelvic floor disorders, the etiology of POP is multifactorial.[4] The biggest risk factor remains childbirth injury, but contributing factors include conditions of chronic increased intraabdominal pressure: chronic pulmonary disease (asthma, COPD), chronic constipation, obesity, and occupations involving heavy lifting.[5] The natural history of POP remains to be described; it is known that some women have significant levels of pelvic organ descent without symptoms, and that some women have identifiable defects that do not progress over many years of observation. Several large-scale studies have included a physical assessment for POP with some surprising findings; 30% of parous women in the Heart and Estrogen/Progestin Replacement study had a cystocele to within 1 cm of the hymenal ring, and of these, only 30% were symptomatic. Thus, an anatomical finding of descent from "normal" pelvic support may have no clinical significance unless the woman has associated symptoms.

Examination

A vaginal examination with a speculum is necessary for the evaluation of POP. A bivalve Graves or a Sims speculum is used to identify the cervix or vaginal cuff after hysterectomy, and is withdrawn while the woman bears down until the point of maximum descent is reached (i.e., the cervix or cuff descends no further). All distances are measured in centimeters relative to the hymenal ring, the remnants of which can be identified in all women. Although the vaginal apex (cervix or cuff) is optimally found at 8–10 cm above the hymenal ring, there can be considerable descent toward the hymen without symptoms. The posterior vaginal wall is retracted with a Sims speculum or the posterior half of a Graves speculum, while watching the anterior wall descend to its maximum point, and is used again to retract the anterior wall while watching the posterior wall descend. Finally, the genital hiatus (distance from midurethra to posterior fourchette) and perineal body (distance from posterior fourchette to midanal opening) are measured. All measurements are taken with the patient while straining/Valsalva; examining a patient in the supine lithotomy position demonstrates less prolapse than examining in a birthing chair at a 45 degree sitting position or in the standing position.[6] Some clinicians examine in the standing position as well if the supine or sitting positions fail to reproduce the woman's symptoms of bulging.

The International Continence Society devised the POP-Q system in 1994 in an effort to standardize the measurement of POP.[7] The system records two points in the anterior wall, two points in the posterior wall, one to two points in the apex, total vaginal length, and the measurements of the genital hiatus and perineal body (Fig. 1.6.1). The leading edge of the prolapse defines the overall stage, with stage II being defined as the leading edge being within 1 cm of the hymenal ring (above or below), stage IV being complete eversion of the vagina, stage III being prolapse greater than stage II but not complete eversion,

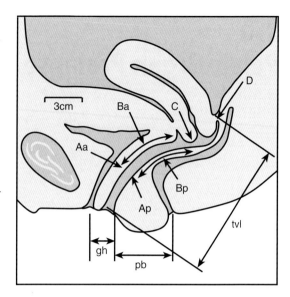

FIGURE 1.6.1. The pelvic organ prolapse quantification system for description of pelvic organ prolapse. In POP-Q, all measurements are relative to the hymen. The two anterior vaginal wall measurements are Aa and Ba. Aa is the point 3 cm cephalad (internal) from external urethral meatus, meant to represent the location of the bladder neck. Ba is the point in the anterior wall which descends most. The apical measurements are C, D, and total vaginal length or TVL. C measures the position the uterine cervix or in the case of hysterectomy, the vaginal cuff, while D measures the posterior fornix when the uterus is present. The two posterior vaginal wall measurements are Ap and Bp. Ap is the point 3 cm cephalad from the posterior forchette (at the hymen), and Bp is that point in the posterior wall which descends most. Additionally, genital hiatus (gh) and perineal body length (pb) are recorded.[7]

stage 0 being no prolapse, and stage I being prolapse less than stage II (not to within 1 cm of the hymenal ring.) Although some specialist gynecologists would prefer a system with more detail that includes vaginal capacity, bladder neck mobility, and the presence of a lateral wall defect, the POP-Q system may be too detailed for other clinicians. One useful modification is the Schüssler diagram (Fig. 1.6.2), devised by Schüssler as part of the original POP-Q system.[8] In this diagram, the vagina is represented diagrammatically with an obtuse representation of the introitus, and an outline of the leading edge of the prolapse with including measurements. Unless precise measurements are needed for research purposes, the measurements for prolapse, including a Schüssler diagram, should take no more than a few minutes.

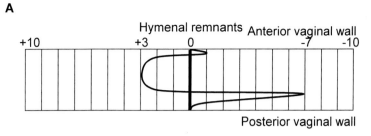
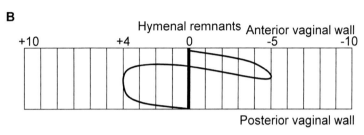

FIGURE 1.6.2. Schüssler diagram demonstrating **(A)** a stage III anterior vaginal wall prolapse (Aa + 3 and Ba + 3) and **(B)** a stage III posterior vaginal wall prolapse (Ap + 3 and Bp + 4).

Descending Perineal Syndrome

Perineal descent refers to the movement of the perineal body (the area between the posterior fourchette of the vagina and the anus) with valsalva, especially during evacuation. At rest, the perineal body should be several centimeters above the ischial tuberosities, which are bony prominences that are easily palpated in supine lithotomy. Terminal branches of the pudendal nerve insert into the perineal body. In childbirth, the fibromuscular wall of the rectovaginal septum may become detached from the perineal body; thus, with any straining the perineal body descends several centimeters below or beyond the ischial tuberosities, further injuring these nerves as they stretched with the movement of their insertion site.[9] It has been suggested that the clinical finding of perineal descent may actually represent a hernia of the levator ani.[10] Some surgeons have referred to a "perineal rectocele" and have advocated repair by pulling bulbocavernosus tissue to the midline in a surgical repair and reattaching the fibromuscular wall to its insertion site on the perineal body.[11] Loss of pelvic floor muscle support may lead to more than POP. Defecation can become disordered as the stool bolus follows the path of least resistance and pushes the rectovaginal septum and perineal body in a caudal direction. The pudendal nerve endings on the perineal body undergo significant stretching during such straining to move the stool bolus back to the anal sphincter, thus, causing a chronic nerve damage in addition to that seen with childbirth injury. Such perineal descent may be compensated by instructing patients to manually splint the perineum or rectovaginal septum during defecation. The Mayo Clinic reviewed their experience with pelvic floor education for descending perineum syndrome;[12] many patients still experienced constipation or excessive straining two years after their regimen for pelvic floor training, especially where the perineal body move more than 4 cm (relative to the ischial spines) with straining.

POP and the Pelvic Floor

Does pelvic floor muscle dysfunction cause POP? This has been the subject of much conjecture, but little definitive research. The hypothesis is well founded; as long as the muscles remained intact, pelvic organs were supported without tension on their suspensory ligaments. This idea originated as long ago as the beginning of the twentieth century with Paramore, who suggested that the support of pelvic organs was similar to a boat in drydock (Fig 1.6.3), where the water (pelvic floor muscles) supports the boat (pelvic structures), while the moorings (connective tissue supports

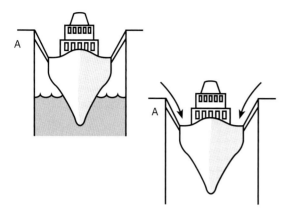

FIGURE 1.6.3. Boat in drydock concept for the development of pelvic organ prolapse.

such as ligaments) help to suspend and stabilize the boat. If the water is withdrawn to work on the underside of the boat, much strain is placed on the moorings, which now support the entire boat through its suspensions. Similarly, loss of some or all of the pelvic floor support may place excessive strain on the connective tissue suspensions in the pelvis. Although good pelvic floor function would be expected to preserve pelvic organ support, it is unknown whether such muscle training could be expected to improve support, once damage (stretching or breaking) of the suspensory components has occurred.

Although it may be difficult to show a cause and effect relationship for pelvic floor dysfunction and POP, a related question is whether pelvic floor muscles prevent POP. The American College of Obstetrics and Gynecology's patient education bulletin on pelvic relaxation recommends that patients employ Kegel exercise to improve their POP, although no published studies exist to confirm this. If the pelvic floor prevents undue burden on the pelvic suspensory/connective tissue components, then one might hypothesize that strengthening the pelvic floor would have a very positive impact on prolapse. However, appreciation of a damaged pelvic floor may come late in the process; if the injury to the muscles is at the time of childbirth and the woman is able to compensate somewhat to achieve normal function, the lack of support may begin the process of POP. Teaching these muscles to contract better after the prolapse has developed is unlikely to reverse the process of POP.

References

1. Olsen A, Smith V, Bergstrom J, et al. Epidemiology of surgically managed pelvic organ prolapse and urinary incontinence. Obstet Gynecol. 1997;89:501–506.
2. Luber K, Boero S, Choe J. The demographics of pelvic floor disorders: current observations and future projections. Am J Obstet Gynecol. 2001; 184:1496–1501.
3. Barber MD, Kuchibhatla MN, Pieper CF, et al. Psychometric evaluation of 2 comprehensive condition-specific quality of life instruments for women with pelvic floor disorders. Am J Obstet Gynecol. 2001;185(6):1388–1395.
4. Bump R, Norton P. Epidemiology and natural history of pelvic floor dysfunction. Obstet Gynecol Clin North Am. 1998;25(4):723–46.
5. Spence-Jones C, Kamm M, Henry M, Hudson C. Bowel dysfunction: a pathogenic factor in utero-vaginal prolapse and urinary stress incontinence. Br J Obstet Gynaecol. 1994;101(2):147–152.
6. Barber MD, Lambers A, Visco AG, et al. Effect of patient position on clinical evaluation of pelvic organ prolapse. Obstet Gynecol. 2000;96(1):18–22.
7. Viereck V, Peschers U, Singer M, Schuessler B. Metrische Quantifizierung des weiblichen Genitalprolapses: Eine sinnvolle Neuerung in der Prolapsdiagnostik? GeburtshFrauenheik. 1997;57: 177–182.
8. Bump R, Mattiasson A, BO K, et al. The standardization of terminology of female pelvic organ prolapse and pelvic floor dysfunction. Am J Obstet Gynecol. 1996;175:10–17.
9. Henry M, Parks A, Swash M. The pelvic floor musculature in the descending perineum syndrome. Br J Surg. 1982;69:470–472.
10. Gearhart SL, Pannu HK, Cundiff GW, et al. Perineal descent and levator ani hernia: a dynamic magnetic resonance imaging study. Dis Colon Rectum. 2004;47(8):1298–1304.
11. Richardson A. The anatomic defects in rectocele and enterocele. J Pelvic Surg. 1995;4:214–221.
12. Harewood GC, Coulie B, Camilleri M, et al. Descending perineum syndrome: audit of clinical and laboratory features and outcome of pelvic floor retraining. Am J Gastroenterol. 1999;94(1): 126–130.

1.7
Anal Incontinence, Constipation, and Obstructed Defecation

Abdul H. Sultan and A. Muti Abulaffi

Key Message

The definitions and prevalence of anal incontinence and constipation are discussed. The anatomy, physiology, and pathophysiology of bowel motility and sphincter control are reviewed. The relationship between posterior compartmental prolapse, rectal prolapse, and uterovaginal prolapse is discussed.

Introduction

Anal incontinence is defined as the loss of normal control of bowel action leading to the involuntary passage of flatus and feces. Incontinence can range from a minor leakage of feculent fluid, occasional leakage of stool during passage of flatus, to a complete loss of bowel control. Constipation, on the other hand, has no universally accepted definition. About 80% of people suffer from constipation at some time during their lives, and brief periods of constipation are normal. Widespread beliefs, such as the assumption that everyone should have a movement at least once each day, have led to overuse and abuse of laxatives. Some of the most frequently used definitions include infrequent bowel movements, typically 2 or less per week, difficulty in defecation (straining at passing stools for more than 25% of bowel movements or a subjective sensation of hard stools), or the sensation of incomplete bowel emptying. Agachan et al.[1] proposed an objective constipation scoring system using a 100-question symptom analysis which correlated well with colon transit, anal manometry, proctography, and electromyography.

Anal incontinence has an estimated prevalence of 0.42%, rising to more than 1% in the over 65 age group.[2] The condition is more common in women than men of young and middle age, although in the elderly, it affects both sexes equally. Constipation is far more common, with an estimated prevalence of 2% with women being affected 3 times more often than men.[3]

Although constipation and incontinence would seem to be at the opposite extremes of the spectrum, there are scenarios where constipation can lead to incontinence (see the following sections).

Anatomy and Physiology

The Large Bowel

The large bowel is 1.5 m long from the ileocecal valve to the anus. It is arbitrarily divided into the cecum, ascending colon, transverse colon, descending colon, sigmoid colon, and rectum. The transverse and sigmoid colon always have a mesentery, but the ascending and descending colons have a mesentery in 12% and 22%, respectively. The function of the large bowel is to mix and propel its contents and to absorb water and electrolytes. Bowel activity (peristalsis) is involuntary and is mediated by a combination of extrinsic and intrinsic nervous systems. The extrinsic innervation is made up of sympathetic and parasympathetic nerves. The intrinsic innervation is made up of the enteric nervous system, which is a network of nerve cells in the bowel wall comprising the myenteric and submucosal plexuses.

The Anal Sphincter and Pelvic Floor

The anal canal is 3–4 cm in length. It is surrounded by two muscles (Fig. 1.7.1):[4] (1) the internal sphincter, a smooth muscle that represents the expanded distal portion of the circular smooth muscle of the rectum and is innervated by the autonomic nerves, and (2) the external sphincter, a striated muscle that is situated around the internal sphincter and extends superiorly to blend with the puborectalis, which is an important constituent of the levator ani. It is divided descriptively into three parts; subcutaneous, superficial, and deep. The innervation of the external sphincter is from the pudendal nerves (S2, S3, S4), and that of the puborectalis is from direct branches of S3 and S4.

The internal sphincter is almost always contracted and contributes up to 70% of the resting tone. Relaxation occurs in response to certain stimuli such as rectal distension (rectosphincteric reflex), and then only for a short time. The external sphincter, puborectalis, and levator ani are constantly active, even during sleep. This activity in the external sphincter further contributes to the resting tone of the anal canal. The activity in these muscles is increased by voluntary squeezing, but usually only for short periods (up to 60 seconds) and is lowered during defecation (initial straining effort).

The anorectal angle is created and maintained by the constant active contraction of the levator ani and its most central portion, the puborectalis, which passes around the rectum in a U-shaped fashion, thus, pulling it forward. This allows the anterior wall of the rectum to cover the top of the anal canal as a flap valve. Any increase in the intraabdominal pressure causes the flap valve to close more tightly, preventing rectal contents from entering the anal canal.

Continence and Defecation

Normal continence and defecation is achieved by the interplay between a host of sensory and motor inputs. The rectum maintains low intraluminal pressure (rectal compliance) and accommodates slowly to the fecal contents, which are propagated slowly down the colon by colonic peristalsis.

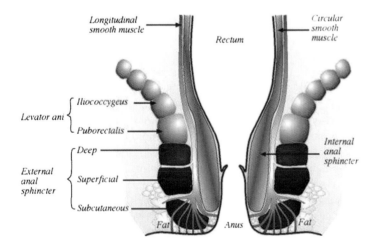

FIGURE 1.7.1. Diagrammatic representation of the anal sphincter mechanism. (Thakar and Sultan, 2003.)

FIGURE 1.7.2. The mechanism of maintaining continence and defecation. (Sultan and Nugent, 2004.)

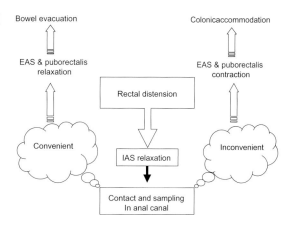

Once a certain volume is reached (usually 200 ml), a sensation of rectal fullness occurs, which is mediated by sensory stretch receptors in the puborectalis and levator ani. The rectosphincteric inhibitory reflex (anorectal inhibitory reflex) is activated, causing relaxation of the internal anal sphincter. The fecal bolus descends towards the anal canal, but is prevented from progressing further by the voluntary contraction of the external sphincter. At this point, sensory sampling by the profuse sensory receptors of the anal canal permits the individual to distinguish between flatus, liquid, or solid (Fig. 1.7.2). When socially convenient, the external sphincter and puborectalis are relaxed and combined with straining, the anorectal angle straightens, allowing the rectal contents to be expelled. If the timing is inconvenient, the external sphincter and puborectalis remain contracted, returning the luminal contents back to the rectum, and the contents are then reaccommodated in the colon.[5]

Pathophysiology

Anal Incontinence

Most of the resting anal pressure is contributed by the internal anal sphincter (up to 70%), the anal cushions (up to 15%), and the external sphincter (the remaining percentage of pressure).

In general, passive soiling and/or flatus incontinence is associated with internal sphincter dysfunction. By contrast, urgency and/or urge incontinence is associated with external sphincter dysfunction. However, more frequently dysfunction involves both sphincters, giving rise to mixed symptoms.

When anal incontinence is attributed to anal sphincter dysfunction, it is usually caused by either mechanical disruption or neuropathy, but sometimes both may coexist. Obstetric trauma is a major etiological factor, and there is a double peak incidence of anal incontinence; the first is in the postpartum period and the second is in the perimenopausal years. After the advent of anal endosonography it has been shown that approximately one third of primiparous women develop anal sphincter injury that is not recognized during vaginal delivery.[6] However, even when it is recognized and repaired, the outcome is suboptimal, as one third continue to suffer im-paired continence.[7] This may be related to concurrent neuropathic damage or inadequate repair. However, there are other factors, such as the effect of ageing, collagen weakness, progression of pelvic neuropathy, estrogen deficiency, concurrent irritable bowel syndrome (IBS), primary degeneration of the internal anal sphincter, severe constipation, and uterovaginal/rectal prolapse that may contribute to a deterioration in anorectal function.

TABLE 1.7.1. Causes of Constipation

Causes of constipation		
Colonic constipation	Organic/Anatomic	Diverticular disease
		Colonic tumors
		Benign colonic strictures
	Functional	IBS
	Colonic (slow transit)	Idiopathic
		Drugs
		Endocrine causes
		Psychiatric
		Environmental
Anorectal constipation	Organic/Anatomic	Rectal Cancer
		Fissure in ano
		Extrinsic compression of rectum
		Rectal intussusception and prolapse
		Rectocele
		Enterocele
	Functional causes	Diminished sensation
		Anismus

Constipation

There may be several, possibly simultaneous, factors leading to constipation, including inadequate fiber and fluid intake, a sedentary lifestyle, psychosocial issues, history of sexual abuse, and environmental changes. Constipation may be aggravated by travel, pregnancy, or a change in diet. In some people, it may result from repeatedly ignoring the urge to have a bowel action.

Constipation results from disorders of the large bowel and anorectum and, for the purposes of this chapter, can be classified into colonic and anorectal constipation. Each can then be subdivided into organic and functional causes[8] (Table 1.7.1).

Colonic Constipation

Organic Causes

In these situations, there is a physical or a structural abnormality in the colon that hinders or interferes with the passage of stools, causing constipation. Tumors of the colon and rectum, diverticular disease, and benign strictures are but a few good examples.

Functional Causes

In these situations, the colon is structurally normal looking during a contrast examination or a colonoscopy. However, the problem lies with disordered motor activity of the colon caused by factors and conditions affecting both the extrinsic and intrinsic nervous systems. Perhaps one of the most common conditions belonging to this group is IBS. In these patients, there is colonic hypermotility producing nonpropulsive haustral waves that hinder transport of feces rather than promote it. In contrast, colonic hypomotility or colonic inertia can also be a cause of severe constipation, such as in idiopathic slow-transit constipation. Other factors affecting motor activity include endocrine, metabolic, environmental, psychiatric, neuromuscular, and drugs.

Anorectal Constipation (Outlet Obstruction)

Organic Causes

As with colonic causes, there is an associated anatomical abnormality. This includes condi-

tions in the pelvis outside the anorectum, such as large fibroids, uterine tumors, endometriosis, and pregnancy pressing on the rectum and causing luminal narrowing. Within the anorectum, evacuatory difficulties occur with tumors of the lower rectum, large hemorrhoids, rectal prolapse, rectocele, enterocele, and anal fissures.

In rectal prolapse, part of the rectum protrudes through the anus. The prolapse may be made of rectal mucosa only (mucosal prolapse) or the full thickness of the rectal wall including the muscularis (full-thickness prolapse). In few instances, the prolapse is internal or occult and is only seen on sigmoidoscopy or during defecation proctography.

Intussusception is a type of a full-thickness internal rectal prolapse that involves one segment of the rectal circumference and is only seen during sigmoidoscopy or defecation proctography.

A rectocele is a herniation of the rectum creating a bulge in the posterior vaginal wall. Rectoceles are described in detail in Chapter 4.8. Many such posterior wall defects are asymptomatic and are often identified during concurrent examination of prolapse in other compartments. Symptoms include a vaginal bulge (lump), constipation, a feeling of incomplete emptying, and the need to perform rectal or vaginal digitations or digital splinting in order to facilitate bowel emptying.[9]

Other posterior compartment prolapse, such as an enterocele (herniation of the small intestine into the vagina), may occur concurrently; therefore, careful evaluation is essential before embarking on surgery. Rectoceles may be a secondary manifestation of outlet obstruction caused by intussusception; therefore, a defecating proctogram should be performed under these circumstances. In a study of 69 patients with an enterocele, Mellgren et al[10] found that 55% had concurrent rectal intussusception, and 38% had rectal prolapse.

Functional Causes

In this type of constipation, the colonic transit is normal, but the abnormality is within the anorectal mechanism responsible for expelling the stools. The individual feels the urge to open the bowels, but is unable to do so even though the stools are soft. It should be noted, however, that there is a small group of patients with anorectal constipation who rarely feel the urge to defecate because of diminished sensation in the anorectal zone.

Relationship Between Rectal Prolapse and Advanced Uterovaginal Prolapse

There is no evidence to support a clear association between rectal prolapse with complete uterine prolapse. Reports in the literature are limited to case reports and small series.[11,12] The largest series is by Goligher,[13] where only 8 out of 83 patients with rectal prolapse had a significant uterine prolapse requiring surgery. Unlike fecal incontinence, the prevalence of rectal prolapse does not appear to be lower in nulliparous women. In one series of 183 women treated for rectal prolapse, 40% were nulliparous. It would therefore appear that although uterovaginal prolapse and rectal prolapse may present simultaneously, there are different mechanisms in their pathogenesis.

Anismus is a condition where the aforementioned normal defecator process is lacking. In this condition, the anal sphincter muscles (external and puborectalis) fail to relax and, in some instances, there is paradoxical contraction during straining. The condition can develop at any age and frequently there is an associated psychological element (see Chapter 1.8).

Functional Causes

Abnormal motility occurs in anismus and Hirschsprung's disease. Abnormal sensitivity occurs in cerebrovascular accidents, lesions of the ascending nerves, as in diabetes, and after damage to pudendal nerves.

Relationship Between Constipation and Fecal Incontinence

There are several scenarios where dysfunction can lead to either constipation or incontinence (Fig. 1.7.3). Indeed, constipation itself can lead to incontinence (Fig. 1.7.4). Patients with fecal

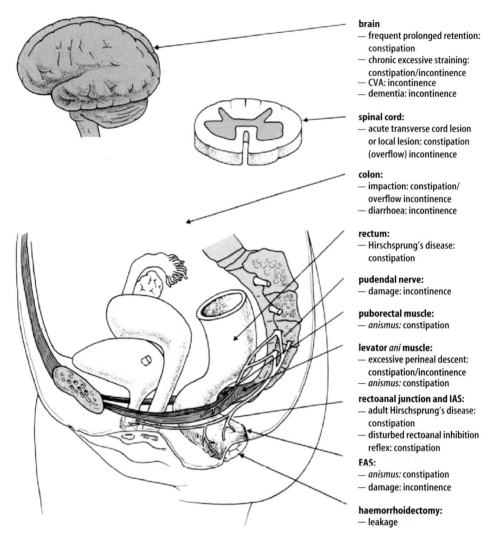

FIGURE 1.7.3. Pathogenesis of constipation and/or incontinence. (Smout and Akkermans, 1994.)

impaction can develop paradoxical diarrhea caused by overflow, and this can be profuse and severe enough to leak through the anus even though the sphincter function is normal. In another much debated condition, it is suggested that patients with chronic constipation and habitual straining develop damage to the pudendal nerves as a result of traction neuropathy (stretch injury), and that this in turn causes weakness of the anal sphincter and pelvic floor muscles leading to incontinence (Fig. 1.7.5).

Descending perineal syndrome is a term introduced by Sir Alan Parks[14] to describe a group of patients with various anal symptoms who showed marked descent of the perineum when asked to strain. One of the findings in this condition is an anterior rectal mucosal prolapse into the upper anal canal that is interpreted as bowel contents, resulting in excessive straining. It is suggested that the pelvic descent in these groups of patients may lead to traction neuropathy and, ultimately, incontinence (Fig. 1.7.4).

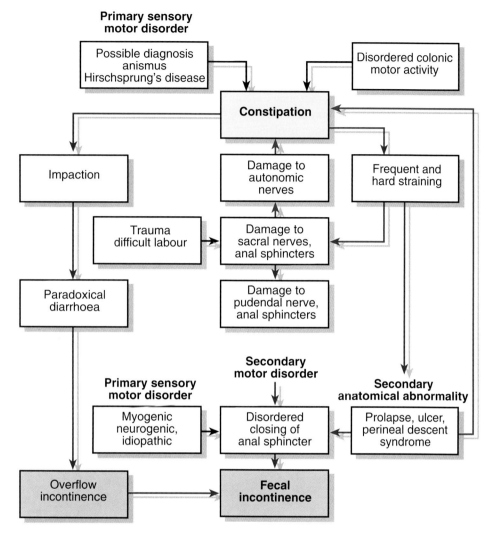

FIGURE 1.7.4. Interrelationship and pathophysiology of constipation and anal incontinence. (Smout and Akkermans, 1994.)

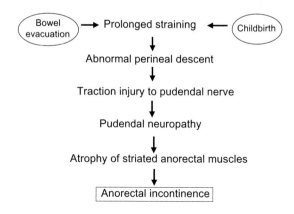

FIGURE 1.7.5. The mechanism of neurogenic anal incontinence. (Sultan and Nugent, 2004.)

Conclusion

The mechanisms that maintain continence are complex and the cause is often multifactorial. Constipation, obstructed defecation, prolapse, and anal incontinence are disorders that may be interrelated; therefore, a careful history and evaluation is essential. When dysfunction involves multiple compartments, a multidisciplinary approach should be adopted.[15]

References

1. Agachan F, Chen T, Pfeifer J, et al. A constipation scoring system to simplify evaluation and management of constipated patients. Dis Colon Rectum. 1996;39:681–685.
2. Thomas TM, Egan M, Walgrove A, et al. The prevalence of faecal and double incontinence. Community Med. 1984;6(3):216–220.
3. Sonnenberg A, Koch TR. Physician visits in the United States for constipation. 1958 to 1986. Dig Dis Sci. 1989;34:606–611.
4. Thakar R, Sultan AH. Management of obstetric anal sphincter injury. Obstet Gynaecol. 2003;5(2):72–78.
5. Sultan AH, Nugent K. Pathophysiology and non-surgical treatment of anal incontinence. Br J Obstet Gynaecol. 2004;111(Supp 1):84–90.
6. Sultan AH, Kamm MA, Hudson CN, et al. Anal sphincter disruption during vaginal delivery. N Engl J Med. 1993;329:1905–1911.
7. Sultan AH, Thakar R. Lower genital tract and anal sphincter trauma. Best Pract Res Clin Obstet Gynaecol. 2002;16(1):99–116.
8. Smout AJP, Akkermans LMA. Normal and disturbed motility of the gastrointestinal tract. Petersfield, UK: Wrightson Biomedical; 1994:175–189.
9. Karlbom U, Graf W, Nilsson S, et al. Does surgical repair of rectocele improve rectal emptying? Dis Colon Rectum. 1996;39:1296–1302.
10. Mellgren A, Johansson C, Dolk A, et al. Enterocele demonstrated by defecography is associated with other pelvic floor disorders. Int J Colorect Dis. 1994;9:121–124.
11. Azpuru CE. Total rectal prolapse and total genital prolapse: a series of 17 cases. Dis Colon rectum. 1974;17:528–531.
12. Dekel A, Rabinerson D, Ben Rafael ZY. Concurrent genital and rectal prolapse: two pathologies – one joint operations. BJOG. 2000;107:125–129.
13. Goligher JC. Prolapse of the rectum. In: Goligher JC, Duthie HL, Nixon HH, editors. Surgery of the anus, rectum and colon. Fifth Edition. London: Baillere Tindall; 1984:246–284.
14. Parks AG, Porter NH, Hardcastle J. The syndrome of the descending perineum. Proc R Soc Med. 1966;59:477.
15. Sultan AH, Abulafi MA. The role of the gynaecologist in faecal incontinence. In: Sturdee R, Olah K, Keane D, editors. Yearbook of Obstetrics and Gynaecology. Volume 9. London: RCOG press; 2001:170–87.

1.8
Overactive Pelvic Floor Muscles and Related Pain

Wendy F. Bower

Key Message

In this chapter, the condition of overactive pelvic floor (OAPF) is defined, and its resultant symptom of pain and its effect on bladder and bowel function are explained. The clinical findings are reviewed. The causation is multifactorial.

Introduction

Overactive pelvic floor is a term synonymous with a dysfunctional/spastic/short pelvic floor, puborectalis syndrome, paradoxical sphincter reaction, and nonrelaxing/hypertonic pelvic floor. By definition, the patient with OAPF has impaired use of the pelvic skeletal muscles in the absence of neurological disease. The entire pelvic floor, rather than one muscle group, is in a state of sustained contraction.[1] Overactive pelvic floor may also be described as the absence or partial presence of relaxation after a pelvic floor contraction.[2] It is unlikely that OAPF is a primary disorder occurring in isolation, but rather part of a cascade of symptoms precipitated by dysfunction within the pelvis and reactive changes in muscle fibers.

Symptoms

The symptoms of OAPF that may be encountered in clinical practice fall into three categories, but can occur in concert: pain, effect on the bladder, and consequences for bowel function.

Pain

Local

Patients with OAPF experience aching or cramping pain in the coccygeal region and a heavy feeling in the rectal or vaginal area (Fig. 1.8.1).[3,4,5] Pain is often triggered or exacerbated by sitting, intercourse, or defecation, which are activities that stretch the pelvic floor musculature.[6]

Other patients will complain of a nonspecific tenderness. Urethral pain or hyperalgesia may be reported, and they are likely caused by local mucosal ischemia, secondary to forceful closure of the urethral sphincter.[7]

Referred

Pain may be felt in structures that are distant from the pelvic floor muscles but innervated by the same spinal nerve. Referred OAPF pain is commonly reported in the perineum and labia, and, less frequently, in the rectum, suprapubic region, abdomen, lower limbs, or lower back.[8] The site of referral may become hyperalgesic and display trophic changes such as edema, altered blood flow, subcutaneous thickening, or muscle atrophy.

Bladder

The effect of an OAPF on the urinary tract is dynamic. Symptoms include hesitancy or difficulty initiating a void, slow or intermittent urine flow, pain with micturition, significant residual urine after voiding, urgency, frequency, and

TABLE 1.8.1. Urinary tract morbidity associated with OAPF

- OAB dry
- OAB wet
- Straining to void
- Recurrent urinary tract infections
- Hydroureteronephrosis
- Vesicoureteric reflux
- Grossly trabeculated large capacity bladder
- Detrusor decompensation

urethral sphincter hyperalgesia (Table 1.8.1).[9,5,10] A significant number of patients also have a diagnosis of interstitial cystitis.[3]

Bowel

Patients with OAPF perceive appropriate perineal fullness and fecal urge, but have trouble passing stools. Because the puborectalis may fail to relax during attempted evacuation, defecation may be painful and mechanically difficult, often requiring external assistance, and rectal emptying may be incomplete.[11] Constipation and obstructed defecation are common in patients with OAPF and can precipitate further anorectal pathology, such as anal fissures and hemorrhoids. In some patients, irritable bowel syndrome and slow-transit constipation may be noted.[12]

Causes

It is likely that the OAPF is both a consequence of pelvic dysfunction and a cause of pain and discomfort. Effects may be mechanical, inflammatory, or histochemical. Repetitive, acute, or sustained muscle overload, such as occurs when the pelvic floor is maintained in a shortened (contracted) position, can injure the motor end plate. Tonic contraction results in decreased local blood flow, heat, and accumulation of potassium ions, lactic acid, histamine, and bradykinin.[7] High local concentrations of these metabolites are known to generate pain. Over time, the muscle becomes habitually shortened and may develop foci of tethering, or trigger points. These structural changes within the muscle fiber anchor regions prevent pelvic floor lengthening and generate local and referred pain. In up to 88% of female patients with pelvic pain, such trigger points can be identified within the pelvic floor muscle group, as well as in the abdominal wall and hip girdle.[3]

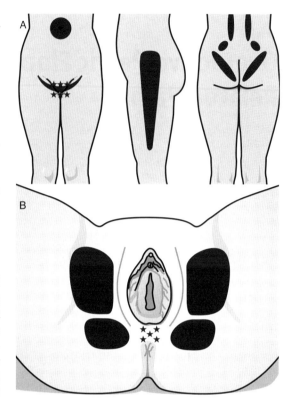

FIGURE 1.8.1. Areas of tissue abnormality in patients with OAPF. (Reprinted from Fitzgerald and Kotarinos, 2003.)

OAPF may be secondary to local inflammation, disease, introital or vaginal infection, or abscess.[4,6,13] Trauma to the perineum or pelvic floor from falls, sexual abuse, vaginal delivery, episiotomy, or surgical scarring can also precipitate the OAPF.[3,4,14] Pudendal nerve entrapment and musculoskeletal limitations may be implicated.[4,15]

Heightened activity in the pelvic floor muscle group and urethral and anal sphincters suggests an underlying dysfunction of neural control. Various theories exist. Mechanosensitive units may develop either a lower threshold of firing or stronger discharge suprathreshold. Cross communication, convergence within the dorsal horn of the spinal cord, or the presence of dichotomizing neurons could explain referred pain. Alternatively the release of excitotoxic neurotransmitters in the pelvic floor could facilitate a change in central drive that allows a drift toward memorized high-pressure activity in pelvic skeletal muscle.[16] Increased excitability of central neurons may be maintained independent of initiating peripheral input, e.g., by cross-system viscerosomatic interaction, and precipitate change in visceral behavior.

It has been shown that emotions affect sphincter tension, and that patients with disordered timing of the puborectalis have significantly higher anxiety, depression, and panic scores than controls.[17] There is clearly a psychological component to OAPF, but whether or not it is causative is contentious.

Findings of Pelvic Floor Pain Associated with Pelvic Floor Muscle Dysfunction

Chronic pelvic floor dysfunction is often associated with lower urinary tract dysfunction, and it is possible that both are maintained by an up regulation of the sacral reflex arc.[10] Motor planning and performance is altered by the expectation of pain. The tight and often painful pelvic floor muscle displays obvious change in recruitment patterning, timing, proprioception, and resting muscle length. Functional use of the pelvic floor is also modified by memory of pain. For example, the alteration of micturition after urinary tract infection or dyssynergic defecation after constipation illustrates the ability of the higher centers to override earlier learning.

Palpation of the normal pelvic floor should not produce pain; however, OAPF patients frequently report moderate to severe tenderness on examination of the pelvic floor muscle group, particularly when trigger points are present. Pressure from the examiner's finger is interpreted as painful, and pain is often noted to spread to the hips and lower abdomen.[3,4] Palpation of muscle spasm often results in pain.[9,18]

Other muscles of the pelvic girdle may compensate for limitations of the pelvic floor in neutralizing intraabdominal pressure rises, stiffening the sacroiliac joint, allowing backward rotation of the sacrum, and stabilizing the spine.[19,20] However, these compensatory muscles are not intended to provide primary support or movement and are themselves vulnerable to imbalance and pain. It is common to find obturator internus, psoas, gluteal, abdominal, and piriformis muscle abnormality and pain.[3,4,6,20]

Findings of Pelvic Floor Overactivity

Patients with OAPF have short and weak pelvic floor muscles. When visualizing the perineum, the genital hiatus may appear small and the perineal body may be displaced anteriorly. On a request to contract the pelvic floor muscles, the OAPF patient is likely to demonstrate poor proprioception, minimal or absent movement, descent in place of elevation, and early fatigue.[3,4,8,14,21] Accessory muscle activity is frequently seen in the thigh adductor muscles, hips, buttocks, and thorax, as patients attempt to locate the pelvic floor muscles.

One of the most common findings of OAPF is a high level of pelvic floor muscle activity at rest, and marked difficulty returning the muscle to a relaxed state.[4,14] Patients may be unable to sense pelvic floor relaxation or take several efforts before tension is released. Successful treatment not only ameliorates pain but also results in posttreatment reduction of vaginal resting pressure.[22]

A less commonly reported finding in OAPF patients is the thickening of subcutaneous connective tissues around perineal or suprapubic trigger points or their regions of pain referral. These trophic changes are identifiable on palpation as altered tissue bulk, contour, elasticity, and temperature, along with variation in color.[3]

Association with Micturition

Initially slow or nonrelaxation of pelvic floor skeletal muscle is most problematic during voiding, creating a hesitancy and resistance to the outflow of urine. The flow of urine may be intermittently slowed, stop and start, or the detrusor may cease contracting prematurely (Table 1.8.2). Associated residual urine after voiding is common, and patients report a need to double void.

TABLE 1.8.2. Patterns of voiding seen in patients with OAPF

Void type	Urethral relaxation	Valsalva evident	Pelvic floor activity	Detrusor contraction	Flow curve shape
1 = Normal	Yes	No	No	Yes	Bell
2 = Abnormal	Yes	No	Yes	Partial	Staccato/steep rise
3 = Abnormal	Yes	Yes	Yes	No	Intermittent

Functional obstruction to complete emptying of the bladder generates escalating detrusor pressures. The feed-forward loop triggers further pelvic floor activity to "brake" additional detrusor pressure rise. Eventually, the micturition reflex becomes destabilized during bladder filling, evidenced on urodynamic investigation as an overactive bladder and on a frequency volume chart as frequent low-volume voids.[23]

Pelvic floor contraction is a common method used to inhibit urgency; however, in the OAPF patient with a shortened muscle very little pressure may be generated. Prescription of pelvic floor exercises to increase urethral closure pressure and maximize detrusor inhibition is inappropriate, as it exacerbates both pain and muscle imbalance and is minimally effective.

Association with defecation

Smaller, more acute straining angles are seen in OAPF patients than in normal defecation attempts; this is a direct consequence of the slow or nonrelaxing pelvic floor.[24,25] Anal manometry is essentially normal for both resting and squeeze pressure.[1] On straining, there is obvious perineal descent and marked EMG activity in place of relaxation in puborectalis and the external anal sphincter.[11] There is delayed initiation of evacuation, prolonged duration of passage of stool, and incomplete emptying of rectum in OAPF patients, compared with control subjects.[1]

In some cases flawed inhibitory gating within the central nervous system may have been laid down during the toilet training process.[16] Children who experience large, hard stools learn to respond to fecal urge with a rise in pelvic floor activity. Withholding of stool, although initially a response to the fear of pain, becomes a learned response to rectal distension. Over a period of time, the call to stool is perceived at higher volumes. Thus, it is probable that in some individuals with OAPF the ability to relax puborectalis and funnel stool through an appropriate aperture was never acquired.

Pontine projections between the reticular formation and nucleus retroambiguus allow the abdominal and pelvic floor muscles to influence each other, making it possible for abdominal pain to lead to dysfunction of the pelvic floor and vice versa.[17]

Association with Sexual Function or Sexual Abuse

Sexual pain and orgasm are the components of female sexual dysfunction most likely to be influenced by OAPF.[26] Pain associated with sexual activity may be either dyspareunia or vaginismus (see Chapter 4.4), whereas orgasmic dysfunction includes infrequent or nonorgasmic response.[27]

There is a higher prevalence of prior sexual abuse among patients with functional disorders in comparison to patients with organic disease.[28] Genitourinary and gastrointestinal symptoms are the most common somatic complaints.[29] Although there is no specific data in respect to patients with OAPF, both women with pelvic pain and patients with dysfunctional voiding have high prevalence rates of childhood or other sexual abuse.[28-31] Clinicians should be aware of this association when treating women with OAPF (see Chapter 4.4, text box #3).

References

1. Halligan S, Bartram CI, Park HJ, et al. Proctographic features of anismus. Radiol. 1995;197:679–682.

2. Messelink B, Benson T, Berghmans B, et al. Standardization of terminology of pelvic floor muscle function and dysfunction: report from the pelvic floor clinical assessment group of the International Continence Society. Neurourol Urodyn 2005;24:374–380.
3. Fitzgerald MP, Kotarinos R. Rehabilitation of the short pelvic floor: Background and patient evaluation. Int Urogynecol J. 2003;14:261–268.
4. Shelly B, Knight S, King P, et al. Aetiology of pelvic floor muscle pain syndromes. In: Laycock J and Haslam J, editors. Therapeutic management of incontinence and pelvic pain. London: Springer-Verlag; 2002:167–170.
5. Meadows E. Treatment for patients with pelvic pain. Urolog Nurs. 1999;19(1):33–35.
6. Segura JW, Opitz JL, Greene LF. Prostatosis, prostatitis or pelvic floor tension myalgia. J Urol. 1979; 122:168–169.
7. Everaert K, Devulder J, De Muynck M, et al. The pain cycle: Implications for the diagnosis and treatment of pelvic pain syndromes. Int Urogynecol J. 2001;12:9–14.
8. Schmidt RA, Vapnek JM. Pelvic floor behavior and interstitial cystitis. Semin Urol. 1991;9(2):154–159.
9. Weiss JM. Pelvic floor myofascial trigger points: manual therapy for interstitial cystitis and the urgency–frequency syndrome. J Urol. 2001;166(6):2226–2231.
10. Zermann DH, Manabu I, Doggweiler R, et al. Neurological insights into the etiology of genitourinary pain in men. J Urol. 1999;161:903–908
11. Bleijenberg G, Kuijpers HC. Treatment of the spastic pelvic floor syndrome with biofeedback. Dis Col Rec. 1987;50(2):108–111.
12. Walker EA, Gelfand AN, Gelfand MD, et al. Chronic pelvic pain and gynecological symptoms in women with irritable bowel syndrome. J Psychosom Obstet Gynaecol. 1996;17(1):39–46.
13. Chung AK, Peters KM, Diokno AC. Epidemiology of the dysfunctional urinary sphincter. In: Corcos J and Schick E, editors. The urinary sphincter. New York: Marcel Dekker 2001:183–191.
14. Glazer HI, Rodke G, Swencionis C, et al. Treatment of vulvar vestibulitis syndrome with electromyographic biofeedback of pelvic floor musculature. J Reprod Med. 1995;40(4):283–290.
15. Antolak SJ, Hough DM, Pawlina W, et al. Anatomical basis of chronic pelvic pain syndrome: the ischial spine and pudendal nerve entrapment. Med Hypotheses. 2002;59(3):349–353.
16. Zermann DH, Manabu I, Schmidt RA. Pathophysiology of the hypertonic sphincter or hypertonic urethra. In: Corcos J and Schick E, editors. The urinary sphincter. New York: Marcel Dekker; 2001:201–218.
17. Zermann DH, Manabu I, Schmidt RA. Management of the hypertonic sphincter or hyperpathic urethra. In: Corcos J and Schick E, editors. The urinary sphincter. New York: Marcel Dekker; 2001:679–686.
18. Hetrick DC, Ciol MA, Rothman I, et al. Musculoskeletal dysfunction in men with chronic pelvic pain syndrome type III: a case-control study. J Urol. 2003;170(3):828–831.
19. Pool-Goudzwaard A, Hoek Van Dijke G, Van Gurp M, et al. Contribution of pelvic floor muscles to stiffness of the pelvic ring. Clin Biomech. 2004; 19(6):564–571.
20. Sapsford RR, Hodges PW. Contractions of the pelvic floor muscles during abdominal maneuvers. Arch Phys Med Rehabil. 2001;82(8):1081–1088.
21. Ab E, Schoenmaker M, van Empelen R, et al. Paradoxical movement of the pelvic floor in dysfunctional voiding and the results of biofeedback training. BJU Int. 2002;89(Suppl 2):1–13.
22. Jarvis SK, Abbott JA, Lenart MB, et al. Pilot study of botulinum toxin type A in the treatment of chronic pelvic pain associated with spasm of the levator ani muscles. Aust N Z J Obstet Gynaecol. 2004;44(1):46–50.
23. Yeung C. Pathophysiology of bladder dysfunction. In: Pediatric urology. Gearhart JP, Rink R, Mouriquand P, editors. WB Saunders: Philadelphia; 2001.
24. Lee HH, Chen SH, Chen DF, et al. Defecographic evaluation of patients with defecation difficulties. J Formos Med Assoc. 1994;93(11–12): 944–949.
25. Kuijpers HC, Bleijenberg G. The spastic pelvic floor syndrome. Dis Col Rec. 1985;28(9):669–672.
26. Munarriz R, Kim NN, Goldstein I, et al. Biology of female sexual function. Urol Clin N Am. 2002; 29:685–693.
27. Bachmann GA, Phillips NA. Sexual dysfunction. In: Steege JF, Metzger DA, Levy BS, editors. Chronic pelvic pain. Philadelphia: WB Saunders; 1998:77–90.
28. Jacob MC, De Nardis MC. Sexual and physical abuse and chronic pelvic pain. In: Steege JF, Metzger DA, Levy BS, editors. Chronic pelvic pain. Philadelphia: WB Saunders; 1998:13–24.

29. Drossman DA. Physical and sexual abuse and gastrointestinal illness: What is the link? Am J Med. 1994;97:105.
30. Ellsworth PI, Merguerian PA, Copening ME. Sexual abuse: another causative factor in dysfunctional voiding. J Urol. 1995;153:773–776.
31. Salonia A, Zanni G, Nappi RE, et al. Sexual dysfunction is common in women with lower urinary fract symptoms and urinary incontinence: Results of a cross-sectional study. Euro Urol. 2004;45:642–648.

Part II
Evaluation of the Pelvic Floor

2.1
Clinical Evaluation of the Pelvic Floor Muscles

Diane K. Newman and Jo Laycock

Key Message

Clinical examination is the basis of diagnosis of urogynecological disorders. It is important that this examination is performed by a well-trained person with the appropriate skills. The patient should actively participate in the examination and be able to carry out pelvic floor muscle (PFM) contractions in a coordinated way when required. This will form the basis of subsequent pelvic floor exercises and is how the woman will learn the different types of muscle contraction which are integral to this. Digital self examination is an important part of pelvic floor re-education, and a woman should be able to do this herself. There are various grading methods, and, in particular, the P.E.R.F.E.C.T. scheme is an important assessment technique, and necessary when planning a treatment program. The Knack is a useful maneuver that can be taught to guard against stress incontinence. Mechanical devices, such as a perineometer with the addition of electromyographic (EMG) recording and vaginal cones, can help in the retraining process. It is important to set a program of exercises and re-evaluate them after a reasonable time period, aided by a urinary diary.

Introduction

Women who present with pelvic floor disorders should undergo a pelvic evaluation to determine abnormalities, the patient's ability to contract and relax the PFMs and to determine the presence of a muscle hypertonus.

This evaluation or examination is performed by a variety of health care clinicians: doctors, nurses, and physical therapists that specialize in the area of pelvic floor dysfunction. Physicians who treat pelvic floor disorders include urologists, gynecologists, fellowship-trained urogynecologists and female urologists, geriatricians, family practitioners, and other interested clinicians. Nurses' background and skills are unique because they encompass the full spectrum of evaluation and treatment, especially in the area of comprehensive pelvic examination. Physiotherapists or physical therapists (PTs) have long played a part in incontinence and pelvic floor care, and involvement in this area is often through association with women's hospitals or obstetric departments, rather than as part of a general physiotherapy practice. As such, they tend to be highly motivated and enthusiastic.

Beginning the Evaluation of the Pelvic Floor

Before examining the pelvic floor, the clinician should explain the procedure to prevent embarrassment and the woman should empty her bladder. Several positions are used, with the lithotomy position considered ideal. This position requires that the patient be supine, with hips and knees bent and abducted; most clinicians' choose this position because PFM assessment may be part of a bimanual speculum pelvic examination. Ideally, the legs should be supported, with one leg against the examiner and the other leg supported on a pillow, or in stirrups. If appropriate and acceptable, the examiner should consider the use of a mirror to demonstrate the findings to the patient.

The vulva and vagina may be examined for irritation from urine or perineal pads, urethral caruncle, atrophic vaginitis, and pelvic organ prolapse (POP). Excuses for not performing the examination include discomfort associated with atrophic vaginitis, patient and clinician embarrassment with the examination, and procedural difficulty because of the need for assistance with confused or elderly patients. A speculum exam is optimal, but in older frail women, a digital exam may be more feasible, especially in those women residing in a nursing home.

Introital laxity is usually obstetrical in origin, and is accompanied by disturbed architecture of the musculature responsible for an enlarged vagina. This defect is common after childbirth. The vagina may be torn away from its intrapelvic attachments, with subsequent loss of the superior vaginal sulcus. There may also be direct attenuation of the vaginal wall itself, which is manifested by loss of vaginal rugae and a thin appearance. Stretching and tearing of the levator ani (LA) muscles results in a longer, wider levator hiatus.

Pelvic Floor Muscle Assessment

The PFM is a group of muscles that extend from the front (anterior) to the back (posterior) of the pelvis, forming a sling that supports the pelvic organs (Fig. 2.1.1). Normal function of the PFM is essential in maintaining appropriate function of the pelvic viscera. The PFM (or circumvaginal muscle [CVM]) made up of the levator ani group and include the pubococcygeus, puborectalis, and ileococcygeus (Table 2.1.1). The PFM not only provides support for all the organs of the pelvis, but because it is a postural muscle, it differs from other skeletal muscles in that it has a higher resting tone. Evaluation of the PFM is made through manual intravaginal (or transrectal) examination and may include the use of measuring devices for complementary assessment.

Palpation

Manual evaluation of the PFM provides information about the woman's ability to relax and the duration and symmetry of contractions. This is an important part of the pelvic examination in patients with pelvic floor disorders, because in addition to evaluating the woman's ability to produce a voluntary contraction, it also provides information about the ability of the woman to maintain that contraction during stressful situations.[1] Palpation of the PFM can be performed per vagina or per rectum.

Transvaginal (per vagina) palpation, with the woman initially positioned supine (as previously described) is the preferred route for assessing women. This technique allows the examiner to evaluate muscle bulk, resting tone, contractile strength, and reflex response to cough or valsalva. In addition, reduced sensation, tenderness, and/or pain should be identified.

Transvaginal assessment can be performed using either one or two digits (Fig. 2.1.2). The PFM of a parous woman is usually assessed using two fingers (Fig. 2.1.3), but this technique may be unsuitable for nulliparous women because of their narrower vaginas, or for elderly women with atrophic vaginitis and vaginal stenosis. The index finger is placed over the middle finger and inserted into the vagina. Normally, the pubococcygeus muscle (PCM) is felt as a distinct 1- to 2-cm band that can be palpated on the lateral vaginal wall when the examiner's finger is introduced to a depth of 3 to 5 cm beyond the vaginal introitus and rotated 360 degrees while palpating the perivaginal muscles (Fig. 2.1.4). A weak or attenuated PCM may be indistinguishable from the surrounding tissues. The woman is asked to contract her PFMs as hard as possible, and then to squeeze the examining fingers together and draw them into the vagina.[2] This test can then be repeated with the index and middle fingers spread laterally, thus, palpating the right and left LA simultaneously; it is not uncommon to find asymmetry in a parous woman.

The strong LA muscles of a young woman will be felt as a thick, firm band of muscle, compared with the weak LA muscles of an elderly woman, which may feel thin and may have reduced tone that renders them indistinguishable from the

2.1. Clinical Evaluation of the Pelvic Floor Muscles

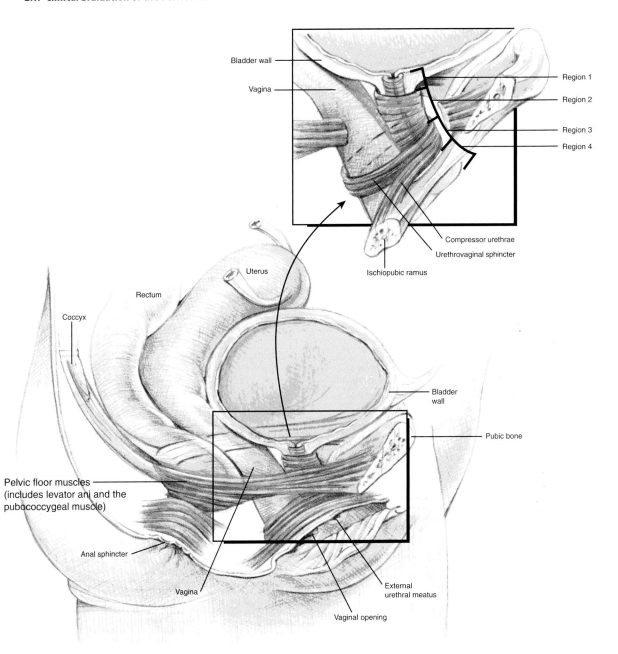

FIGURE 2.1.1. Pelvic floor muscle. (*Source*: reproduced from Newman DK, 2003.)

TABLE 2.1.1. The pelvic floor musculature (PFM)[26,28,29]

The PFM can be viewed in three layers:

First Layer (Endopelvic Fascia) The first layer is called the endopelvic fascia. It is a lining made of a mesh of smooth fibers, ligaments, nerves, blood vessels, and connective tissue. It supports and cors the bladder, the inner organs such as the intestine, and the uterus and upper vagina in women. Some of the ligaments of the endopelvic fascia connect to the lumbar spine and the symphysis pubis. Even though this layer cannot be exercised, training the pelvic floor muscles of the second layer (pelvic diaphragm) can improve back pain by increasing support of the bladder and uterus from below and decreasing the strain on the ligaments upon which the support otherwise depends. Strong pelvic floor muscles can help support the bladder, uterus, and rectum if a woman has sustained tears of the endopelvic fascia through difficult childbirth or other injury.

Second Layer (Pelvic Diaphragm) The pelvic diaphragm consists of the LA, the most important muscle forming the pelvic floor. It is the main muscular support for the pelvic organs and is of crucial importance for pelvic floor muscle rehabilitation or training. The LA is a paired muscle that stretches hammock-like between the pubic bones in the front and coccyx behind. It is attached along the lateral pelvic wall to a thickened band in the obturator fascia, the arcus tendineus.

The LA muscle has three principal parts:

1. *The puborectalis muscle* arises from the inner surface of the pubic bones, proceeds backward along the edge of the genital hiatus in contact with the side of the vagina and loops around the rectum, pulling it forward during contraction and assisting with providing continence.
2. *The PCM* lies in the middle of the vagina and feels like a ridge above the hymeneal ring and can be felt during a pelvic examination. The muscle originates in the inner surface of the pubic bones, passes backward along the puborectalis muscle and inserts into the anococcygeal and the superior surface of the coccyx. Its fibers create a slinglike band (often referred to as a hammock). This muscle provides essential support to the proximal bladder neck and urethra. *The pubovaginalis muscle* (in women only) loops around the vagina. These fibers run in the direction of front to back.
3. *The iliococcygeus muscle* arises from the arcus tendineus LA, running posteromedially to insert on the coccyx (tail bone) and the anococcygeal raphe, which is clinically called the levator plate. Some of these muscle fibers run from one side to the other; some in a more diagonal direction. This muscle does not participate in lifting the anus. *The coccygeus muscle* lies adjacent to the iliococcygeus muscle and arises from the ischial spine and the sacrospinous ligament. It passes backward against the posterior border of the iliococcygeus muscle and inserts into the coccygeus and lower segment of the sacrum. It can influence stability of the sacroiliac joint.

Third Layer (Urogenital Diaphragm) The outer layer of the pelvic floor consists of several muscles. The urethra, vagina and rectum pass within this area. Sometimes the third layer is described as two different layers because the deep transverse perineal muscle lies deeper than the other muscles of the urogenital diaphragm. The deep transverse perineal muscle (sometimes referred to as the perineal membrane) is very important for continence and supports the function of the LA. The other muscles of this layer are important for sexual function. The muscles of the third layer do not support the organs of the pelvis.

4. *The deep transverse perineal muscle* provides additional fibers, such as the sphincter, loops around the urethra in men and women, and assists with continence. It is under voluntary control.
5. *The superficial transverse perineal muscle* reinforces the action of the deep transverse perineus.
6. *The bulbocavernosus muscle* connects contracts during orgasm, erecting the clitoris.
7. *The ischiocavernosus muscle* has little function in women. The muscle fibers run in a diagonal direction.
8. *The anal sphincter muscle* loops around the anus like a ring and provides continence. It comprises the internal and external anal sphincters and the puborectalis muscle.

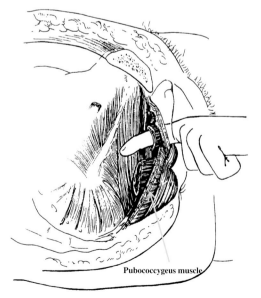

FIGURE 2.1.2. Lateral view of digital palpation of the pelvic floor muscles using one finger. (*Source*: modified from Laycock J, 1994.)

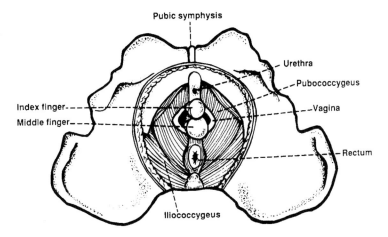

FIGURE 2.1.3. PFM measurement using 2 digits. (*Source*: reproduced from Laycock J, 1994.)

surrounding tissues. Many patients with pelvic floor disorders exhibit low tone muscle dysfunction (Table 2.1.2).

Posterior pressure palpates the rectum, and rectal contents can be detected. An active voluntary PFM contraction is felt as a tightening, lifting, and squeezing action around the examination finger. If possible, this muscle test should also be assessed while standing, since women will perform their PFM exercises in standing, during activities of daily living.

Transrectal (per rectum) palpation of the puborectalis and external anal sphincter at rest and during a muscle contraction is best carried out with the patient in the left-lateral position. The puborectalis swings around the anorectal junction, and at rest, it pulls the junction forward to create an acute angle between the anal canal and the rectal ampulla. The well-lubricated index finger is introduced 3 to 4 cm through the anus to the rectum, and the puborectalis muscle is palpated with the distal pad of the index finger; the effect of a PFM contraction is felt as tension on the examining finger as the sling-like puborectalis pulls the anorectal junction anteriorly. A contraction of the external anal sphincter is detected on withdrawing the finger to a position 2 cm inside the anal canal. Both the puborectalis and external anal sphincter should relax during defecation, and this is evaluated by instructing the patient to bear down, as if emptying the bowels.

Either assessment should note any discomfort or pain that can occur in patients with chronic pelvic pain, interstitial cystitis/painful bladder syndrome, or vulvodynia, as many of these patients have high tone muscle dysfunction as described in Table 2.1.2.[3] Pain should be graded using a scale similar to the one shown in Table 2.1.3.

Elevation of the urethrovesical junction during a PFM contraction should be evaluated (Fig. 2.1.5). In addition, palpation of the periurethral

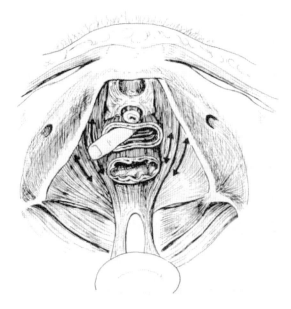

FIGURE 2.1.4. Transvaginal palpation of the perivaginal muscles from above, noting rotation of assessment. (*Source*: reproduced from Laycock J, 1994.)

TABLE 2.1.2. Identification of pelvic floor dysfunction[26]

- Low-tone pelvic floor dysfunction (LTPFD) refers to the examination findings of an impaired ability to isolate and contract the PFM in the presence of weak and atrophic musculature and is seen in women with to POP, UI, vaginal laxity, or fecal incontinence. LTPFD may be encountered in patients with partial pelvic floor denervation as a result of parturition, ageing, or some combination[30,31]
- High-tone pelvic floor dysfunction (HTPFD) refers to the clinical condition of hypertonic, spastic PFM with resultant impairment of muscle isolation, contraction, and relaxation[32]. Urological and gynecological disorders include interstitial cystitis, voiding dysfunction, overactive bladder symptoms of urinary frequency and urgency, and pelvic pain, sexual dysfunction with dyspareunia. All of the syndromes contributing to HTPFD are characterized by tender, spastic PFM manifesting as pain localized to the suprapubic area, coccyx and lower sacrum, rectal pain, or generalized pelvic discomfort. A musculosketal etiology for HTPFD has been suggested as PFM spasm with poor posture and prolonged sitting which can lead to overflexion of the coccyx. Some have described a "typical pelvic pain posture," which is characterized by exaggerated lumbar lordosis, anterior pelvic tilt, and thoracic kyphosis, causing sacroiliac pathology. The lumbar and pelvic changes cause the PFM to be stretched or compressed thus preventing the maintenance of normal resting tone, This can lead to trigger point formation and hypertonicity (Lukban, 2002). The presence of sacroiliac dysfunction with or without the contribution of poor posture may reasonably serve as a trigger for the development of HTPFD[3] Specific presentations of HTPFD include:
    i. Coccydynia which is characterized by pain localized to the coccyx, but can include LA and coccygeus muscle spasm. Symptoms include pain localized to the lower sacrum and coccyx.
    ii. Tension myalgia of the pelvic floor can cause low back pain and heaviness in the perineum or pelvis aggravated. Prolonged sitting will aggravate this condition.
    iii. Coccygeus-levator spasm syndrome has been seen in patients with pelvic floor spasm and "rectal" pain with complaints of tenderness at muscular sites adjacent to the ischial spines and coccyx, which can include involvement of musculofascial, ligamentous, and tendinous structures.
    iv. Levator ani spasm is used to describe patients exhibiting PFM spasm and tenderness.[33] A predominant symptom is discomfort. In patients with interstitial cystitis, levator ani syndrome is reported as pain in the area of the bladder and also in the regions of the sacrum, coccyx, and anus. It is felt that these muscle spasms are the result of bladder pathology, with increased PFM tone appearing in response to afferent autonomic impulses emanating from the bladder wall.

muscles centrally (12 o'clock) and each side of the urethra (11 o'clock and 1 o'clock), will determine their contractility and symmetry.

Assessment of a PFM Contraction

It is important to realize that when asked to contract the PFMs, a woman may use the wrong muscles, strain down, or perform a Valsalva maneuver, or fail to activate all layers of the pelvic musculature. It is known that most women are largely unaware that the PFM exist, and simple instruction in technique may not be adequate preparation. To be effective, PFM exercises must be performed correctly, which requires careful and specific training. Without adequate instruction, 30% to 50% of women perform them incorrectly, potentially worsening the very condition they are intending to prevent or treat.[7]

Instruct the woman on how to contract the PFM by asking her to "pull in" or "lift up" the floor of her vagina or to imagine that she is trying to control passing wind or pinching off a stool. Confirm by digital palpation that she's performing the contraction correctly and can maintain it while breathing in and out. The same protocol is performed for both the right and left PFM. A very important part of the physical examination in women is to determine the strength of the PFM. The woman is asked to contract the PFM around the examiner's fingers with as much force and for as long as she is able. Women are taught the followingthree different types of muscle contractions: (1) Strong maximal contractions, (2) short, one-second maximum contractions (sometimes referred to as "quick flicks"), and (3) submaximal contractions, which may be sustained for 5 to 10 seconds. It is recommended that both maximal and submaximal PFM contractions should be practiced as part of a home exercise program.[16,25,26]

TABLE 2.1.3. Pelvic floor muscle tenderness (hypertonus) scale

0 – no pressure or pain associated with exam
1 – comfortable pressure associated with exam
2 – uncomfortable pressure associated with exam
3 – moderate pain associated with exam, intensifies with contraction
4 – severe pain associated with exam; patient is unable to perform contraction maneuver because of pain

Source: Whitmore K et al., 1998.

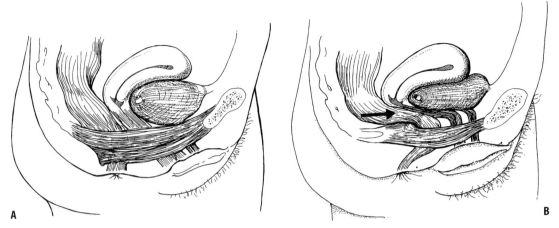

FIGURE 2.1.5. (A) Palpation of the urethra. (B) Assessing elevation of the urethra during a pelvic floor contraction. (*Source*: Reproduced from Laycock J, 1994.)

Observations During a PFM Contraction

Educating women regarding the anatomy and workings of the lower urinary and lower gastrointestinal tracts, including the PFM and sphincters, is important. This is followed by a brief lesson in contracting the PFM before muscle contractility is assessed. During a strong contraction, the main observation is a puckering and cephalad drawing-in of the vaginal introitus, anus, and perineum; a weak contraction may only demonstrate a slight puckering, and some women are unable to produce any movement of the perineum. Observation of a downward movement suggests that the woman is straining and not producing a correct PFM contraction.

The impact of a cough on a healthy, strong PFM produces little or no movement, either at the vaginal introitus or the perineum as a whole. In some cases, an anticipatory precough PFM contraction is noted. On the other hand, a woman with a very weak pelvic floor may demonstrate perineal descent, possibly below the level of the ischial spines,[5] and the vaginal introitus may bulge and gape. There may also be caudal movement of any prolapse, and urine loss may be observed. These observations should be relayed to the woman so that she is aware of the bulging perineum on coughing and the possible damage repeated coughing may cause. At this time, the women should be instructed to contract the PFM and repeat the cough, to test whether she can minimize the perineal descent and urine loss.

Observations of Extraneous Muscle Activity

Muscles rarely work in isolation, and the PFMs are no exception. Research has shown that the abdominal muscles, in particular the transversus abdominis (TrA) muscles, are always recruited (contracted in unison) during a PFM maximum voluntary contraction (MVC).[6] These muscles (PFM and TrA) work in conjunction with the lumbar multifidi muscles and the respiratory diaphragm to stabilize the lumbar spine.

In light of the difficulty some patients have initiating a voluntary PFM contraction,[7] and the fact that when TrA is contracted the PFMs are recruited,[8] it has been suggested that patients can be taught to contract the TrA so as to recruit the PFMs.[9] However, more research is needed before traditional PFM exercises are discontinued.[10] Furthermore, muscles are not recruited if their activation causes pain, or there is pain in the region.[11] Consequently, pain in the pelvis, pelvic floor, pelvic joints, or abdomen may inhibit PFM activity.

Contraction of TrA is characterized by a drawing-in of the lower abdomen. When learning this technique, patients are instructed to avoid using other abdominal muscles. Many women may have difficulty coordinating the TrA without contracting the entire abdomen. To avoid holding the breath while learning this technique, TrA and PFM contraction is initially taught on expiration.

Observations in Standing

The postural role of the PFM to counteract the pull of gravity and the increase in intraabdominal pressure is often overlooked, and weak muscles may be unable to give adequate support in the upright position over a period of time. Consequently, because stress urinary incontinence (SUI) and POP are more likely to be provoked in the standing position, it is recommended that the vulva is observed and palpated at rest during a PFM contraction and in supine and standing positions on coughing.

Digital Self-Assessment

Digital self-assessment is an important part of a pelvic floor re-education regimen and should be taught to any woman who is willing and able to perform it, as it can be a useful feedback technique for women who are unsure of the correct muscle action. The woman can be taught transvaginal palpation with one or two fingers while sitting or standing in the bath or shower, with water as a lubricant (gloves are not needed). Some women find it easier if the thumb is used; the distal pad of the thumb is placed over the posterior vaginal wall, just inside the vagina. Furthermore, this technique may help in understanding pelvic floor activity, as well as in monitoring the progress of a treatment program.

Measurement of PFM Strength

Appropriate training includes the initial examination, to determine muscle strength and place the woman at an appropriate level of exercise, and a graduated strength-training protocol supported by periodic assessment of muscle strength.[4]

When contracting the PFMs, three criteria of muscle strength should be noted: strength, duration, and alteration in position. The amount of pressure or strength of the muscle contraction can range from imperceptible to a firm squeeze. Duration involves the number of seconds that the examiner feels the muscle contraction. In a woman with a well-supported pelvic muscle, the muscle contraction can lift the base of the examiner's fingers.

Assessment of Muscle Coordination

Coordination is ascertained by monitoring the ability to contract and relax quickly and slowly; special attention is paid to the ability to relax quickly and completely. A sluggish response indicates poor coordination, and this test completes the digital assessment. The time to maximum contraction, which can be described as brisk, fair, or sluggish, not only gives a measure of PFM coordination, it also provides further information on the fast-twitch fiber capability. This is important because it is thought that the reflex response of the fast fibers to coughing is the mechanism that maintains continence, by lifting the proximal urethra and increasing the urethral occlusive pressure.

Use of Scales or Assessment Tools

Digital evaluation of the PFM has been described by several authors, and many have attempted to quantify this assessment through the use of grading scales or assessment tools. The early ones are from nurse researchers in the U.S. Worth and colleagues[12] described the "CVM Rating Scale," a one-to-three scale using a one-finger assessment of pressure, duration, muscle "ribbing" (tone during contraction), and position (degree of displacement) of the examiner's finger during examination. This scale was adapted from scales to assess orgasmic dysfunction and was tested on 25 women (Table 2.1.4). Interobserver and test–retest reliability for this technique was proved. Brink and colleagues[2] reported on the "Pelvic Muscle Rating Scale," which is a 1–4 scale using a two-finger assessment of lateral and anteroposterior pressure, time or duration, and vertical displacement. This scale was tested in

2.1. Clinical Evaluation of the Pelvic Floor Muscles

TABLE 2.1.4. The circumvaginal musculature rating scale

Grade	1	2	3
Pressure	Slight	Moderate	Firm
Duration	None	3 seconds or less	4 seconds or more
Ribbing	Muscle feels soft; no ribbing	Muscle feels different from surrounding tissue but not very ribbed	Muscle feels distinct. Like rings of ribbing or ribbed muscle tissue
Position of examiner's fingers	Can easily slip out	Remains in the same position	Forcibly gripped or expelled, finger in upward position

Source: Worth AM et al., 1986.

338 noninstitutionalized women with UI. Inter-rater and test–retest reliability for this scale was also proved. (Table 2.1.5). Confirmation of PFM strength was also measured using a vaginal EMG perineometer. There were significant positive correlations between the perineometer and digital scores of PFM strength, but also significant relationships with such variables as ability to control urine stream and age. Table 2.1.6 is a scale modified from the Oxford grading system, which is an internationally accepted muscle grading method. However, none of the above grading systems is recognized by the International Continence Society document on standardization of terminology of the PFM, which recommends the following less subjective grading: absent, weak, normal, strong.[13]

Laycock[14] developed a notable PFM digital assessment tool that uses a 0–5 scale that is reproducible and assesses pressure and displacement

TABLE 2.1.5. Pelvic Muscle rating scale

Dimensions	1	2	3	4
Pressure	No response; cannot perceive on finger surface	Weak squeeze felt as flick at various points along finger surface, not all the way around	Moderate squeeze; felt all the way around finger surface	Strong squeeze; full circumference of fingers compressed; override
Duration	None	<1 second	>1 but <3 seconds	>3 seconds
Duration	None	Fingertips may move anteriorly (pulled up by muscle bulk) <1 second	Whole length of fingered move anteriorly	Whole fingers move anteriorly, are gripped and pulled in

Source: Brink et al., 1989.

TABLE 2.1.6. Pelvic floor muscle assessment of strength; the modified "Oxford Scale"

Grade	Description
0	**Nil** – lack of any discernible response in the peri-vaginal muscles.
1	**Flicker** – "flicker" describes a fluttering/quivering of the muscle.
2	**Weak** – contraction which is not fluttering.
3	**Moderate** – increase in pressure, compressing the examiner's fingers and incorporating a small degree of lift, as the fingers are moved in a cranial direction.
4	**Good** – contraction is firm causing lifting of the PFM up and in against resistance.
5	**Strong** – implies a very strong grip of the examiner's finger and positive movement in a cranial direction against strong resistance.

TABLE 2.1.7. The "P.E.R.F.E.C.T"

Scale	Measurement	Description
P	**P**ower (measurement of strength)	Graded from 0 (no movement) to 5 (strong movement) and is based on the Oxford grading system. Both slow and fast twitch muscle fibres contribute to power.
E	**E**ndurance measured in seconds	Duration of time (in seconds) that a maximum vaginal contraction can be maintained before a drop of 50% of the strength (up to 10 s) and reflects the activity of the slow-twitch muscle fibres.
R	Number of **R**epetitions that a woman is able to achieve	Another way to determine endurance of the muscle is to assess the repetitions of the muscle contractions. This is done by instructing the patient to contract the muscle for as long as possible – say 5 seconds, then repeat the contraction as many times as possible with 4 seconds rest in between each contraction. If the power of the contraction is reduced by 50% or the holding time decreases, the assessment stops and the number of repetitions is recorded. Ideally, the prescription for the home exercise program would be based on the number of repetitions that are performed during the PFM assessment.
F	Number of **F**ast – 1 to 2 second contractions that can be performed	The contractility of the fast-twitch fibres is determined after two minutes rest, by recording the number of fast or quick MVCs (1 to 2 seconds in length) that the woman can perform.
E	**E**very	This reminds the examiner to monitor progress by timing all contractions.
C	**C**ontraction	This method equates with general muscle assessment and re-education;
T	**T**imed	for example, after knee surgery, a woman is encouraged to strengthen her quadriceps muscles, being taught to increase the hold-time and number of repetitions of a "straight-leg raise."

The P.E.R.F.E.C.T. is a mnemonic of an assessment technique that can be used when performing a PFM examination as it supplies data regarding muscle contractility and coordination which can be used when planning a patient-specific exercise programme.[14]
Source: Adapted from Laycock and Jerwood, 2001.[15]

(Table 2.1.7). The P.E.R.F.E.C.T. scheme of PFM assessment has been validated to verify that the increase in vaginal pressure is caused by contraction of the perivaginal muscles, and not a manifestation of transmitted abdominal pressure caused by concomitant abdominal muscle contraction. Furthermore, interexaminer and test–retest studies have shown that this digital technique is reliable and reproducible.[15] This methodology may be criticized for its complexity, but it is the only reported method of PFM assessment that enables the planning of a patient-specific exercise program. In its favor, this method requires no expensive equipment and is quick and easy to perform. Table 2.1.8 shows examples.

The use of an assessment tool for documentation on the patient's medical record is advised,

TABLE 2.1.8. Examples of the P.E.R.F.E.C.T. assessment techniques

Example	Assessment	Explanation	Comment
1	P/E/R//F 4/7/5//6 – brisk response	P = 4, good contraction E = 7, held for 7 s R = 5, repeated 5 times F = 6, 6 fast contractions	This woman can perform five (good) contractions lasting 7 s, followed by six fast contractions. Every contraction is timed (E.C.T.) and this will be her exercise regimen. "Brisk response" describes the level of coordination.
2	P/E/R/F 2/5/2//3 – sluggish response	P = 2, weak contraction E = 5, held for 5 s R = 2, repeated 2 times F = 3, 3 fast contractions	This woman can perform two contractions lasting 5 s, followed by three fast contractions. Every contraction is timed (E.C.T.) and this is her initial exercise regimen. "Sluggish response" describes the poor coordination when asked to contract quickly.

2.1. Clinical Evaluation of the Pelvic Floor Muscles

TABLE 2.1.9. Clinical scale for grading digital evaluation of muscle strength

Patient Name _____ Date _____

(CHECK ONE) Scale	☐ VAGINAL EXAM Grade	☐ RECTAL EXAM Description
NONE	0	NO DURATION OF MUSCLE CONTRACTION, PRESSURE, DISPLACEMENT
		RECRUITMENT OF LARGE MUSCLE GROUP (E.G. GLUTEALS, ADDUCTORS, ABDOMINALS
TRACE	1/5	SLIGHT BUT INSTANT CONTRACTION: <1 SECOND
		RECRUITMENT OF LARGE MUSCLE GROUP (E.G. GLUTEALS, ADDUCTORS, ABDOMINALS
WEAK	2/5	WEAK CONTRACTION: WITH OR WITHOUT POSTERIOR ELEVATION OF FINGERS, HELD FOR >1 SECOND BUT ≤3 SECONDS
MODERATE	3/5	MODERATE CONTRACTION: WITH OR WITHOUT POSTERIOR ELEVATION OF FINGERS, HELD FOR AT LEAST 4–6 SECONDS, REPEATED 3 TIMES
GOOD	4/5	STRONG CONTRACTION: WITH POSTERIOR ELEVATION OF FINGERS, HELD FOR AT LEAST 7–9 SECONDS, REPEATED 4–5 TIMES
STRONG	5/5	UNMISTAKABLY STRONG CONTRACTION WITH POSTERIOR ELEVATION OF FINGERS, HELD FOR AT LEAST 10 SECONDS, REPEATED >5 TIMES

Evaluation – muscle hypertonus/spasm

CIRCLE ONE:	0	NO PRESSURE, TENDERNESS OR PAIN
	1	COMFORTABLE PRESSURE, SLIGHT TENDERNESS BUT BEARABLE
	2	UNCOMFORTABLE PRESSURE, SLIGHT PAIN
	3	MODERATE PAIN THAT INTENSIFIES WITH MUSCLE CONTRACTION
	4	SEVERE PAIN, WOMAN UNABLE TO PERFORM MUSCLE CONTRACTION DUE TO PAIN

Comments _____

Adapted with permission from Newman.[16]

especially in the U.S. Table 2.1.9 is a Clinical Scale for Grading Digital Evaluation of PFM Strength that has been used in clinical practice by this author.[16] It was developed by combining scales used in clinical research. It also allows the grading of women with pelvic pain and possible muscle spasms, and it is easy to use in a busy clinical practice.

There is a relative paucity of data regarding digital scoring systems in the evaluation of high-tone pelvic floor dysfunction. Most scales seem to address a woman's ability to contract her PFM without an assessment of tenderness or impaired relaxation.

Initiating a PFM Exercise Program

Before initiating a course of PFM exercises, clinical evaluation should be designed to assess the integrity of the PFM and their contractility. Questions the examiner should consider during the examination involve the functional integrity of the PFM and include the following:

1. Are the muscles morphologically symmetric?
2. Are there any defects in the muscles, such as hernias or tears from obstetrical trauma?
3. Is there any scarring?
4. What is the strength and bulk (volume) of the muscles?
5. Is there voluntary symmetric contraction?
6. Does contraction evaluate the bladder neck and anorectal angle?
7. Is there tenderness and pain during the examination? With a contraction?

The examiner uses both inspection and palpation to answer these questions. During inspection, when the LA muscle group is healthy, voluntary contraction of these muscles results in a puckering and drawing-in of the vaginal introitus, anal sphincter, and perineal body. Coughing should produce little or no descent of the perineum, in either the supine or standing position. In the woman whose LA muscle function is compromised, voluntary contraction may produce minimal puckering or no movement at all. Coughing when levator muscle function is compromised produces perineal descent and gaping of the vaginal introitus with accompanying POP. The standing position may be necessary to demonstrate the full extent of prolapse and perineal descent.

Other aspects of PFM function that can be evaluated include coordination and reflex response. A woman with a healthy functioning pelvic floor should be able to contract and relax the pelvic floor both quickly and slowly on command. An increase in intraabdominal pressure should result in a reflex contraction of the PFM that can be observed at the introitus and palpated digitally.[22]

PFM Rehabilitation or Training

The actual effects of PMEs on lower urinary tract function are not completely understood. Some studies show a relationship between changes in various measures of pelvic floor strength, such as anal sphincter strength or increased urethral closure pressure and resistance, all of which will prevent urine leakage. Kegel introduced PFM exercises through the implementation of a comprehensive program of progressive or maximal contractions of the LA muscle that incorporated biofeedback technology and was under direct supervision of a trained nurse.[17] The proposed mechanisms of action for PMEs are that:

1. A strong and fast pelvic muscle contraction closes the urethra and increases urethral pressure to prevent leakage during sudden increase in intraabdominal pressure (e.g., during a cough).[18,19] Urethral compression can be maximized by timing the muscle contraction at the exact moment of intraabdominal force (called the "Knack").[20-22] It refers to the skill of consciously timing a PFM contraction just before and during the intraabdominal pressure rise associated with a predictable stressful activity, such as a cough. A woman should practice the Knack by contracting her PFM about 1 second before a hard cough and maintaining the contraction throughout the cough.

2. Rising intraabdominal pressure (e.g. during coughing, laughing, or sneezing) exerts a downward (caudal) pressure or force on the bladder and urethra. Contraction of the LA exerts a counterbalancing upward (cephalic) force by lifting the endopelvic fascia, on which the urethra rests, and pressing it upward toward the pubic symphysis, creating a mechanical pressure rise.

3. Muscle contraction causes a pelvic muscle "reflex" contraction that precedes increased bladder pressure and may inhibit bladder overactivity. The aim is to acquire learned reflex activity.

Use of Instrumentation

In order to re-educate a muscle or muscle group, it is first necessary to evaluate the state of the muscle at rest and during a maximum voluntary contraction (MVC). Generally, muscle evaluation involves (1) inspection, (2) palpation and then (3) testing the muscles, with some method of standardized recording to allow comparison of clinical data before and after treatment; this methodology also applies to the PFM.

Instrumentation that objectively measures the PFM can be used to validate the digital examination and was first described by Kegel, who developed and used a vaginal perineonmeter that utilized pressure (manometry) for measurement. Traditionally, this device consisted of an airfilled vaginal pressure probe connected to a manometer that registered changes in pressure caused by a contraction of the perivaginal muscles,[17,24] and it is used to measure the strength of the PFM, as well as for biofeedback. The perineometer can, thus, provide objective information regarding the strength of the PFM. However, care must be taken to ensure that the pressure increase is caused by a PFM contraction and not transmitted abdominal pressure. This is most easily done by observing a drawing-in of the perineometer during a PFM contraction and palpating the abdominal muscles.

Most clinicians who specialize in this field prefer EMG measurements.[25] Electromyography is the recording of the electrical activity of a muscle and is a practical indicator of muscle activity which has been defined as:

- The study of electrical potentials generated by the depolarization of muscle.
- A monitor of bioelectrical activity correlating to motor unit activity; it does not measure the

muscle contractility itself, but the electrical correlate of the muscle contraction.
- An indicator of the physiological activity.

The advantage of EMG over manometric pressure is that, provided the machinery is of sufficient sophistication with adequate filtering, the EMG apparatus can engage the use of the newer types of electrodes that are lightweight and designed to stay in place, hence, allowing more functional positions during assessment and treatment.[26]

Weighted cones may also be used as an adjunct assessment tool, being the measure of the weight a subject can retain in the vagina while standing. It is generally accepted that as the muscle strength increases, heavier cones can be retained, thus improving the symptoms of SUI.[27] The disadvantage of this method lies in the varying vaginal diameters and amount of mucous present, which will influence the initial cone assessment.

Other Assessment Methods

Further methods of assessment of incontinence and pelvic floor strength include the stress test; SUI is a condition in which increased intra-abdominal pressure transmitted to the bladder (e.g., during coughing or jumping) exceeds intra-urethral pressure, resulting in urine loss. It is generally accepted that urinary leakage is more likely to occur with increasing volumes of urine in the bladder. The stress test is devised as a self-test for patients to monitor their own progress and to set targets for improvement. An example would be to aim to perform five coughs or five jumping jacks (jumping and landing with the arms and legs apart) with a full bladder (assuming this to be impossible at the beginning of treatment). The patient starts by carrying out five coughs (or jumping jacks) 10 minutes after emptying the bladder, and gradually (over several weeks) increases the time between voiding and coughing (or jumping jacks), providing there is no urine loss, until she can reach her target. The numbers/activities will need modification for individual patients, to provide a challenge without setting impossible goals. The "Stop Test," instructing patients to stop and start the flow of urine, is discouraged because of possible harmful effects.[7]

Monitoring Therapy

As a pelvic floor re-education program may last up to six months, it is important to monitor the effect of therapy by regular reassessment of both the PFM and the severity and frequency of pelvic floor symptoms. However, many patients are unable to attend a hospital/clinic on a regular basis because of work and family commitments. Monthly completion of a bladder diary (frequency/volume chart) and encouragement to continue with the exercise regimen should be tailored to resources and patient's needs.

References

1. Bourcier AP, Juras JC, Villet RM. Office evaluation and physical examination. In: Bourcier AP, McGuire EJ, and Abrams P, editors. Pelvic Floor Disorders. Philadelphia: Elsevier Saunders, 2004; 133–148.
2. Brink C, Sampselle C, Wells T, et al. A digital test for pelvic muscle strength in older women with urinary incontinence. Nurs Res. 1989;38(4):196–199.
3. Lukban JC, Whitmore KE. Pelvic floor muscle re-education treatment of the overactive bladder and painful bladder syndrome. Clin Obstet Gynecol. 2002;45:273–285.
4. Miller J, Kasper C, Sampselle C. Review of muscle physiology with application to pelvic muscle exercise. Urol Nurs. 1994;14(3):92–97.
5. Parks AG, Porter NM, Hardcastle JD. The syndrome of the descending perineum. Proc Royal Soc Med. 1966;59:477–482.
6. Neumann P, Gill V. Pelvic floor and abdominal muscle interaction: EMG activity and intra-abdominal pressure. Int Urogynecol J. 2002;13: 125–132.
7. Bump RC, Hurt WG, Fantl A, et al. Assessment of Kegel pelvic muscle exercise performance after brief verbal instruction. Am J Obstet Gynecol. 1999;165:322–329.
8. Hulme, JA. Research in geriatric urinary incontinence: Pelvic muscle force field. Top Geriatric Rehabil. 2000;16(1):10–21.
9. Sapsford R. The pelvic floor. A clinical model for function and rehabilitation. Physiotherapy. 2001; 87:620–630.
10. Bo K. Pelvic floor muscle training is effective in treatment of stress urinary incontinence, but how does it work? Int Urogynecol J. 2004;15(2):76–84.

11. Hodges PW, Richardson CA. Inefficient muscular stabilization of the lumbar spine associated with low back pain. Spine. 21;22:2640-2650.
12. Worth AM, Dougherty MC, McKey PL. Development and testing of the circumvaginal muscles rating scale. Nurs Res. 1986;35(3):166-168.
13. Messelink B, et al. The standardisation of terminology of pelvic floor muscle function and dysfunction. Report from the pelvic floor clinical assessment group of the International Continence Society (ICS). Neurourol Urodyn 2005:24;374-380.
14. Laycock J. Pelvic muscles exercises: physiotherapy for the pelvic floor. Urol Nurs. 14(3);136-140.
15. Laycock J, Jerwood D. Pelvic floor assessment; the PERFECT scheme. Physiotherapy. 2001;87(12): 631-642.
16. Newman DK. Managing and treating urinary incontinence. Baltimore: Health Professions Press; 2002.
17. Kegel AH. Physiologic therapy for urinary incontinence. JAMA. 1951;146:915-917.
18. Theofrastous JP, Wyman JF, Bump RC, et al. Effects of pelvic floor muscle training on strength and predictors of response in the treatment of urinary incontinence in women. Neurourol Urodyn. 2002;21:486-490.
19. Bo K, Talseth T. Change in urethral pressure during voluntary pelvic floor muscle contraction and vaginal electrical stimulation. Int Urogynecol J. 1997;8:3-7.
20. Miller JM. Criteria for therapeutic use of pelvic floor muscle training in women. JWOCN. 2002; 29(6):301-311.
21. Miller JM, Perucchini D, Carchidi L, DeLancey JOL, Ashton-Miller JA. (2001) Pelvic floor muscle contraction during a cough and decreased vesical neck mobility. Obstet Gynecol. 97(2):255-260.
22. Miller J, Aston-Miller J, DeLancey J. (1996) The Knack: Use of precisely-timed pelvic muscle contraction can reduce leakage in SUI. Neurol Urodyn. 15:302-393.
23. Brubaker L, Benson JT, Bent A, et al. (1997) Transvaginal electrical stimulation for female urinary incontinence. Am J Obstet Gynaecol. 177:536-540.
24. Shepherd AM, Montgomery E, Anderson RS. Treatment of genuine stress incontinence with a new perineometer. Physiotherapy. 1983;69:13.
25. Newman DK. Behavioral treatments. In: Vasavada SP, Appell RA, Sand PK, et al., editors. Female urology. Urogynecology and voiding dysfunction. New York: Marcel Dekker; 233-266.
26. Newman DK. Clinical Manual for Pelvic Muscle Rehabilitation. Dover, NH. Prometheus, Inc; 2003: 89-98.
27. Peattie AE, Plevnik S, and Stanton SL. Vaginal cones; a conservative method of treating genuine stress incontinence. Br J Obstet Gynaecol. 1988; 95:1049-1053.
28. Peschers UM, DeLancey JOL. Anatomy. In: Laycock J, Haslam J, editors. Therapeutic management of incontinence and pelvic pain. London: Springer-Verlag; 2002:7-16.
29. Ashton-Miller JA, Howard D, DeLancey JOL. The functional anatomy of the female pelvic floor and stress continence control system. Scand J Urol Nephrol 2001;207:1-7.
30. Allen RE, Hosker GL, Smith ARB, Warrel DW. Pelvic floor damage and childbirth: A neurophysiological study. Br J Obstet Gynaecol. 1990;97:770-779.
31. Smith ARB, Hosker GL, Warrell DW. The role of partial denervation of the pelvic floor in the aetiology of genitourinary prolapse and stress incontinence of urine. A neurophysiological study. Br J Obstet Gynaecol. 1989;96(1):24-28.
32. Oyama IA, Rejba A, Lukban JC, et al. Modified thiele massage as therapeutic intervention for female patients with interstitial cystitis and high-tone pelvic floor dysfunction. Urol. 2004;64(5):862-865.
33. Lilius HG, Oravisto KJ, Valtonen EJ. Origin of pain in interstitial cystitis. Scand J Urol Nephrol. 1973; 7:150-152.
34. Whitmore K, Kellog-Spradt S, and Fletcher E. Comprehensive assessment of pelvic floor dysfunction. Issues in Incontinence. 1998:1-10.

2.2
Examination of Patients with Pelvic Organ Prolapse

Ursula M. Peschers

> Examination of POP should aim at reproducing a patient's own observation. Assessment should separate between quality (which organ is prolapsed?) and quantity (what is the extent?) of POP. The ICS prolapse staging system is a validated and necessary tool for any studies on the treatment of POP. POP stage I and II are very common in women after vaginal childbirth and very rarely are the origin of symptoms.

Pelvic organ prolapse (POP) might cause a variety of urogynecological problems. While some women simply have the feeling of a bulge protruding out of the vagina and causing discomfort while sitting others cannot empty the bladder or the bowel completely or have symptoms of urinary or anal incontinence. It has to be kept in mind, however, that stages less than grade II in the ICS prolapse staging system normally do not cause specific symptoms.[1]

Patients with prolapse have defective pelvic support allowing the vagina and the uterus to descent below their normal position. The organs lying directly adjacent to the virgina (the bladder and the rectum) follow the vagina downwards. Traditionally the organ that is lying next to the part of the vagina that has prolapsed names the prolapse, e.g. in case of anterior vaginal wall prolapse this is called a cystocele. Qualitative (which is the prolapsed organ behind the vaginal wall?) and quantitative (how much prolapse is it?) assessment are necessary in order to come to a definitive prolapse diagnosis allowing for specific treatment. Qualitative assessment is based on carefully clinical examination, followed by imaging techniques (see Ultrasound, MRI) and in some cases final diagnosis reveals only during surgery.

A variety of classifications of POP are in use. To precisely describe the degree of POP the International Continence Society published a system of quantitative description of POP.[2] This systems allows to measure the degree of prolapse exactly and to give an exact descriptions of its amount. The interobserver and intraobserver reliability of the classification system has been shown.[3,4]

Systematic classification is important as it allows to measure and to compare the success of different treatment options including physiotherapy for POP.

Conditions of Examination

The examiner should see and describe the maximum protrusion noted by the individual during her daily activities. If in doubt the patient herself may be the best witness to decide on the maximal descent by the help of a handhold mirror. Usually the prolapse descends more in the standing position. The bladder should be empty because otherwise the patient might hesitate to strain for fear of leaking urine. A pair of separated vaginal specula should be used to investigate one compartment of the vagina while the others are reduced by the speculum.

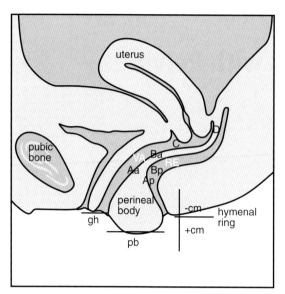

FIGURE 2.2.1. Quantative description of pelvic organ position. gh = genital hiatus; VA = vagina; RE = rectum; pb = perineal body. The description for the measurement points Aa, Ba, C, D, Bp, and Ap is found in Chapter 1.6.

Quantitative Description of Pelvic Organ Position (Fig. 2.2.1)

The vagina contains three compartments. Each compartment has one or two points whose position has to be measured. The maximum protrusion of each compartment has to be evaluated while the other compartments are held back with speculum:

- Anterior compartment (Point Aa and Ba): anterior vaginal wall with the adjacent organs urethra, bladder neck and the bladder. Aa corresponds to the bladder neck and is usually easily identified by the urethrovesical crease on the anterior vaginal wall, 3 cm from the outer meatus. By definition Aa is always located in a range from −3 cm to +3 cm from the hymen. Ba represents the most distal part of the anterior vaginal wall between Aa (corresponding to the bladder neck) and the cervix or the vaginal cuff. In a patient without prolapse Ba is located at −3 cm.
- Middle compartment: (Point C and D): cervix or vaginal cuff (after hysterectomy), posterior fornix. In patients with a uterus Point D corresponds to the insertion of the uterosacral ligaments. If the cervix is elongated a large difference between C and D will be found. Point D is not applicable for patients after hysterectomy.
- Posterior compartment: (Point Ap and Bp): posterior vaginal wall. Point Ap corresponds to the position of Aa but there is no anatomical landmark on the posterior vaginal wall. By definition Ap is located at −3 cm in a patient without prolapse of the posterior vaginal wall and its maximum protusion is +3 cm. Point Bp is the most distal part of the posterior vaginal wall between Ap and the cervix or the vaginal cuff.

Additionally the total vaginal length (tvl), and the size of the genital hiatus (gh) and of the perineal body (pb) are measured. The measurement of the lengths do not have minus or plus attached.

The ICS recommends recording the position of the points and the lengths in a simple 3 by 3 box. However, introducing each measured point into a special grid has rendered both: Simple and easy communicative. (see Figs. 2.2.2–2.2.5).

Stages are assigned according to the most severe portion of the prolapse when the full extend to the protrusion has been demonstrated.

Stage 0: No prolapse.

Stage I: The most distal part of prolapse is >1 cm above the level of the hymen. From a clinical point of view stage I prolapse is not a pathologic condtion.

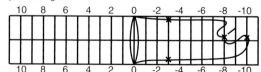

FIGURE 2.2.2. From Viereck et al., Geburtsh. u. Frauenheilk. 1997; 57:177–182, Abb. 2a, p. 179.

2.2. Examination of Patients with Pelvic Organ Prolapse

Stage II: The most distal part of prolapse is located between −1 cm and +1 cm. Clinically stage II leads to symptoms very rarely.

Stage III: The most distal part of prolapse is located at more than +1 cm beyond the hymen but protrudes no further than the total vaginal length minus 2 cm.

Stage IV: Prolapse further than stage III.

The following letter qualifiers name the leading part of the prolapse:

a = anterior vaginal wall
p = posterior vaginal wall
C = vaginal cuff
Cx = Cervix

Even though the ICS classification for POP seems to be complicated it has been shown that it is easy to use after a while. Studies regarding the treatment of POP should always include the ICS classification.

The following four examples demonstrate how to apply the staging system:

Patient # 1: Normal anatomy, no prolapse. The points Aa, Ba, Ap, and Bp are all located at −3 cm because there is no prolapse of the anterior and the posterior vaginal wall. The cervix is located 8 cm above the hymen (C = −8) and Point D (insertion of the sacrouterine ligaments 10 cm above the hymen (D = −10). The vaginal length is 10 cm, the genital hiatus 4 cm and the perineal body 3 cm. Staging: Grade 0.

Patient # 2: A patient presents with total protrusion of the vaginal cuff after hysterectomy. Point Aa and Ap are located at their maximum position at +3 cm. The most distal parts of the anterior and posterior vaginal wall (Points Ba and Bp) and the vaginal cuff (Point C) are found 8 cm beyond the hymen (+8). The total vaginal length (tvl) is 9 cm. Tvl (9) minus maximum prolapsed point (8) results into 1. Staging: Grade IV.

Patient # 3: A patient presents with prolapse of the anterior vaginal wall after hysterectomy. The vaginal cuff does not prolapse beyond the hymen. Aa is located at +3 cm. Ba is protruding further to +5 cm and is the leading part of the prolapse. C is found inside the vagina at −2 cm. The total vaginal length is 7.5 cm.

tvl (7.5) minus Ba (5) = 2.5. Staging: Grade IIIa.

Patient # 4: Prolapse of the posterior vaginal wall after hysterectomy. The anterior vaginal wall (Aa and Ba at −3) and the vaginal cuff (C at −6) do not descend. The leading point of the prolapse is Bp 4 cm beyond the hymen. The total vaginal length is 4 cm.

tvt (7) minus Bp (4) = 3. Staging: Grade IIIp.

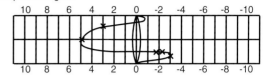

FIGURE 2.2.4. From Viereck et al., Geburtsh. u. Frauenheilk. 1997; 57:177–182, Abb. 3a, p. 180.

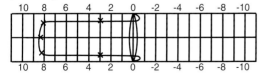

FIGURE 2.2.3. From Viereck et al., Geburtsh. u. Frauenheilk. 1997; 57:177–182, Abb. 2b, p. 179.

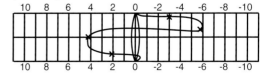

FIGURE 2.2.5. From Viereck et al., Geburtsh. u. Frauenheilk. 1997; 57:177–182, Abb. 3b, p. 180.

References

1. Swift S, Woodman P, O'Bogle A, Kahn M, Valley M, Bland J, Wang W, Schaffer J. Pelvic organ support study (POSST): The distribution clinical definition and epidemiologic concition of pelvic organ support defects. Am J Obstet Gynecol 2005;1992:795–802.
2. Bump RC, Mattiasson A, Bo K, Brubaker LP, DeLancey JO, Klarskov P, et al. The standardization of terminology of female pelvic organ prolapse and pelvic floor dysfunction. Am J Obstet Gynecol 1996;175:10–17.
3. Hall AF, Theofrastous JP, Cundiff GW, Harris RL, Hamilton LF, Swift SE, Bump RC. Interobserver and intraobserver reliability of the proposed International Continence Society, Society of Gynecologic Surgeons, and American Urogynecologic Society pelvic organ prolapse classification system. Am J Obstet Gynecol 1996;175:1467–1470; discussion 1470–1471.
4. Viereck V, Peschers U, Singer M, Schuessler B. Metrische Quantifizierung des weiblichen Genitalprolapses: Eine sinnvolle Neuerung in der Prolapsdiagnostik? Geburtsh Frauenheilk 1997;57:177–182.

2.3
Urodynamics

Ursula M. Peschers

Key Message

Not every woman with urinary incontinence needs urodynamic testing before physiotherapy. Nevertheless, a physiotherapist should be familiar with the purpose and technique of urodynamics for several reasons:

1. To understand normal function and pathophysiology of the lower urinary tract based on urodynamic investigation.
2. To understand when urodynamic studies are necessary.
3. To understand that urodynamic results may help to improve physiotherapy in certain patients.

This chapter summarizes the indications for urodynamic evaluation, introduces the basics of this technique, and shows typical patient profiles and their corresponding urodynamic traces.

Indications

Many patients present to the urogynecology clinic with unclear symptoms. Ideally, urodynamics should reproduce the patient's symptoms under controlled conditions. However, reality shows that this is not always possible. Therefore, urodynamic investigations cannot substitute for a thorough patient history and a good physical examination.

Patients should undergo urodynamics

- if history and physical examination do not lead to a diagnosis
- if voiding problems are present
- if at least 2 months conservative treatment did not improve symptoms
- before any planned continence surgery
- after failed incontinence surgery.

Urodynamic Techniques: Definitions and Technical Performance

The International Continence Society has published several papers to standardize definitions and measurement techniques.[1–3]

Urodynamics can be used to evaluate lower urinary tract symptoms and signs of lower urinary tract dysfunction.

Lower urinary tract symptoms include the following:

- storage symptoms (increased daytime frequency, nocturia, urgency, stress urinary incontinence, urge urinary incontinence, mixed urinary incontinence, noctural enuresis, and other types of incontinence and bladder sensation)
- voiding symptoms (slow stream, splitting and spraying, intermittent stream, hesitancy, straining, terminal dribble) and post micturition symptoms (incomplete emptying, post micturition dribble)

Conventional urodynamic studies are performed in the urodynamic laboratory. They involve artificial filling of the bladder via a catheter with a specified liquid at a specified filling rate and temperature (Fig. 2.3.1, A and B). In contrast,

Urodynamic studies provide information about the storage function of the bladder (filling cystometry or cystometrogram [CMG]), the voiding function (uroflowmetry, pressure-flow studies), and the function of the urethra (urethral pressure measurement/profilometry, leak point pressure measurements). Bladder filling with contrast medium allows for simultaneous fluoroscopic imaging of bladder filling and voiding. Electromyographic (EMG) studies of the pelvic floor muscles (with surface electrodes) or of the urethral sphincter muscle (with needle or wire electrodes) can be added.

Uroflowmetry

Voiding is the result of opening of the outlet (relaxation of the urethra and the pelvic floor muscles) and an increase of pressure from above (the detrusor). An obstructed outlet or a weak detrusor pressure or contraction of the pelvic floor leads to pathological voiding patterns.

Uroflowmetry is a simple, noninvasive test that measures and plots the amount of voided urine (y axis) per time unit (x axis) in milliliters per second while the patient sits on a micturition chair. The voided volume, the voiding time, the average flow rate, and the maximum flow Q_{max} should be recorded (Fig. 2.3.2). For women, a peak flow rate (Q_{max}) of 15–20 ml/s is regarded as normal. As Q_{max} is dependent on the voided volume, a minimal bladder volume of 150 ml is recommended.

Even in normals, flow curves differ individually, as well as in the same person, depending on several factors (e.g., bladder filling, degree of abdominal pressure, contractile status of the pelvic floor muscle).

A normal detrusor contraction combined with a normal bladder outlet results in a smooth, arc-shaped flow rate curve with high amplitude. A flattened asymmetric curve with a slow-declining end part might be caused by obstruction (e.g., after continence surgery or in women with a neuropathic pelvic floor) or from a weak detrusor. If women void by straining only, a typical interrupted flow pattern is resulting (see patient example # 4 and Fig. 2.3.7).

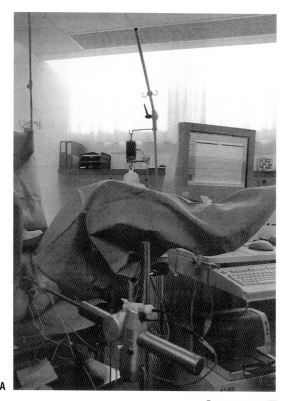

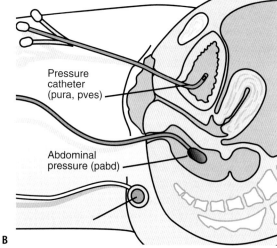

FIGURE 2.3.1. (A) Urodynamic lab with a patient in supine position under investigation. (B) The pressure lines are in bladder/urethra and a rectal catheter for abdominal pressure measurement as well as a surface electrode for pelvic floor muscle recording.

ambulatory urodynamics studies the lower urinary tract during natural filling, while the patient is engaged in every day activities. To date, this test is not adequately validated and rarely used and, thus, will not be described in this chapter.

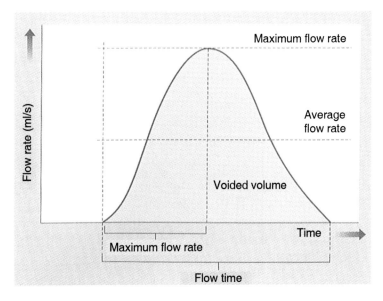

FIGURE 2.3.2. Urine flow curve and its description terminology. (*Source*: Cardozo and Staskin, 2001.)

Filling Cystometry and Pressure Flow Studies

Filling cystometry and pressure-flow studies are invasive procedures that involve artificial filling of the bladder. During filling, the detrusor pressure (y axis) is constantly monitored and plotted against the filled volume (x axis).

As the pressure generated by the detrusor muscle cannot be measured directly, the intravesical pressure (p_{ves}) is measured with an intravesical probe, intraabdominal pressure is measured with an intrarectal probe, and the detrusor pressure (p_{det}) is calculated by subtracting p_{abd} from p_{ves} (Fig. 2.3.1 B). During voiding, the flow curve is plotted parallel to the detrusor pressure.

Pressure can be measured either by fluid filled lines connected to external pressure transducers or by pressure transducers mounted on a catheter (microtip transducers). The equipment should include at least three measurement channels (two for pressure, one for flow), a display (monitor or printer), and secure storage of three pressures (p_{abd}, p_{ves}, p_{det}) and flow (Q) as tracings against time.

Several landmarks are necessary for the interpretation of results, as follows:

- First sensation, which occurs when the patient first becomes aware of the filling.
- First desire to void, which is the feeling that the patient would pass urine if conveniently possible, but that voiding could still be postponed.
- Strong desire to void, which is defined as a persistent desire to void without fear of leakage.
- Maximum cystometric capacity, which is reached when the patient has a very strong urge to void and can no longer delay micturition.
- Bladder compliance, which describes the relationship between change in bladder volume and change in detrusor pressure (ml/cm H_2O). Compliance is reduced if the bladder is unable to stretch during filling (e.g., after radiation or radical pelvic surgery). This leads to a steady increase in intravesical pressure.
- Normal detrusor function, which is present when there is no or little increase in bladder pressure during bladder filling. No involuntary phasic contractions should occur despite provocation tests (e.g., coughing).

Spontaneous or provoked involuntary detrusor contraction during the filling phase is defined as detrusor hyperactivity. If the patient leaks urine during a sudden increase of abdominal pressure in the absence of the detrusor contraction (e.g., a cough) this is defined as urodynamic stress incontinence.

Pressure-flow studies allow examiners to determine detrusor function during voiding

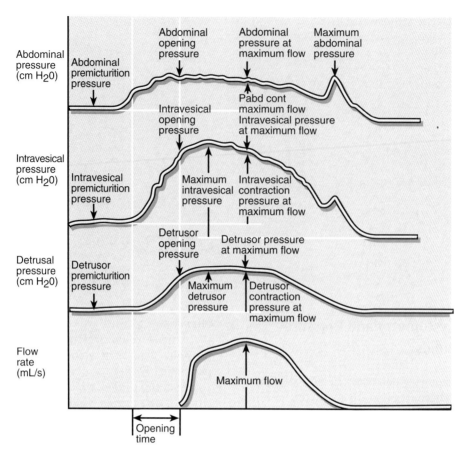

FIGURE 2.3.3. Example of a pressure/flow recording during filling and micturition. (Nomenclature as recommended by the International Continence Society.) (*Source*: adaptation from Bourcier AP, McGuire EJ, Abrams P: Pelvic floor disorders. Philadelphia: Elsevier Saunders; 2004: 205. With permission from Elsevier.)

by measuring flow and voiding pressure at the same time.

The same equipment is used as for uroflow and filling cystometry. After filling the bladder to maximum capacity, the patient attempts to void with the intravesical and the intrarectal catheter still in place. Two pressures (p_{abd}, p_{ves}) and the flow (Q) are recorded. The detrusor pressure p_{det} is calculated simultaneously (Figs. 2.3.2 and 2.3.3; see Chapter 1.5).

Pressure-flow studies are not required in women with stress urinary incontinence unless pathological micturition (e.g., residual urine or reduced uroflow) is present; this is to distinguish between three main causes:

1. Intravesical obstruction
2. Weak detrusor function (hypocontractile or acontractile detrusor)
3. Detrusor-urethral or pelvic floor dyssynergia, which is uncoordinated detrusor contraction and pelvic floor/urethral muscle relaxation.

Urethral Pressure Measurement

Understanding urethral function is important to the investigation of stress incontinence. Urethral pressure is the pressure needed to open a closed (collapsed) urethra. The measurements are prone to artifacts caused by the catheters (stiffness, direction of the tip) and by the patient (movement, etc.).

The urethral closure pressure profile is a graph indicating the pressure along the length of the urethra (urethral pressure profile) (Fig. 2.3.4, A and B). It is calculated by subtracting the inter-

2.3. Urodynamics

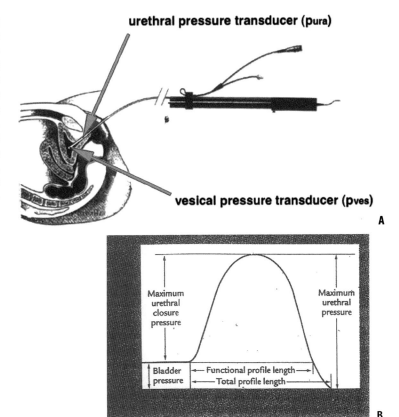

FIGURE 2.3.4. (A) Schematic drawing of the technique for urethral pressure measurement. A puller device is retracting the catheter with the pressure transducer at constant speed through the urethra, registering the pressures alongside the urethra while at the same time intravesical pressure is measured. Subtracting p_{ves} from p_{ura} is the urethral closure pressure p_{clos} (A). (*Source*: reprinted from Cardozo and Staskin, 2001.) (B) Urethral pressure profile trace at rest and its description terminology according to ICS definition.

vesical pressure (p_{ves}) from the intraurethral pressure (p_{ura}). Both pressures are measured by a transducer, one constantly lying in the bladder (p_{ves}), while the other is withdrawn with constant speed (1 mm/s) alongside the urethra (p_{ura}), either at rest or during physical stress (stress profile). The maximum urethral closure pressure (MUCP) is age dependent and its normal range is between 50 and 80 cm H_2O. A MUCP of <20 H_2O is regarded as a low-pressure urethra.[4] The functional urethral length is the length of the urethra, along which urethral pressure exceeds intravesical pressure.

In women with normal pelvic floor anatomy, the urethra is connected to the levator ani muscle by connective tissue (see chapter 1.1). Contraction of the levator ani muscle leads to an increase, and relaxation of the levator ani to a decrease, in urethral pressure.[5] A constant pelvic muscle contraction results in an abnormally high urethral pressure. If the urethral connection to the pelvic floor muscles is damaged (e.g., by childbirth), contraction of these muscles are unable to increase the intraurethral pressure.

During coughing, the pressure transmission ratio can be determined; it is the increment in urethral pressure on stress as a percentage of the simultaneously recorded increment in intravesical pressure. Stress urinary incontinence is ideally represented when the closure pressure equals 0 cm H_2O during every cough alongside the urethra (see Patient # 2; Fig. 2.3.6 B). It has to be understood that the stress profile is very often subjected to artifacts.

Leak Point Pressure

The abdominal leak point pressure is the minimum intravesical pressure at which leakage occurs because of increased intravesical pressure in the absence of a detrusor contraction. It is commonly performed by letting the patient cough or strain with a full bladder (300 ml) until leakage occurs. Although the examination technique is widely used, there is still no standardization on how to measure this pressure.

As it is known that the cough leak point pressure is significantly higher than the Valsalva leak point pressure, it should be stated which maneuver is used to provoke leakage.[6] Additionally the size of the intravesical catheter,[7] its position (intravesical or intravaginal), and the bladder volume[8,9] have a significant influence on the results. Also, it is likely that the leak point pressure is higher if the patient contracts the pelvic floor muscles.

Typical Case Examples

Patient # 1: The Continent Woman with Normal Voiding

A 20-year-old nulliparous woman with no incontinence symptoms and is able to contract her pelvic floor muscles voluntarily. There is no pelvic organ descent, nor prior pelvic surgery.

The patient is able to void normally (Fig. 2.3.5 A). The flowchart is bell-shaped with final spikes (arrows), indicating additional abdominal strain.

Her maximum flow rate (Q_{max}) is 25.7 ml/s, and her voided volume is 353 ml. She is able to voluntarily interrupt her urine flow (arrow) during micturition with the help of a pelvic floor muscle contraction (Fig. 2.3.5 A).

Her urethral pressure profile at rest shows a MUCP of 72 cm H_2O (Fig. 2.3.5 B). The high pressure measured indicates the intactness of the intrinsic urethral closure mechanism. However, it could also indicate that her pelvic floor is not completely relaxed because of the urodynamic examination. Active contraction of the PFM increases the pressure (Fig. 2.3.5 B). This shows that the pelvic floor is able to influence the sphincter unit with or without a concomitant contraction of the striated urethral sphincter.[10]

Her cough stress profile (Fig. 2.3.5 C) shows a positive pressure transmission onto the urethra during coughing, indicating that there might be also reflex activity present within the striated sphincter muscle and/or the pelvic floor muscle.

During filling cystometry (Fig. 2.3.5 D) the detrusor pressure remains stable despite provocational coughs. A Valsalva maneuver at a bladder filling of 300 ml with a pressure increase of 80 cm H_2O does not lead to urinary leakage, also indicating the intactness of the sphincter mechanism. Maximum bladder capacity normally exceeds 400 ml. Compliance is unmeasurably low because of the failure of an increase in detrusor pressure during filling up to 400 ml.

Patient # 2: Stress-Incontinent Woman

A 34-year-old woman who has undergone three vaginal deliveries. No previous pelvic surgery. During examination no significant prolapse can be found. She is hardly able to contract her pelvic floor muscles. She describes symptoms of stress urinary incontinence and denies any urgency. With the bladder almost empty she leaks while coughing and straining in the supine position.

During urodynamics she has a normal voiding pattern and a normal filling cystometry (see patient #1).

Figure 2.3.6 a shows the urethral pressure measurement at rest.

Her MUCP is 14 mm H_2O, her functional urethral length is 23 cm. Active contraction of the PFR does not increase urethral closure pressure, indicating either an inability to contract the pelvic floor muscle or damage to its connection with the urethra. The very low urethral closure pressure (<20 cm H_2O), together with a cough leak point pressure of 43 cm H_2O, indicates intrinsic urethral insufficiency, a condition in which denervation or defects within the sphincter unit, rather than hypermobility of the urethra, is the primary cause of stress urinary incontinence.. This condition is generally difficult to treat with physiotherapy.

The stress profile shows a negative closure pressure along side the length of the urethra (Fig. 2.3.6 B) representing the absence of any closure

2.3. Urodynamics

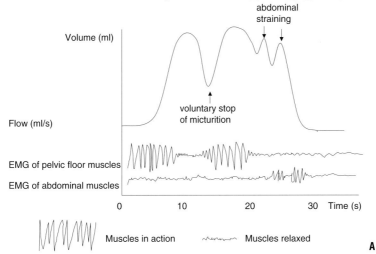

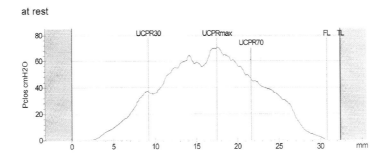

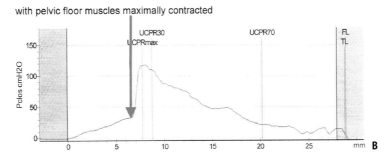

FIGURE 2.3.5. (A) Uroflowmetry. Urine flow pattern in a normal female. Arrows pointing down indicate straining episodes in order to speed up urine flow at the end of micturiton. Voluntary interruption of the urine stream by contraction of the pelvic floor muscle indicated by an arrow pointing up. EMG tracing indicates pelvic floor muscle relaxation to allow normal uroflow and contraction to interrupt midstream. **(B)** Urethral closure pressure profile (at rest): Pelvic floor muscle completely relaxed (upper curve). Maximal pelvic floor muscle contraction during measurements leads to a typical sharp increase in closure pressure at the inner third of the urethra (arrow) clearly exceeding a relaxed closure pressure profile.

Stressprofile

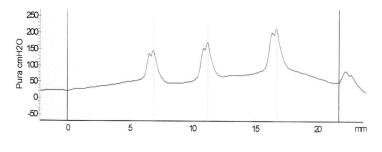

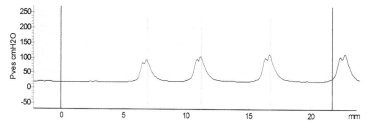

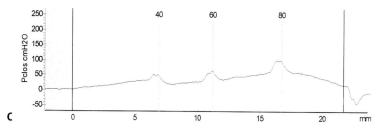

Filling Cystometry

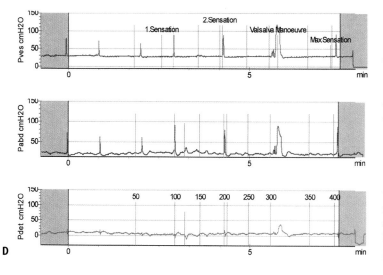

FIGURE 2.3.5. (**C**) Cough stress profile of a continent female. P_{ura} is the pressure measured alongside the urethra; p_{ves} measures the pressure in the bladder at the same time; p_{clos} is the pressure calculated by subtracting p_{ves} from p_{ura}. It stands for the closure pressure of the urethra during coughing. Positive pressure transmission during the coughs indicates reflex contractions of the pelvic floor muscles. (**D**) Normal filling cystometry of a continent female without symptoms of OAB. P_{ves} is the pressure measured in the bladder during constant filling; p_{abd} is the pressure in the rectum which stands for the intraabdominal pressure; p_{det} is the detrusor pressure which is calculated by subtracting p_{abd} from p_{ves}.

Hypotonic Urethra

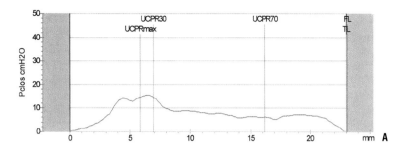

Cough stress profile

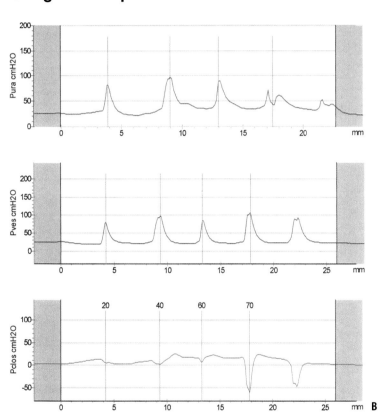

FIGURE 2.3.6. **(A)** Urethral closure pressure profile (at rest) of a stress incontinent female with a hypotonic urethra (intrinsic sphincter deficiency). Max. urethral closure pressure: 18 cm H_2O. **(B)** Cough stress profile of the same patient.

barrier of the urethra when the patient starts coughing.

Patient # 3: Voiding Dysfunction

A 54-year-old woman presenting with voiding difficulties after abdominal colposuspension for stress urinary incontinence two years ago. She does not have incontinence. During pelvic examination there is no prolapse. The anterior vaginal wall and the bladder neck are immobile during straining. There is 80 ml of residual urine, and there is no urinary infection. Despite a hyperactive contraction of the detrusor muscle, the urine flow is low and intermittent. Additional flow spikes are the result of concomitant abdominal straining (Fig. 2.3.7).

Patient # 4: Overactive Bladder Syndrome

A 63-year-old woman presents with symptoms of urgency and urge incontinence. She complains about the need to void every hour during the daytime and three to four times during the night.

When feeling the urge to void, she leaks while trying to get to the toilet. A prolapse of the anterior vaginal wall is found, but no urinary tract infections are evident.

Filling cystometry reveals uninhibited detrusor hyperactivity, starting early after the bladder is filled above 100 ml and immediately leading to a complete loss of stored urine (Fig. 2.3.8). The very high uninhibited detrusor contractions are indicative of a neurological disease as the origin, rather than an idiopathic cause.

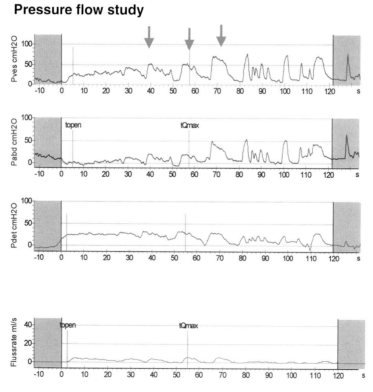

FIGURE 2.3.7. Pressure flow study of a female with voiding difficulties after incontinence surgery. Although detrusor pressure with 30 cm H_2O is quite high, resulting urine flow rate is low (8 ml/s) thus indicating intravesical obstruction. Arrows indicate additional straining efforts to further press out intravesical urine.

Filling Cystometry

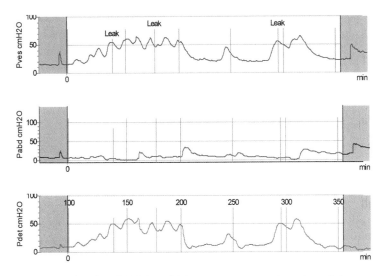

FIGURE 2.3.8. Filling cystometry in a female with OAB / wet (high pressure motor urge incontinence). (*Source*: reprinted with permission from Wall LL, 1993.)

References

1. Abrams P, Cardozo L, Fall M, et al. The standardisation of terminology of lower urinary tract function. Neurourol Urodyn. 2002;21:167–178.
2. Lose G, Griffiths DJ, Hosker GL, et al. Standardisation of urethral pressure measurement: report from the Standardisation Sub-Committee of the International Continence Society. Neurourol Urodyn. 2002;21:258–260.
3. Schaefer W, Abrams P, Liao L, et al. Good urodynamic practices: uroflowmetry, filling cystometry, and pressure-flow studies. Neurourol Urodyn. 2002;21:261–274.
4. Wall LL, Norton PA, DeLancey JO. Practical urodynamics. In: Wall LL, Norton PA, DeLancey JO, editors. Practical urogynecology. Baltimore: Williams & Wilkins; 1993.
5. Baessler K, Miska K, Draths R, et al. Effects of voluntary pelvic floor contraction and relaxation on the urethral closure pressure. Int Urogynecol J Pelvic Floor Dysfunct. 2004;16:187–190.
6. Peschers UM, Jundt K, Dimpfl T. Differences between cough and Valsalva leak-point pressure in stress incontinent women. Neurourol Urodyn. 2000;19(6):677–681.
7. Bump RC, Elser DM, Theofrastous JP, et al. Valsalva leak point pressures in women with genuine stress incontinence: reproducibility, effect of catheter caliber, and correlations with other measures of urethral resistance. Continence Program for Women Research Group. Am J Obstet Gynecol. 1995;173(2):551–557.
8. Miklos JR, Sze EH, Karram MM. A critical appraisal of the methods of measuring leak-point pressures in women with stress incontinence. Obstet Gynecol. 1995;86(3):349–352.
9. Theofrastous JP, Cundiff GW, Harris RL, et al. The effect of vesical volume on Valsalva leak-point pressures in women with genuine stress urinary incontinence. Obstet Gynecol. 1996;87:711–714.
10. Miller JM, Umek WH, DeLancey JO, et al. Can women without visible pubococcygeal muscle in MR images still increase urethral closure pressures? Am J Obstet Gynecol. 2004;191:171–175.

2.4
Applying Urodynamic Findings to Clinical Practice

Christopher K. Payne

Key Message

This chapter depicts how urodynamic studies are helpful in deciding whether or not surgery and physiotherapy are appropriate treatment/strategies for stress, urge, and mixed incontinence, as well as voiding dysfunction. Carefully selected practically based opinions are, given as there is a lack of evidenced-based data thus far.

Introduction

As described in previous chapters, urodynamic studies are used to evaluate women with incontinence and lower urinary tract symptoms in two common clinical settings:

1. Preference for surgical therapy in women with clinically predominant stress incontinence symptoms
2. Unsatisfactory response after empirical application of conservative measures.

Urodynamic studies are also employed initially, when the complexity of the clinical situation prevents making a provisional diagnosis. Some clinicians prefer to test all patients before any therapy is administered. In any case, once the examination has been completed, the clinician will need to interpret the findings and counsel the patient about the relevant treatment options. Unfortunately, there is considerable disagreement on the meaning and significance of many urodynamic tests and little evidence on which to base opinions. The Urodynamics Committee of the 3rd International Consultation on Incontinence found mostly low level grades 4 and 5 evidence (case series and expert opinion) to support the use and interpretation of urodynamic tests in the majority of clinical situations.[1] The Committee on Research Methodology pointed out the need for large-scale prospective studies to determine the predictive value of urodynamic tests in determining optional treatment for patients with incontinence.[2] Thus, although the urodynamic diagnosis may suggest a preferable course of action, the ultimate decision must incorporate the goals and values of the patient as well as the locally available resources.

It is rare that only one form of treatment is acceptable, and common that available data do not clearly define the "best" or "optimal" therapy.

Despite the limitations, urodynamic studies remain the primary method of evaluating lower urinary tract complaints. Interpreting urodynamic studies and counseling the patient about therapy calls for a combination of clinical expertise and empathetic understanding of the patient's condition and objectives. The most common urodynamic scenarios are discussed below and are, in most cases, reflective of expert opinion.

Pure Stress Incontinence

When conservative measures are preferred, urodynamic studies are not typically performed, yet the option of conservative therapy should not be ignored when counseling patients after urodynamic testing. First, there is no evidence that specific urodynamic findings (e.g., severe incontinence as manifested by low leak pressure or presence of urethral sphincter dysfunction as shown by a low urethral closure pressure), or clinical criteria (e.g., age, obesity, prior incontinence surgery) define patient groups less likely to respond to pelvic floor rehabilitation.[6] Therefore, no patient should be primarily excluded from pelvic floor muscle training (PFMT) if they are willing to train. Expert opinion suggests that patients with a positive stress test and a good voluntary pelvic floor muscle contraction, which is not coordinated during a cough, represent a particularly attractive group for PFMT. The chance for successful conservative treatment is very high, given that leakage is controlled after instructions to perform both cough and contraction simultaneously. This could be performed on a "see and treat" basis in connection with the urodynamic assessment.[7] The authors believe that conservative therapy with pelvic muscle exercises and/or biofeedback should be strongly recommended when pelvic floor strength is poor and the objective severity of stress urinary incontinence (SUI) is low, as there is substantial room for improvement.

There are many groups that experts feel are less likely to respond to conservative measures. For example, patients who show no urethral mobility on clinical examination or perineal ultrasound despite a sustained voluntary contraction. Some experts also identify patients who are able to perform a sustained contraction in which either the urethra is not adequately lifted upwards or is not kept in this position (e.g. during a cough) as poor candidates for training. In this case, there is presumably a poor connection of the urethra to the puborectalis muscles. This entity is easily encountered during digital examination or perineal ultrasound. Others have suggested that patients with chronic bronchitis or athletes in endurance sports are not good candidates.

Urodynamic studies may help identify the patients most appropriate for conservative therapy by highlighting poor voluntary function of the pelvic floor. To define that weakness, the urodynamic examination must isolate and assess voluntary and reflex pelvic floor activity. Perineal patch electrodes and/or direct urethral pressure measurement can be used to establish baseline function—a simple survey could include evaluation of voluntary contraction strength and duration, "quick flicks," and response to cough and bulbocavernosus stimulation. When preoperative urodynamic studies indicate an increased chance of postoperative retention (hypocontractile bladder) a concerted effort to manage the problem with conservative measures should be considered before surgical intervention.

The following three urodynamic criteria define the ideal candidate for surgical treatment of stress incontinence:

1) There is significant and objective stress leakage
2) There is no or minimal associated detrusor overactivity
3) Voiding dynamics are normal.

Such patients are highly likely to have favorable outcomes with surgical treatment and should be encouraged to undergo surgery if so inclined.

The presence of detrusor overactivity does not in any way preclude surgical treatment of stress incontinence (see section on mixed incontinence), but it does call for different counseling. When the filling phase is stable, the patient is simply informed about the risk of de novo overactivity/urge incontinence as appropriate to the chosen procedure, approximately 0–27% for fascial slings, 1–16% for colposuspension, and probably less for tension-free midurethral slings.[3] In contrast, patients with documented detrusor overactivity are counseled about the likelihood of the resolution of urge symptoms. For all patients with preoperative urgency, symptoms persists in 30–50% of patients, but this rate may be higher in those patients with documented detrusor overactivity, particularly those with high pressure unstable contractions.[4] In some of these cases, urgency can even increase despite resolution of stress incontinence, particularly if the operation creates any voiding dysfunction.

Finally, it is important to demonstrate normal voiding. Urinary retention is one of the most serious complications of stress incontinence surgery and is not always reversible. If there is no clinical reason to suspect voiding problems, a normal uroflow and low postvoid residual will suffice. Patients with abnormally low flow rates, patterns suggesting Valsalva voiding, elevated residual urine, or medical problems correlated with bladder dysfunction (e.g., diabetes or lumbar spine disease) are probably investigated by a pressure-flow voiding study. In the absence of high quality prospective data, it is most appropriate to assume that a patient who depends on abdominal straining to void will be at higher risk for urinary retention after any type of stress incontinence procedure, including injection therapy, and to counsel accordingly.[5] On the other hand, low pressure bladder contractions are not necessarily worrisome. Women with stress incontinence often have very low urethral resistance and, thus, may void without a detectable contraction. If the flow curve is normal, the peak flow rate is normal, and the residual is low, there should be no increased risk of retention.

Pure Detrusor Overactivity

The patient with clinically pure urge incontinence is typically managed with pharmacotherapy, behavioral techniques, and pelvic floor exercises. The primary role for urodynamics is when these simple measures fail and more invasive treatment is being considered. Here, the role of urodynamic studies is to identify other treatable problems, such as associated stress incontinence or bladder outlet obstruction, for which curative therapy exists. Also, patients with low pressure overactivity and leakage should be referred for biofeedback training and/or electrical/magnetic stimulation before considering sacral neuromodulation or reconstructive surgery.

Patients with high pressure overactive contractions must have good pelvic floor function; if they were unable to oppose the overactive contraction appropriately then there would be leakage before high pressures could be generated.

True Mixed Incontinence

The patient with true mixed incontinence, both SUI and detrusor overactivity incontinence on urodynamic testing, presents a challenge in forming a treatment plan. The clinician's goal is to:

- Select patients who are most likely to benefit from surgical therapy
- Identify patients who are most likely to suffer from persistent urge symptoms after surgery.

The urodynamic study gives an assessment of pelvic floor muscle strength to correlate with the physical exam, an objective measure of the severity of the stress incontinence component, and an objective measure of the degree of bladder dysfunction. Those patients with more severe SUI and better pelvic floor function are better candidates for surgery than are patients with less SUI, weaker pelvic floor muscles, and more severe detrusor overactivity. In the latter case, surgery should be employed only if concerted efforts at conservative therapy fail.

It is generally felt that patients with mixed incontinence have poorer outcomes with surgery for stress incontinence because of the possibility of persistent urge symptoms, although data are conflicting and incomplete. Therefore, these patients should usually be offered PMFT as a first-line treatment, with or without concomitant drug treatment (e.g. anticholinergics, Duloxetine).

Dysfunctional Voiding

As with the case of acontractile bladder, dysfunctional voiding is seen much more frequently in pressure-flow urodynamic studies than it is as an actual important clinical issue. The pressure-flow study must be correlated with the clinical symptoms and a free-flow/residual urine determination. Very often, the low flow rate and high EMG activity are only artifacts of the testing situation for the reasons mentioned above. When the clinical complaints, free-flow, and pressure-flow study do all suggest a diagnosis of dysfunctional voiding, consideration should be given to the possibility of neurogenic bladder with detrusor–sphincter dyssynergia. (see Chapter 1.5, Fig. 1.5.2 B). This diagnosis mandates a search for a suprasacral spinal cord lesion. In the absence of a neurogenic cause, the diagnosis is dysfunctional voiding and the treatment options include pelvic floor physical therapy with maximal emphasis on treatment of a hypertonic pelvic floor, which is often found in these patients, biofeedback, and sacral neuromodulation.

References

1. Griffiths D, Kondo A, Bauer S, et al. Dynamic testing. In: Incontinence. Abrams P, Cardozo L, Khoury S, et al., editors. Plymouth, UK: Health Publications, Ltd.; 2005.
2. Payne CK, van Kerrebroeck P, Blaivas J, et al. Research methodology. In: Incontinence. Abrams P, Cardozo L, Khoury S, et al., editors. Plymouth, UK: Health Publications, Ltd.; 2005.
3. Smith ARB, Daneshgari F, Dmokowski R, et al. Surgery for stress incontinence in women. In: Incontinece. Abrams P, Cardozo L, Khoury S, et al., editors. Plymouth, UK: Health Publications, Ltd.; 2005.
4. Schrepferman CG, Griebling TL, Nygard IE, et al. Resolution of urge symptoms following sling cysto-urethropexy. J Urol. 164:2000:1628–1631.
5. Nguyen JK. Diagnosis and treatment of voiding dysfunction caused by urethral obstruction after anti-incontinence surgery. Obstet Gynecol. 2002: 468–475.
6. Wilson P, Berghmans B, Hagen S, et al. Adult conservative management. In: Incontinece. Abrams P, Cardozo L, Khoury S, et al., editors. Plymouth, UK: Health Publications, Ltd.; 2005.
7. Miller JM, Ashton JA, DeLancey JOL. A pelvic muscle precontraction can reduce cough-related urine loss in selected women with mild SUI. J Am Geriatr Soc. 1998;46:870–874.
8. Anger JT, Rodriguez LV. Mixed incontinence: stressing about urge. Curr Urol. Rep 6;2004:427.

2.5
Anorectal Physiology

David Z. Lubowski and Michael L. Kennedy

Key Message

The most import clinical manifestation of abnormal anorectal physiology is anal incontinence, and this, together with its causes, is comprehensively reviewed in this chapter. The physiology includes studies on anorectal and colonic functions, using manometric and electrophysiological studies. The latter includes motor and sensory nerve conduction and electromyography. Measurements of colonic transit rely on radioisotope studies, such as defecation scintigraphy, and dynamic studies, such as defecography.

Endoanal ultrasound using static and rotating probes is widely available and used for anal sphincter morphology.

Anal incontinence is defined as the involuntary loss of stool or flatus and is a distressing condition frequently seen by clinicians. It has previously been estimated that the prevalence of anal incontinence may be as high as 15%[1] in the general community and up to 20% of patients in elderly care institutions.

Introduction

When a patient presents with anal incontinence, a thorough investigation is required to find evidence of neuropathy, muscle weakness, colonic abnormality, or combinations of these. Normal anorectal function requires an intact anatomy and is a complex, multifactorial physiological process. The mechanisms of continence and defecation are interdependent, and neither can function completely normally without the other. Continence and defecation both involve colonic and anorectal function, and the term "anorectal physiology" is misleading because any discussion about anorectal function must now include the colon in order to have clinical relevance.

A variety of tests are available to study anorectal and colonic function. Some are useful in everyday clinical practice, whereas other tests have an important research role. This chapter examines the available physiological tests, and places each in a clinical context. It is important to carefully select those tests which will provide useful clinical information for each individual patient, as each patient has a tolerance level for anorectal tests above which he or she will become less compliant, and exceeding this level with unhelpful tests will be counterproductive.

Anorectal Manometry

The anal canal's high pressure zone is the most important component of the continence mechanism. Anal pressure may be measured at different levels in the anal canal, usually at approximately 1-cm intervals. Several recording devices are available, such as perfusion catheters, microballoons, sleeve catheters, or strain gauge transducers. Perfusion catheter systems are the most commonly used. The maximum diameter of the recording device should be no greater than 5 mm because large-diameter catheters may artificially raise anal pressure.[2]

2.5. Anorectal Physiology

Perfusion Systems

A perfusion catheter is perfused with water at a constant rate, from 0.4 to 0.6 mL/min. A minimum of four pressure channels are generally required. Each channel terminates at a side-hole, the first being 2 cm proximal to the tip for recording rectal pressure. Three centimetres proximal to this is the first of three side-holes placed 1 cm apart for recording anal pressure. One channel opens at the tip to allow inflation of a rectal balloon. The catheter is positioned with the distal recording side-hole in the lowest part of the anal canal, where the basal resting pressure is allowed to stabilize for 1–3 min. The subject is then requested to maximally contract the sphincter for 3–5 s three times at 10-s intervals. The subject is also asked to cough forcefully three times at 10-s intervals. In this way both internal and external sphincter strength are assessed.

Normal Recordings

Resting Anal Pressure

Resting anal pressure is largely caused by the internal sphincter (Fig. 2.5.1). The muscle is in a

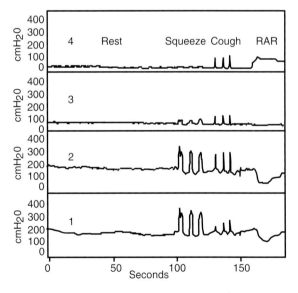

FIGURE 2.5.1. Anorectal manometry using a perfusion system. Channel 1 is recording just within the anal verge, and channels 2 and 3 progressively more proximally in the anal canal. Channel 4 is rectal pressure. Pressures at rest, during muscle contraction and then during coughing are shown. Rectal pressure increases during coughing but not with voluntary contraction.

TABLE 2.5.1. Normal mean anal pressures

	Sun and Read[10]	Loening-Baucke and Anuras[17]	Williams et al.[56]
	cm H$_2$O	mm Hg	mm Hg
Resting pressure			
Mean		60 (13)	
Mean male	91 (5)	63 (12)	62 (5)
Mean female	61 (6)	50 (13)	64 (5)
Squeeze pressure			
Mean		204 (54)	
Mean male	257 (20)	238 (38)	189 (10)
Mean female	107 (13)	159 (45)	142 (7)
Method	Perfusion	Strain gauge	Perfusion

Values in parentheses are standard deviations.

state of continuous intrinsic contraction,[3] which has phasic variation known as slow waves. These are more apparent when resting pressure is elevated, and they occur at a rate of 6–20 per minute with an amplitude 10–25 cm H$_2$O[4]. Additionally, ultra–slow waves are high-pressure waves with a frequency of 1–3 per minute; they are found in 5% of normal subjects,[5] but in approximately 50% when maximum resting pressure is greater than 100 cm H$_2$O.[6] They are frequently present in conditions caused by internal sphincter spasm, such as hemorrhoids[7] and anal fissure.[8] There is no consensus as to which part of the ultra–slow wave should be regarded as the maximal resting pressure, and it is best to document peak and trough pressures, as well as mean pressure. It is important to note that there is wide variation in resting and squeeze pressure, depending on the methodology, and there are also variations in repeatability studies.[9] Resting pressure is lower in women than in men[10] and decreases with age.[11] Generally, a pressure of between 50 and 100 cm H$_2$O is considered normal (Table 2.5.1).

Spontaneous relaxation of the internal sphincter has been shown to occur at a mean of 7 times per hour,[12] thus, a "sampling reflex" allows the mucoas of the upper anal canal to discriminate gas from stool. Spontaneous relaxations occur more frequently than normal in patients with fecal incontinence.[13] This is caused by a rise in rectal pressure due to rectal distension or contraction[14] mediated via an intramural neural pathway[15] and known as the rectoanal inhibitory reflex (Fig. 2.5.2).

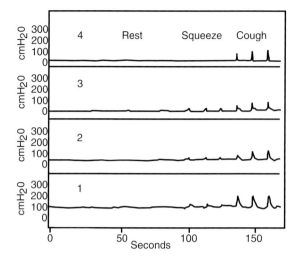

FIGURE 2.5.2. Rectoanal reflex. With distension of the rectal balloon there is a rise in rectal pressure (channel 4). Pressure falls in the upper- and mid-anal canal (channels 3 and 2) due to internal sphincter relaxation, and then returns toward the basal level. Pressure in the lowest part of the anal canal rises (channel 1) due to reflex contraction of the subcutaneous external sphincter.

This reflex is tested by rapid balloon distension of the rectum while simultaneously recording anal pressure. A fully deflated latex balloon, attached to the tip of a perfusion catheter, is inserted into the rectum. Resting anal pressure is allowed to stabilize, and the rectal balloon is then inflated with 20-mL increments of air. A fall in resting anal pressure of 20%, followed by return to the resting level, constitutes a positive reflex.

Voluntary Contraction Pressure

The external sphincter is in a state of tonic partial contraction, contributes about 15% of the resting pressure,[3] and is controlled by the spinal Onuf's nucleus. The external sphincter produces the voluntary contraction pressure[16] and is measured as the increase above the resting pressure (see Fig. 2.5.1). Like the resting pressure, it is lower in women and reduces with age.[17] The external sphincter undergoes reflex and concomitant contraction in response to a rise in abdominal pressure, and this is tested as the "cough pressure." In some cases, a more accurate measure of external sphincter strength is obtained with the cough pressure (Fig. 2.5.1),[18] which should be routinely tested during anal manometry.

Electrophysiology

Manometry gives some indication of sphincter function and can help determine whether sphincter weakness is likely to be the cause of symptoms. However, other tests, such as electromyography and/or ultrasound, may be needed to determine the underlying pathology.

Motor Nerve Conduction Studies

Pudendal Nerve Terminal Motor Latency

For further background see Chapter 2.8. Stretch-induced damage to the pudendal nerve may be measured using motor conduction studies. Electrical stimulation of the pudendal nerve distal to the ischial spine is carried out as described by Kiff and Swash.[19] A disposable electrode (Dantec 13L40, Skovlunde, Denmark) with an adhesive backing is placed on the gloved index finger with a node immediately proximal to the finger-tip (Fig. 2.5.3). The finger is inserted into the rectum with the tip of the finger positioned at one ischial spine. Square-wave stimuli lasting for 0.1 ms are delivered at 1-s intervals, and the amplitude is progressively increased to 20 mA. The fingertip is repositioned until it lies over the pudendal nerve, at which time a reproducible tracing with shortest possible pudendal latency is obtained. The ischial spine cannot be accurately palpated in many subjects and the correct position of the finger is determined by a reproducible tracing with minimum latency. The latency is defined as the time from stimulus to the point of takeoff when the impulse reaches the recording electrode at the external sphincter.

The validity of this test remains in doubt, and some institutes regard it as unreliable. In our practice, it has become clear that the test requires an experienced technician to produce repetitive reliable measurements.

Normal Recordings

Initially, the average of the left and right pudendal latencies was calculated.[19] However, nerve damage is often asymmetrical and

2.5. Anorectal Physiology

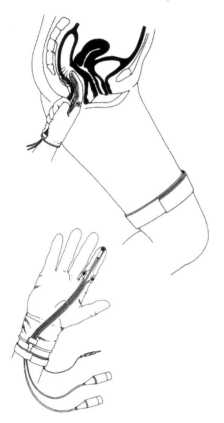

FIGURE 2.5.3. Disposable pudendal electrode and connector (Dantec, Denmark). The electrode is mounted on the right index finger with the anode at the finger-tip, with the recording electrodes at the base of the finger.

TABLE 2.5.2. Normal mean pudendal latencies

Reference	Pudendal motor latency (ms)
Kiff and Swash[19]	2.1 (0.2)
Snooks et al.[57]	1.9 (0.2)
Rogers et al.[40]	1.95 (1.7–2.25)
Beevors et al.[29]	Right 1.9 (0.2); Left 2.0 (0.2)

Values in parentheses are standard deviation or range.

Electromyography

For further background to this subject refer to Chapter 2.8.

Single-Fiber Electromyography

When a stretch-induced nerve injury occurs, action potentials pass down the fastest conducting nerve fibers that remain intact. Therefore, significant nerve damage may occur before motor latencies become prolonged. Examination of the innervated muscle provides evidence of nerve damage at an earlier stage. Single-fiber EMG is the most accurate method of testing for denervation and reinnervation of skeletal muscle, by recording action potentials from individual muscle fibers (Fig. 2.5.4).[31] Denervation of skeletal muscle is accompanied by reinnervation from neighboring axons. Histological studies show the normal random distribution of type 1 and type 2 fibers replaced by type 1 and type 2 grouping.[32] These changes can be measured accurately with single-fiber EMG, which consists of a 25-μm-diam platinum wire insulated in a stainless steel cannula, and terminating end-on at the side of the cannula, 5 mm proximal to the tip (recording surface = 25 μm).

the latencies should be reported separately.[20] (Table 2.5.2) Pudendal nerve terminal motor latency increases with age.[11] The amount of perineal descent correlates with pudendal latency.[21] An acute, severe stretch injury during vaginal delivery,[22,23,24] or a repetitive injury due to chronic straining at stool,[19,25] may result in nerve damage. This is supported by evidence of acute prolongation of pudendal latency during straining in patients with abnormal perineal descent.[26] Prolonged pudendal latencies are found in several disorders associated with difficult vaginal delivery or chronic straining at stool, including fecal incontinence,[27] severe constipation,[25] hemorrhoids,[28] and uterine prolapse.[29] Prolonged pudendal latencies can differentiate minor soiling from major incontinence.[30]

The changes of denervation and reinnervation can be quantified by calculating the fiber density, which is a measure of the number of muscle fibers in a motor unit within the area of uptake of the single-fiber electrode[33] (e.g., the average number of fibers recorded at 20 sites in the muscle). With reinnervation, the number of muscle fibers within this area increases, and the number of recorded spikes in the action potentials increases correspondingly. Normal fiber density in the external anal sphincter is 1.5 (SD 0.16);[34] fiber density

Anal and Rectal Sensation

The anal mucosa is richly innervated with nerve endings and is sensitive to touch, pain, temperature, and movement. Pain sensation extends 0.5–1.5 cm above the level of the dentate line. Sensory fibers travel via the pudendal nerves to the S2–S4 nerve roots.

Mucosal electrosensitivity is one measure of rectal sensation. Current is applied to the anal mucosa and the sensory threshold is quantitatively measured A ring electrode (Dantec, Skovlunde, Denmark) is placed on a Foley catheter with the balloon deflated. The stimulator delivers a current lasting 0.1-ms at a rate of five cycles per second, and the amplitude is increased from zero in 0.1-mA increments until a tingling or tapping sensation is felt by the patient. The test is reproducible[9] and provides an accurate measure of anal sensation. Normal median values and range[39] are:

Lower anal canal (mA)   4.8 (3.0–7.0)
Midanal canal (mA)   4.2 (2.0–6.0)
Upper anal canal (mA)   5.7 (3.3–7.3)

Anal sensation is an important part of the continence mechanism and is reduced in patients with neurogenic incontinence.[40]

Rectal sensation plays an important role in the continence mechanism, and a sensation of rectal filling is also an integral part of normal defecation. The rectum is sensitive to distension but lacks pain receptors. Rectal sensation is tested by balloon distension of the rectum (Table 2.5.3) or with mucosal electrosensitivity.

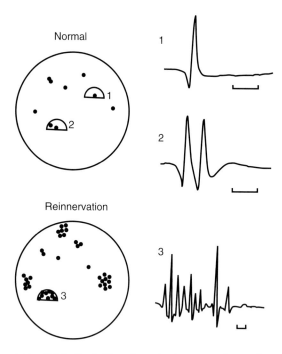

FIGURE 2.5.4. Single-fiber electrode shown superimposed on normal muscle to demonstrate the recording surface of the wire electrode on the side of cannula. The field of electrical uptake involves only a small number of muscle fibers. (b) Recording of muscle fibers in a normal muscle (above) and a denervated muscle (below). The denervated and reinnervated muscle has fiber grouping, resulting in a larger number of muscle fibers in the motor unit, and hence the polyphasic action potentials shown diagrammatically. (c) Single-fiber EMG recording from a reinnervated muscle showing a normal action potential (left) and a polyphasic action potential (right). (*Source*: Stalberg and Trontelj, 1979.)

greater than 2.0 indicates denervation with reinnervation.

Single-fiber EMG is a reproducible test.[35] Fiber density increases with age.[36] It is increased in conditions associated with stretch-induced pudendal neuropathy, including fecal incontinence, rectal prolapse,[37] urinary stress incontinence,[38] and uterine prolapse.[29] Single-fiber EMG is more invasive than other tests of pudendal function. Although it is more accurate, it is often used solely for patients in whom there is clinical evidence of neurogenic muscle weakness, but pudendal latency and mucosal electrosensitivity are normal.

Balloon Distension

A balloon attached to a catheter is inserted into the rectum, with the lower edge of the inflated balloon lying on the pelvic floor. Three endpoints are taken: (1) the volume first felt, which is the rectal sensitivity threshold; (2) the volume producing the urge to defecate; and (3) the volume causing intolerable pain, which is called the maximum tolerable volume. Air and water can be used to inflate the balloon.

The ability to evacuate the rectum may be tested by rectal balloon expulsion. A balloon is inserted into the rectum and inflated with 50 mL warm water. The subject is instructed to expel

2.5. Anorectal Physiology

TABLE 2.5.3. Rectal sensation showing balloon distension volumes in normal subjects

	Mean (ml) (years)	Balloon distension volume			
Reference		RST	Urge to defecate	MTV	Method
Loening-Baucke and Anuras[17]	29	17 (8)	173 (64)		Bolus air
	72	17 (9)	151 (61)		Bolus air
Sun et al.[58]	22	12 (1)	75 (10)	110 (10)	Bolus air
		20 (5)	128 (10)	178 (20)	Continuous air 20 ml/min
		20 (6)	130 (12)	176 (20)	Continuous water 20 ml/min
		43 (10)	167 (7)	230 (21)	Continuous air 100 ml/min
		40 (10)	175 (10)	216 (20)	Continuous water 100 ml/min

Values in parentheses are standard deviations. RST, rectal sensitivity threshold; MTV, maximum tolerable volume.

the balloon, and if this cannot be done then it is inflated to a point that is closer to the volume producing the urge to defecate. However, balloon expulsion is an unphysiological test and should be interpreted with caution. It does form the basis of one form of biofeedback training.

Measurement of Colonic Function

Constipation is a common clinical problem and the objective measurement of colonic function is essential. Symptoms that are seemingly referable to the anorectal region may have their origin higher in the colon in some patients, and anorectal function cannot always be considered in isolation.

Colonic Transit Studies

Current techniques to measure colonic transit time use radioopaque markers or radioisotopes. For marker studies, all laxatives are ceased for at least 48 hours, and 20 markers are swallowed. A single radiograph is taken on day 5 and in normal subjects at least 14 (80%) have been passed. The test is a measure of whole gut transit time, but since colonic transit time forms a large component of this, it is a useful, simple method of assessing colonic transit time.

For radioisotope studies, the isotope is given orally and the abdomen is scanned with a wide-field-of-view gamma camera 6, 24, 48, 72, and 96 hours after ingestion. The colon is divided in right, left, and sigmoid/rectal segments for analysis. Segmental transit is expressed in one of two ways; either the percentage of isotope retained in each segment and total percentage retained, or as the midpoint of the isotope column (mean activity position).[41] Total percentage retention is significantly increased in patients with severe idiopathic constipation (Fig. 2.5.5).[41] The total

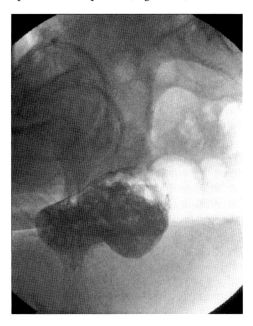

FIGURE 2.5.5. Radioisotope colon transit study. Subject with severe constipation. There is prolonged retention of isotope in the colon to 96 hours.

percentage retained is a simple measure for clinical use and can be compared with normal data.

Defecation Scintigraphy

Recently, colon transit scintigraphy has been adapted as a research tool to assess the rectum and colon during defecation. The impetus for this arose from the hypothesis that in some patients with obstructed defecation there may be an underlying diffuse colonic abnormality. Oral indium-111-DTPA (4MBq) is given as for standard colon scintigraphy, and then on the following day, when the subject develops a normal call to stool, the colon and rectum are screened immediately before, during, and after defecation.[42]

Colonic Manometry

Colonic manometry is still largely a research tool, but is now becoming clinically relevant. There is evidence from ambulatory anorectal manometry that vigorous colonic propulsion may cause fecal incontinence, so that an apparently "anorectal" condition may actually arise from the colon. Similarly, defecation is not only a process of rectal evacuation but also involves propulsive activity in the colon.

Colonic manometry has been studied by placing a catheter colonoscopically into the prepared colon.[43] The catheter is fed manually through the nose and checked fluoroscopically to prevent curling in the stomach. Recording is begun when the catheter tip reaches the rectum.

The motor events surrounding normal defecation are recorded under physiological conditions.[44] In the hour preceding "defecation," a series of up to three propagating waves arise in the proximal colon: the origin of each wave arises progressively more distally. In the 15 minutes before defecation, up to three propagating waves were again seen, but the reverse pattern was seen (each successive propagating wave having arose more proximally). These propagating waves were associated with the urge to defecate at the end of the wave. There was a marked reduction of propagating sequences both in the distal and proximal colon in patients with obstructed defecation including those with anismus, suggesting a more diffuse colonic abnormality rather than an anorectal abnormality alone.

Dynamic Imaging of the Rectum

Several methods are available. Scintigraphic methods provide a quantitative measure of rectal evacuation, but do not demonstrate the anatomy of the anorectum; therefore, defecography using barium has been more widely used.

Defecography (Defecating Proctogram)

Dynamic imaging of the rectum has been used for over 30 years.[45] A variety of anatomical abnormalities and disorders of function have been identified in patients with obstructed defecation. Defecography does not assess defecation under physiological conditions because it ignores the normal sensory responses, as well as the colonic motor events, during defecation.

The barium mixture used varies widely. A mixture which approximates the volume and consistency of stool is optimal, but these parameters are variable in normal subjects. A solid artificial stool may alter the process of evacuation[46] and a standard mixture has not been established. Typically, barium is diluted 1.5:4 in water and mixed with potato starch to form a thick paste. The mixture is instilled while the patient lies in the lateral position. The line of the anal canal is defined radiologically. Lateral radiographs are obtained with the subject sitting on a water-filled balloon ring or plastic commode to eliminate an air interface. X-rays are obtained at rest, while coughing, during a Valsalva maneuver, and during evacuation of the barium. The examination room should be darkened and the number of observers limited.

Normal Evacuation

As straining begins, abdominal pressure produces a slight concavity of the anterior rectal wall (Fig. 2.5.6). The pelvic floor descends, the anorectal angle widens, the anal canal begins to

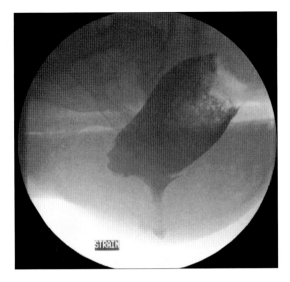

FIGURE 2.5.6. Proctogram showing descent of the perineum from the resting to the straining position. Descent is measured in relation to the line of the ischial tuberosities which are usually clearly visualized.

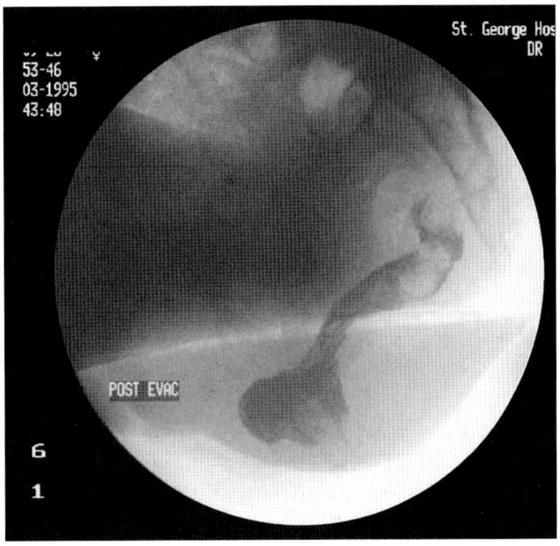

FIGURE 2.5.7. Proctogram showing anterior rectocele which fails to empty after evacuation.

open, shortens, and becomes funnel shaped. Rectal evacuation begins and emptying is completed. A slight degree of rectal wall intussusception may occur normally. There are very few studies in normal controls, particularly in young women, because of the radiation exposure. There is wide variation in normal values.[47]

The anorectal angle is measured at rest, during contraction of the pelvic floor muscles, and during defecation. The angle measures about 90 degrees at rest, is more acute (60–80 degrees) during contraction, and more obtuse (mean 130 degrees, range 120–145 degrees) during evacuation.[48] In an individual subject, the exact measurement is not as important as the change that occurs from the rest to straining positions.

Findings at Defecating Proctography

Small rectoceles are a normal finding in asymptomatic women.[48] The size is measured anteriorly from a line drawn upwards from the anterior wall of the anal canal. Precisely what size rectocele will cause symptoms is not defined, but a bulge greater than 2–3 cm is considered abnormal. Failure of the rectocele to empty is an important finding (Fig. 2.5.7).

Intussusception is the prolapse of the rectal wall that occurs within the rectum (rectorectal), within the anal canal (rectoanal), or there may be external rectal prolapse. Minor infoldings of the rectal wall are seen in up to 45% of asymptomatic subjects[47] and are considered normal. There is debate about the clinical importance of intussusception, particularly in relation to symptoms of obstructed defecation.[49]

Neurogenic incontinence is associated with perineal descent and a widened anorectal angle at rest.[50] There may be leakage of barium during coughing or a Valsalva maneuvre, which is a useful, simple test to help determine whether sphincter weakness is the cause of the incontinence.

Endoanal Ultrasound

The anal sphincters may be seen clearly using an endoanal probe. Two systems are available: a rotating 7 MHz or 10 MHz probe housed within a water-filled plastic cone (Bruel and Kjaer, Gentofte, Denmark) or a static linear 7 MHz probe. The rotating probe provides a 360 degree horizontal cross-sectional image at each level, and is inserted into the rectum and then withdrawn from

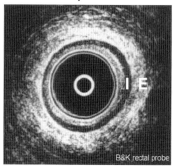 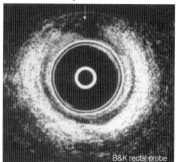

FIGURE 2.5.8. Normal ultrasound, rotating probe. **(A)** Mid-anal canal. Intract internal sphincter (I), external sphincter (E). **(B)** Anterior internal and external sphincter defect. (*Source*: Dr. Peter Stewart, Sydney, Australia.)

the top of the anal canal to the anal verge. The linear probe produces a longitudinal image of the entire sphincter length and recordings are taken at several points in each quadrant.

The internal sphincter is a well-defined hypoechoic layer. Beyond this is the thick echogenic external sphincter ring (Fig. 2.5.8). The internal sphincter ring is absent in the lowest part of the anal canal. In the upper anal canal the puborectalis is seen above the upper limit of the external sphincter.

Abnormal Findings

The normal internal sphincter thickness increases with age from 1 mm at under 20 years to 2–3 mm in the elderly.[51] Therefore a 3-mm internal sphincter in a young patient is abnormal. It is thin in patients with neurogenic incontinence, in keeping with the known muscle atrophy in that condition.[52]

Defects in the internal and external sphincter can be found following fistulotomy, lateral sphincterotomy, and, in some cases, after vaginal delivery.[53,54]

Endoanal ultrasound detects external sphincter defects as accurately as EMG,[55] and has become the procedure of choice for sphincter mapping because it causes much less discomfort than EMG using a needle.

References

1. Lam TCF, Kennedy ML, Chen FC, et al. Prevalence of faecal incontinence: obstetric and constipation-related risk factors; a population-based study. Colorectal Dis. 1999;1:197–203.
2. Gibbons CP, Bannister JJ, Trowbridge GA, et al. An analysis of anal sphincter pressure and anal compliance in normal subjects. Int J Colorectal Dis. 1986;1:231–237.
3. Lestar B, Penninckx F, Kerremans R. The composition of anal basal pressure: An in vivo and in vitro study in man. Int J Colorectal Dis. 1989;4:118–122.
4. Hancock BD, Smith K. The internal sphincter and Lord's procedure for haemorrhoids. Br J Surg. 1975;62 833–866.
5. Hancock BD. Internal sphincter and the nature of haemorrhoids. Gut. 1977;18:651–655.
6. Haynes WG, Read NW. Anorectal activity in man during rectal infusion of saline: a dynamic assessment of the anal continence mechanism. J Physiol. 1982;330:45–56.
7. Sun WM, Read NW, Shorthouse AJ. The hypertensive anal cushion as a cause of the high anal pressures in patients with haemorrhoids. Br J Surg. 1990;77:458–462.
8. McNamara MJ, Percy JP, Fielding IR. A manometric study of anal fissure treated by subcutaneous lateral internal sphincterotomy. Ann Surg. 1990;211:235–238.
9. Ryhammer AM, Laurberg S, Hermann AP. Test-retest repeatability of anorectal physiology tests in healthy volunteers. Dis Colon Rectum. 1997;40:287–292.
10. Sun WM, Read NW. Anorectal function in normal subjects: the effect of gender. Int J Colorectal Dis. 1989;4:188–196.
11. Ryhammer AM, Laurberg S, Sorensen FH. Effects of age on anal function in normal women. Int J Colorectal Dis.1997;2:225–229.
12. Orkin BA, Hanson RB, Kelly KA, et al. Human anal motility while fasting, after feeding and during sleep. Gastroenterol. 1991;100:1016–1023.

13. Sun WM, Read NW, Miner PB, et al. The role of transient internal sphincter relaxation in faecal incontinence. Int J Colorectal Dis. 1990;5:31–36.
14. Naudy B, Planche D, Monges B, Salducci J. Relaxations of the internal anal sphincter elicited by rectal and extra-rectal distensions in man. In: Roman C, editor. Gastrointestinal Motility. London: MTP Press; 1984:451–458.
15. Lubowski DZ, Nicholls RJ, Swash M, et al. Neural control of internal anal sphincter function. Br J Surg. 1987;74:668–670.
16. Keighley MRB, Henry MM, Bartolo DCC, et al. Anorectal physiology measurement: report of a working party. Br J Surg. 1989;76:356–357.
17. Loening-Baucke V, Anuras S. Effects of age and sex on anorectal manometry. Am J Gastroenterol. 1985;80:50–53.
18. Meagher AP, Lubowski DZ, King DW. The cough response of the anal sphincter. Int J Colorectal Dis. 1993;8:217.
19. Kiff ES, Swash M. Slowed conduction in the pudendal nerves in idiopathic (neurogenic) faecal incontinence. Br J Surg. 1984a;71:614–616.
20. Lubowski DZ, Jones PN, Swash M, et al. Asymmetrical pudendal nerve damage in pelvic floor disorders. Int J Colorectal Dis. 1988;3:158–160.
21. Ho YH, Goh HS. The neurophysiological significance of perineal descent. Int J Colorectal Dis. 1995;10:107–111.
22. Snooks SJ, Setchell M, Swash M, et al. Injury to the innervation of the pelvic floor musculature in childbirth. Lancet. 1984;2:546–550.
23. Snooks SJ, Swash M, Mathers SE, et al. Effect of vaginal delivery on the pelvic floor: a 5 year follow-up. Br J Surg. 1990;77:1358–1360.
24. Sultan AH, Kamm MA, Hudson CN. Pudendal nerve damage during labour: prospective study before and after childbirth. Br J Obstet Gynaecol. 1994;101:22–28.
25. Snooks SJ, Barnes PRH, Swash M, et al. Damage to the innervation of the pelvic floor musculature in chronic constipation. Gastroenterol. 1985;89:977–981.
26. Lubowski DZ, Swash M, Nicholls RJ, et al. Increase in pudendal nerve terminal motor latency with defaecation straining. Br J Surg. 1988;75:1095–1097.
27. Kiff ES, Swash M. Normal proximal and delayed distal conduction in the pudendal nerves of patients with idiopathic (neurogenic) faecal incontinence. J Neurol Neurosurg Psychiatry 1984;47:820–823.
28. Bruck CE, Lubowski DZ, King DW. Do patients with haemorrhoids have pelvic floor denervation? Int J Colorectal Dis. 1988;3:210–214.
29. Beevors MA, Lubowski DZ, King DW, et al. Pudendal nerve damage in women with symptomatic utero-vaginal prolapse. Int J Colorectal Dis. 1991;6:24–28.
30. Kafka NJ, Coller JA, Barrett RC, et al. Pudendal neuropathy is the only parameter differentiating leakage from solid stool incontinence. Dis Colon Rectum. 1997;40:1220–1227.
31. Lubowski DZ, Swash M, Henry MM. Neural mechanisms in disorders of defaecation. In Grundy D, Read NW, editors. Clinical Gastroenterology. Gastrointestinal Neurophysiology. London: Baillière Tindall/WB Saunders; 1988:201–203.
32. Parks AG, Swash M, Urich H. Sphincter denervation in ano-rectal incontinence and rectal prolapse. Gut. 1977;18:656–665.
33. Trontelj JV, Stalberg E. Single fiber electromyography in studies of neuromuscular function. Adv Exp Med Biol. 1995;384:109–119.
34. Neill ME, Swash M. Increased motor unit fibre density in the external sphincter muscle in anorectal incontinence: a single fibre EMG study. J Neurol Neurosurg Psychiatry. 1980;43:343–347.
35. Rogers J, Laurberg S, Misiewicz JJ, et al. Anorectal physiology validated: a repeatability study of the motor and sensory tests of anorectal function. Br J Surg. 1989;76:607–609.
36. Laurberg S, Swash M. Effects of ageing on the anorectal sphincters and their innervation. Dis Colon Rectum. 1989;32:734–742.
37. Neill ME, Parks AG, Swash M. Physiological studies of the pelvic floor in idiopathic faecal incontinence and rectal prolapse. Br J Surg. 1981;68:531–536.
38. Smith ARB, Hosker GL, Warrell DW. The role of partial denervation of the pelvic floor in the aetiology of genitourinary prolapse and stress incontinence of urine. A neurophysiological study. Br J Obstet Gynaecol. 1989;96:24–28.
39. Roe AM, Bartolo DCC, Mortensen NJMcC. New method for assessment of anal sensation in various anorectal disorders. Br J Surg. 1986;73:310–312.
40. Rogers J, Henry MM, Misiewicz JJ. Combined sensory and motor deficit in primary neuropathic faecal incontinence. Gut. 1988;29:5–9.
41. Smart RC, McLean RG, Gaston-Parry D, et al. Comparison of oral iodine-131-cellulose and Indium-111 DTPA as tracers for colon transit scintigraphy: analysis by colon activity profiles. J Nucl Med. 1981;32:1668–1674.
42. Lubowski DZ, Meagher AP, Smart RC, et al. Scintigraphic assessment of colonic function during defaecation. Int J Colorectal Dis.1995;10:91–93.

43. Furukawa Y, Cook IJ, Panagopoulos V, et al. Relationship between sleep patterns and human colonic motor patterns. Gastroenterol. 1994;107:1372–1381.
44. Dinning PG, Kennedy ML, Lubowski DZ, et al. Manometric findings in the unprepared proximal colon in patients with severe obstructed defaecation. Gastroenterol. 1996;110:657.
45. Burhenne HJ. Intestinal evacuation study: A new roentgenologic technique. Radiol Clin. 1964;33:79–84.
46. Bannister JJ, Davison P, Timms JM, et al. Effect of stool size and consistency on defecation. Gut. 1987;28:1246–1250.
47. Shorvon PJ, McHugh S, Diament NE, et al. Defecography in normal volunteers: normal results and implications. Gut 1989;30:1737–1749.
48. Finlay IG (moderator). Symposium: Proctography. Int J Colorectal Dis.1988;3:67–89.
49. van Tets WF, Kuijpers JHC. Internal rectal intussusception – fact or fancy? Dis Colon Rectum. 1995;38:1080–1083.
50. Pinho M, Yoshioka K, Keighley MRB. Are pelvic floor movements abnormal in disordered defaecation? Dis Colon Rectum. 1991;34:1117–1119.
51. Burnett SJ, Bartram CI. Endosonographic variations in the normal internal anal sphincter. Int J Colorectal Dis. 1991;6:2–4.
52. Swash M, Gray A, Lubowski DZ, et al. Ultrastructural changes in internal sphincter in neurogenic faecal incontinence. Gut 1988;29:1692–1698.
53. Sultan AH, Kamm MA, Hudson CN, et al. Anal-sphincter disruption during vaginal delivery. N Engl J Med 1993;329:1905–1911.
54. Sultan AH, Kamm MA, Bartram CI, et al. Anal sphincter trauma during instrumental delivery. Int J Gynaecol Obstet. 1993;43:263–270.
55. Sultan AH, Kamm MA, Talbot IC, et al. Anal sphincter endosonography for identifying external sphincter defects confirmed histologically. Br J Surg. 1994;81:463–465.
56. Williams N, Barlow J, Hobson A, et al. Manometric asymmetry in the anal canal in controls and patients with fecal incontinence. Dis Colon Rectum. 1995;38:1275–1280.
57. Snooks SJ, Swash M, Henry MM, et al. Risk factors in childbirth causing damage to the pelvic floor innervation: a precursor of stress incontinence. Int J Colorectal Dis. 1986;1:20–24.
58. Sun WM, Read NW, Prior A, et al. Sensory and motor responses to rectal distension vary according to rate and pattern of balloon inflation. Gastroenterol. 1990;99:1008–1015.
59. Stalberg E, Trontelj J. Single Fibre Electromyography. Surrey: Mirvalle Press; 1979.

2.6
Ultrasound Imaging

Kaven Baessler and Heinz Kölbl

Key Messages

- Perineal ultrasound has an important role in the clinical evaluation of pelvic floor dynamics in women with stress and urge incontinence, and pelvic organ prolapse, as follows:
  - dynamic assessment of pelvic floor muscle activity during functional tasks such as coughing and straining
  - assessment of efficacy of pelvic floor re-education
  - pre and postoperative assessment
  - search for pathologies, especially in women with overactive bladder
- Use as a Biofeedback tool in pelvic floor re-education
- Scientific evaluation of pelvic floor dynamics to compare women with and without pelvic floor disorders as well as pre and post partum.

The Scope of Ultrasound Imaging of the Pelvic Floor

Ultrasound scanning has gained popularity and importance for nearly all body parts and functions. In urogynecology, bladder neck imaging has become routine practice and has provided another piece in the jigsaw puzzle of pelvic floor assessment.

The advantages of ultrasound in the assessment of pelvic floor disorders are numerous: there is no radiation; it is painless and may be noninvasive, easily applied supine and standing, and easily learned, performed, and taught; it is accepted and easily understood by women; and it is commonly available, is carried out by gynecologists, and does not require the involvement of a radiologist. Because of the lack of radiation, ultrasound permits prolonged imaging of pelvic floor function, especially when used as a feedback method for pelvic floor rehabilitation. Fluid-filled structures can be visualized without the use of contrast medium. Perineal ultrasound (translabial or introital ultrasound) has almost completely replaced lateral cystourethrography because of its advantages and similar results.[1,2] It is the instrument of choice to image the dynamics of the pelvic floor during coughing and straining. Although MRI can give very precise pictures, it is not universally accessible and is expensive. With the recent advance of real-time three-dimensional ultrasound, pelvic floor images similar to those produced by an MRI can be obtained.

Perineal ultrasound is also a noninvasive and painless alternative to the Q-tip or cotton-swab test, where a Q-tip is inserted into the urethra and angles are measured during straining, coughing, and pelvic floor contraction.[3] Simultaneous imaging during urodynamics studies is also possible.[4]

Although abdominal ultrasound can give reasonably static and dynamic images of the filled bladder, and also, indirectly, of a pelvic floor contraction, by visualizing bladder base elevation,[5] perineal ultrasound is usually preferred because of its superior quality and more direct imaging. Obesity, suprapubic incisions, and advanced pelvic organ prolapse can make accurate and meaningful abdominal scanning impossible. It is, however, the approach of choice for the noninvasive observation of the concomitant or primary action of the transversus abdominis muscle.[6]

Three-dimensional ultrasound of the urethra, bladder, anal sphincter, and pelvic floor is primarily a research tool, but is becoming more used for perineal scanning. Intraurethral ultrasound two- and three-dimensional ultrasound scanning also remains mainly a research tool. However, endoanal ultrasound is the gold standard in the assessment of the anal sphincter in women with fecal incontinence.

This chapter will describe the use of ultrasound in the clinical assessment of the pelvic floor and as a research tool in the evaluation of the physiology and pathophysiology of pelvic floor function. In the first part of the chapter, the different ultrasound techniques are depicted while the second part informs about the functional findings.

Applications, Equipment, and Pelvic Floor Structures

Perineal and Introital Ultrasound

The use of a curved linear array scanner with frequencies between 3.5 and 5 (7.5) MHz for perineal (synonym translabial) ultrasound or an endovaginal probe with (5) 7–10 MHz placed on the introitus is recommended. For a perineal ultrasound, the probe is placed on the perineum in the midsagittal plane with plenty of ultrasound gel to allow for good contact. The labia may have to be parted to obtain a satisfactory image. In contrast to endovaginal ultrasound, perineal ultrasound does not affect the topography of the bladder as long as there is no prolapse beyond the introitus and not too much pressure on the perineal probe, as this can displace the bladder neck cranially.[7] Introital ultrasound with an endovaginal probe may interfere with the exact assessment of bladder neck mobility, pelvic organ prolapse, and pelvic floor contraction. This may not be important in the clinical setting, but can be relevant when used in research. Different bladder volumes do not seem to affect bladder neck mobility in a clinically significant way,[7–9] but bladder neck funneling is usually better observed at higher volumes of approximately 300 ml.[7,9]

Depending on the purpose of the ultrasound the probe can be applied in the supine, sitting, or standing position. The bladder, bladder wall, urethra, bladder neck, pubic symphysis, vagina, uterus, rectum, anorectal junction, puborectalis sling, anal sphincter, and ureteric jets can be visualized. With perineal ultrasound, the urethra appears as hypoechoic because of the smooth muscle orientation of the urethra. The striated urethral muscle is not easily differentiated with perineal ultrasound, probably because of the amount of connective tissue and fiber orientation giving off a similar echogenicity as the surrounding structures. Because of the substantial improvement of ultrasound equipment with better resolution, the insertion of a urethral catheter to demarcate the urethra and the bladder neck with the filled catheter balloon is no longer necessary. The vagina is hyperechoic, but is not persistently identifiable. The insertion of ultrasound gel into the vagina can be helpful if required. The position and the descent of the uterus, bladder, rectum, and an enterocele can be determined,[10–12] although quantification of pelvic organ prolapse is limited, especially when the prolapse extends beyond the introitus. Stool and gas can seriously disturb the images, particularly after repeated straining.

The German Urogynecology Association has advocated a standardization of perineal ultrasound to assist with clinical assessment and facilitate research.[13] Figure 2.6.1 explains the different measurements that can be obtained. The pubic symphysis is used as the reference structure. Its central axis represents the x axis and the perpendicular line at the inferior edge of the symphysis represents the y axis, thus, allowing the calculation of the coordinates of the bladder neck position and its mobility vectors.

According to the primary focus, the probes can be tilted more towards the urethra and bladder or towards the anal sphincter and rectum (Fig. 2.6.2). To visualize the anal sphincter muscle, the curved probe has to be orientated on the perineum

2.6. Ultrasound Imaging

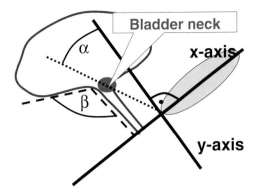

FIGURE 2.6.1. The pubic symphysis serves as the reference structure with the X-axis running along its longitudinal axis and the Y-axis crossing perpendicularly at the inferior border of the symphysis. The rotational or angle alpha is measured between the Y-axis and a line determined by the bladder neck position and the inferior border of the pubic symphysis. Angle alpha is the posterior urethrovesical angle.

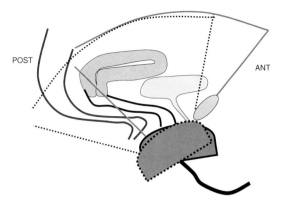

FIGURE 2.6.2. Sagittal scan/plan. Orientation of the ultrasound probe according to the focus of the examination.

perpendicular to the axis of the anal canal (Fig. 2.6.3). Although currently inferior to endoanal ultrasound, perineal ultrasound can be used to assess anal sphincter integrity and allows visualization of the mucosa, internal, and external anal sphincter without distortion of the anatomy (Fig. 2.6.3).[14,15]

Paravaginal (lateral) support of the anterior vaginal wall can be imaged by placing the ultrasound probe transversely onto the perineum and vaginal introitus and by orientating it ventrally (Fig. 2.6.4).[16] In a transverse image, a vagina with normal paravaginal support has a "butterfly" shape (Fig. 2.6.3), which has been described as "tenting" in three-dimensional ultrasound.[17]

Bladder wall thickness can be measured using perineal, introital, or vaginal ultrasound. An increased bladder wall thickness of more than 5 mm has been reported to be associated with detrusor overactivity (Fig. 2.6.5).[18] Urethral diverticula, suburethral cysts, fibroids, tumors, or foreign bodies may also be detected.[11] Imaging the urethral and bladder neck mobility after a bladder neck suspension or suburethral tape procedure can help in the assessment of postoperative voiding dysfunction (see DVD).[19]

Abdominal Ultrasound

To visualize the bladder base and endopelvic fascia, suprapubic placement of the scanner with

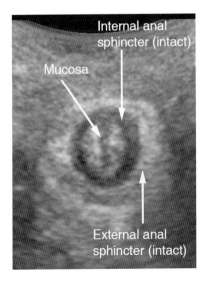

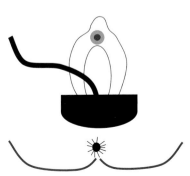

FIGURE 2.6.3. Coronal scan. Placement of the ultrasound probe transverse to the axis of the anal canal to visualise the mucosa and the internal and external sphincter muscles without distortion.

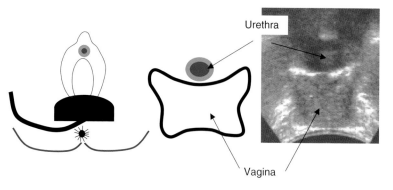

FIGURE 2.6.4 Coronal scan. With placement of the ultrasound probe transverse on the perineum and introitus, the vagina with its lateral support (paravaginal attachment) can be visualised. The vagina has a "butterfly" shape when imaged coronally.

lower frequencies (3.5–5 MHz) is required. Higher frequencies (5–10 MHz) are recommended to image the transversus abdominis muscle in the horizontal (transverse) plane on the lateral abdominal wall.[6] Abdominal ultrasound is very limited in the investigation and visualization of the pelvic floor, particularly of the urethra and bladder neck. There is no standardization of the assessment although measurements of the bladder displacement during a pelvic floor contraction have been reported to be reproducible.[5,20]

Imaging of paravaginal support has been attempted with transabdominal ultrasound. The methods cannot easily be standardized because of the lack of reference points and difficult delineation of the endopelvic fascia and vagina.[16,21] Results are conflicting and vaginal examination remains the gold standard.

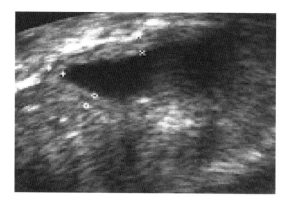

FIGURE 2.6.5. Perineal ultrasound of the minimally filled bladder, sagittal plane. Bladder wall thickness measured at three different sites. Mean measurements above 5 mm are indicative of detrusor overactivity.

Intraurethral, Endoanal, and Three-Dimensional Ultrasound

The equipment for an intraurethral ultrasound ranges from 7.5- to 10-MHz mechanically rotated endoprobes with catheter sizes between 6.2–9 French. Intraurethral ultrasound is feasible and has been shown to correlate well with histological examinations of the urethra.[22] The striated urethral sphincter assumes an omega-shape caused by the loss of signal at the 6 o'clock position.[22,23] For endoanal ultrasound, there are sector scanners that cover nearly 360 degrees, and there are mechanically rotated, 360-degree endoprobes. Frequencies between 7.5 and 10 MHz are currently used, with the higher frequency producing superior images. (see Chapter 1.7.)

There are several different scanners with 3D facilities available, mainly curved probes, but also endoprobes for intraurethral and endoanal ultrasound. The use of a 3D endoanal probe to explore the urethra at rest and during pelvic floor contraction has been shown to be feasible.[24] The perineal 3D application can demonstrate the relationship between urethra, suburethral tape, and vagina.[17] It holds great promise for the assessment of the pelvic floor muscle before and after delivery as puborectalis avulsions from the pubic bone can be identified.[17]

Sonographic Evaluation of Pelvic Floor Function

Valsalva, Straining, and Coughing

Imaging of the position of the bladder neck at rest and during coughing, straining, and pelvic floor contraction and assessment of bladder neck mobility are currently the cornerstones of perineal ultrasound scanning. Position and mobility of the bladder neck can reliably be measured using a coordinate system with the pubic symphysis as a reference point.[7,8]

The bladder neck position at rest is significantly lower in the upright position compared with the supine position.[7,25,26] Bladder neck descent during straining or a Valsalva maneuver in young, nulliparous, continent women has been reported between 0 and 40 mm.[27-30] To distinguish between normal bladder neck mobility and bladder neck hypermobility, a cutoff value of 5 mm[29,31] or 14 mm[32,33] has been used. However, bladder neck hypermobility is not clearly defined. Confounding factors include the difficult standardization of a Valsalva maneuver (forced expiratory effort against a closed glottis to increase intrapleural pressure, resulting in bradycardia and hypotonia) and that it is a nonphysiological test that might result in concomitant pelvic floor contraction instead of relaxation in some women.[28] Table 2.6.1 presents bladder neck mobility measurements during Valsalva and coughing obtained in different studies.

In a normal subject, the bladder neck is usually stabilized during coughing as part of the abdominal/pelvic floor muscle coactivation.[34] Therefore, bladder neck mobility is not as extensive during coughing as during straining.[34-36] Furthermore, bladder neck mobility is significantly reduced when a pelvic floor contraction is performed before coughing ("the Knack").[36] In continent nulliparous women, without the Knack, the bladder neck was displaced by a median of 4.6 mm (up to 19.5 mm); with the Knack, the median displacement was zero (up to 17 mm). This effect was not as manifest in stress incontinent women who might have sustained damage to the pelvic floor, its innervation, or the fascial support structures.[36]

TABLE 2.6.1. Measurements of bladder neck mobility and posterior urethrovesical angle in different populations

Bladder neck mobility on Valsalva	Author	Continent controls	Incontinent subjects	P
60 cmH$_2$O	Peschers[34]	young NP, n = 39 14 ± 9 mm (2–31)		
	Dietz[30]	young NP, = 106 17 ± 9 mm	young NP, SUI, n = 10 18 ± 10 mm	0.7
40 cmH$_2$O	Reed[29]	premenopausal NP, n = 48 6.3 ± 0.42 mm (0–18.7)		
	Howard[35]	young continent NP, n = 17 12.4 ± 4.7 mm young continent PP, n = 18 14.5 ± 7.0 mm	9–12 months pp, VD, PP, SUI, n = 23 14.8 ± 6.4 mm	0.42
maximal straining	Chen[25]		various complaints, no POP, n = 78 Supine: 18.1 ± 4.6 mm Standing: 16.1 ± 3.4 mm	<0.001
Bladder neck mobility on Coughing	Peschers[34]	n = 39 9 ± 6 mm (4–32)		
	Howard[35]	young continent NP, n = 17 8.2 ± 4.1 mm young continent PP, n = 18 9.9 ± 4.0 mm	9–12 months pp, VD, PP, SUI, n = 23 13.8 ± 5.4 mm	0.001
	Miller[36]	young continent NP, n = 11 median 5.4 mm, range 20 with "Knack" median 2.9 mm, range 18.3		<0.001
Posterior urethrovesical angle	Alper[42]	continent age-matched, n = 50 Rest: 93 ± 5° (85–110) Valsalva: 96 ± 8 (85–130)	SUI, n = 50 Rest: 95 ± 11 (70–117) Valsalva: 103 ± 16 (84 – 142)	<0.001 <0.001

NP = nulliparous, PP = primiparous, SUI = stress urinary incontinence, pp = post partum, VD = vaginal delivery.

After vaginal delivery, there was a significant loss of bladder neck support at rest and increased bladder neck mobility during straining. The position of the bladder neck at rest and during straining was lower postpartum compared with nulliparous controls and with women who had an elective cesarean section.[31,34,37]

Bladder neck funneling during coughing and straining is often visible in stress incontinent and parous women, but has not been demonstrated in continent nulliparous women.[27,38] Intravesical contrast medium can help delineate the bladder neck and potential funneling during straining or coughing.[39,40] Color Doppler has successfully been used to demonstrate urethral leakage, although the stress test probably remains the gold standard.[41]

The posterior urethrovesical angle (Fig. 2.6.1) significantly increases during straining in both continent and stress incontinent women (Table 2.6.1).[42] Stress incontinent women displayed a significantly greater angle at rest and while straining than continent women.[42] The posterior urethrovesical angle was significantly increased six weeks after childbirth (67 vs. 57 degrees prepartum).[43] Nulliparous controls had a smaller angle compared with postpartum women (44 vs. 67 degrees).

The bladder neck–symphysial angle (Fig. 2.6.1) has been suggested to identify women with stress urinary incontinence better than bladder neck mobility. The angle was significantly greater at rest and Valsalva in stress incontinent women compared with continent women of similar age and other demographics.[8] The angle correlated reasonably with digital pelvic floor contraction assessment and vaginal perineometry using an air-filled pressure transducer.[44]

Pelvic Floor Contraction

Provided that the subject can perform a sufficient pelvic floor contraction and the attachment of the anterior endopelvic fascia to the puborectalis muscle is intact, the cranial and ventral elevation of the bladder neck elevation during a pelvic floor contraction is clearly visible on perineal ultrasound (see DVD). This pattern can be used for pelvic floor education, as the images with the pelvic floor structures are easily understood by the women. The displacement of the pubocervical fascia during a pelvic floor contraction can also be measured transabdominally[5] and has been reported between 8.2 mm and 17 mm, respectively, depending on primary transversus abdominis or pelvic floor contraction.[20]

Using perineal ultrasound, bladder neck elevation during a voluntary levator ani muscle contraction has been shown to significantly decrease after vaginal delivery and to return to antepartum values after 6–10 weeks in most women.[45]

The thickness of the pelvic floor muscle can be measured perineally, with the ultrasound probe placed paramedially in the sagittal direction.[46] Nulliparous continent women had a significantly thicker pelvic floor muscle at rest (median 0.77 cm) and during contraction (median 0.99 cm) than incontinent nulliparous women (median 0.60 cm and 0.83 cm, respectively).

Micturition and Defecation

Perineal ultrasound imaging during micturition in the physiological sitting position demonstrated that the bladder neck opens in continent volunteers and incontinent patients. In most women, there was also bladder neck descent.[47] Similarly, perineal imaging during defecation (evacuation of ultasonographic gel) in the unphysiological left-lateral position documented a widening of the anorectal angle and descent of the anorectal junction.[10] These investigators also used the symphysis-based coordinate system for their measurements, which correlated well with results obtained with conventional defecography.

Ultrasound as a Biofeedback Tool

Biofeedback via ultrasound can be given to enhance understanding and improve pelvic floor function during (before = Knack) coughing and straining.[48,49] The "Knack," a pelvic floor contraction that is generated before coughing or sneezing to prevent urinary leakage, can be taught using perineal ultrasound.[36,49] The knack has been confirmed to improve the stability of the bladder neck during coughing. Straining and Valsalva are not exactly physiological activities but can be used as a substitute for activities that involve increased intraabdominal pressures like nose-blowing, defecation, bending, or playing a wind instrument. In conjunction with perineal ultrasound, abdominal ultrasound is a valuable instrument in assessing the synergy of the pelvic floor and deep abdominal muscles. It can be used for pelvic floor re-education especially for retraining of functional tasks that result in urinary leakage in the individual subject.[50]

Visual feedback via endoanal ultrasound has been studied in patients with fecal incontinence. Biofeedback with endoanal ultrasound and anal manometry did not prove to be of additional benefit compared to digital feedback.[51]

Conclusion

Ultrasound is an excellent tool to assess pelvic floor function and dysfunction and is suitable as an instrument for biofeedback-directed pelvic floor (re)-education.

References

1. Kolbl H, Bernaschek G, Wolf G. A comparative study of perineal ultrasound scanning and urethrocystography in patients with genuine stress incontinence. Arch Gynecol Obstet. 1988;244:39–45.
2. Troeger C, Gugger M, Holzgreve W, et al. Correlation of perineal ultrasound and lateral chain urethrocystography in the anatomical evaluation of the bladder neck. Int Urogynecol J Pelvic Floor Dysfunct. 2003;14:380–384.
3. Caputo RM, Benson JT. The Q-tip test and urethrovesical junction mobility. Obstet Gynecol. 1993;82:892–896.
4. Schaer GN, Koechli OR, Schüssler B, et al. Can simultaneous perineal sonography and urethrocystometry help explain urethral pressure variations? Neurourol Urodyn. 1997;16:31–38.
5. Thompson JA, O'Sullivan PB. Levator plate movement during voluntary pelvic floor muscle contraction in subjects with incontinence and prolapse: a cross-sectional study and review. Int Urogynecol J Pelvic Floor Dysfunct. 2003;14:84–88.
6. Sapsford RR, Hodges PW. Contraction of the pelvic floor muscles during abdominal maneuvers. Arch Phys Med Rehabil. 2001;82:1081–1088.
7. Schaer GN, Koechli OR, Schüssler B, et al. Perineal ultrasound: determination of reliable examination procedures. Ultrasound Obstet Gynecol. 1996;7:347–352.
8. Pregazzi R, Sartore A, Bortoli P, et al. Perineal ultrasound evaluation of urethral angle and bladder neck mobility in women with stress urinary incontinence. BJOG. 2002;109:821–827.
9. Dietz HP, Wilson PD. The influence of bladder volume on the position and mobility of the urethrovesical junction. Int Urogynecol J Pelvic Floor Dysfunct. 1999;10:3–6.
10. Beer-Gabel M, Teshler M, Schechtman E, et al. Dynamic transperineal ultrasound vs. defecography in patients with evacuatory difficulty: a pilot study. Int J Colorectal Dis. 2004;19:60–67.
11. Tunn R, Petri E. Introital and transvaginal ultrasound as the main tool in the assessment of urogenital and pelvic floor dysfunction: an imaging panel and practical approach. Ultrasound Obstet Gynecol. 2003;22:205–213.
12. Dietz HP, Haylen BT, Broome J. Ultrasound in the quantification of female pelvic organ prolapse. Ultrasound Obstet Gynecol. 2001;18:511–514.
13. Tunn R, Schaer G, Peschers U, et al. Updated recommendations on ultrasonography in urogynecology. Int Urogynecol J Pelvic Floor Dysfunct. 2005;16:236–241.
14. Peschers UM, DeLancey JO, Schaer GN, et al ultrasound of the anal sphincter: normal anatomy and sphincter defects. Br J Obstet Gynaecol. 1997;104:999–1003.
15. Cornelia L, Stephan B, Michel B, et al. Transperineal versus endo-anal ultrasound in the detection of anal sphincter tears. Eur J Obstet Gynecol Reprod Biol. 2002;103:79–82.
16. Martan A, Masata J, Halaska M, et al. Ultrasound imaging of paravaginal defects in women with stress incontinence before and after paravaginal defect repair. Ultrasound Obstet Gynecol. 2002;19:496–500.
17. Dietz HP, Steensma AB, Hastings R. Three-dimensional ultrasound imaging of the pelvic floor: the effect of parturition on paravaginal support structures. Ultrasound Obstet Gynecol. 2003;21:589–595.
18. Khullar V, Cardozo LD, Salvatore S, et al. Ultrasound: a noninvasive screening test for detrusor instability. Br J Obstet Gynaecol. 1996;103:904–908.
19. Vierhout ME, Hol M. Vaginal ultrasound studies before and after successful colposuspension and in continent controls. Acta Obstet Gynecol Scand. 1998;77:101–104.
20. Bo K, Sherburn M, Allen T. Transabdominal ultrasound measurement of pelvic floor muscle activity when activated directly or via a transversus

20. abdominis muscle contraction. Neurourol Urodyn. 2003;22:582–588.
21. Nguyen JK, Hall CD, Taber E, et al. Sonographic diagnosis of paravaginal defects: a standardization of technique. Int Urogynecol J Pelvic Floor Dysfunct. 2000;11:341–345.
22. Schaer GN, Schmid T, Peschers U, et al. Intraurethral ultrasound correlated with urethral histology. Obstet Gynecol 1998;91:60–64.
23. Fischer JR, Heit MH, Clark MH, et al. Correlation of intraurethral ultrasonography and needle electromyography of the urethra. Obstet Gynecol. 2000;95:156–159.
24. Umek WH, Obermair A, Stutterecker D, et al. Three-dimensional ultrasound of the female urethra: comparing transvaginal and transrectal scanning. Ultrasound Obstet Gynecol. 2001;17:425–430.
25. Chen GD, Lin LY, Gardner JD, et al. Dynamic displacement changes of the bladder neck with the patient supine and standing. J Urol. 1998;159:754–757.
26. Dietz HP, Clarke B. The influence of posture on perineal ultrasound imaging parameters. Int Urogynecol J Pelvic Floor Dysfunct. 2001;12:104–106.
27. Brandt FT, Albuquerque CD, Lorenzato FR, et al. Perineal assessment of urethrovesical junction mobility in young continent females. Int Urogynecol J Pelvic Floor Dysfunct. 2000;11:18–22.
28. Peschers UM, Vodusek DB, Fanger G, et al. Pelvic muscle activity in nulliparous volunteers. Neurourol Urodyn. 2001;20:269–275.
29. Reed H, Freeman RM, Waterfield A, et al. Prevalence of bladder neck mobility in asymptomatic non-pregnant nulliparous volunteers. BJOG. 2004; 111:172–175.
30. Dietz HP, Eldridge A, Grace M, et al. Pelvic organ descent in young nulligravid women. Am J Obstet Gynecol 2004;191:95–99.
31. Reilly ET, Freeman RM, Waterfield MR, et al. Prevention of postpartum stress incontinence in primigravidae with increased bladder neck mobility: a randomised controlled trial of antenatal pelvic floor exercises. BJOG. 2002;109:68–76.
32. Lin LY, Chen SY, Lee HS, et al. Female bladder neck changes with position. Int Urogynecol J Pelvic Floor Dysfunct. 1999;10:277–282.
33. Meyer S, De Grandi P, Schreyer A, et al. The assessment of bladder neck position and mobility in continent nullipara, mulitpara, forceps-delivered and incontinent women using perineal ultrasound: a future office procedure? Int Urogynecol J Pelvic Floor Dysfunct. 1996;7:138–146.
34. Peschers UM, Fanger G, Schaer GN, et al. Bladder neck mobility in continent nulliparous women. BJOG. 2001;108:320–324.
35. Howard D, Miller JM, Delancey JO, et al. Differential effects of cough, valsalva, and continence status on vesical neck movement. Obstet Gynecol. 2000;95:535–540.
36. Miller JM, Perucchini D, Carchidi LT, et al. Pelvic floor muscle contraction during a cough and decreased vesical neck mobility. Obstet Gynecol. 2001;97:255–260.
37. King JK, Freeman RM. Is antenatal bladder neck mobility a risk factor for postpartum stress incontinence? Br J Obstet Gynaecol. 1998;105:1300–1307.
38. Schaer GN, Perucchini D, Munz E, et al. Sonographic evaluation of the bladder neck in continent and stress-incontinent women. Obstet Gynecol. 1999;93:412–416.
39. Schaer GN, Koechli OR, Schüssler B, et al. Usefulness of ultrasound contrast medium in perineal sonography for visualization of bladder neck funneling–first observations. Urology. 1996;47: 452–453.
40. Schaer GN, Koechli OR, Schüssler B, et al. Improvement of perineal sonographic bladder neck imaging with ultrasound contrast medium. Obstet Gynecol. 1995;86:950–954.
41. Dietz HP, Clarke B. Translabial color Doppler urodynamics. Int Urogynecol J Pelvic Floor Dysfunct. 2001;12:304–307.
42. Alper T, Cetinkaya M, Okutgen S, et al. Evaluation of urethrovesical angle by ultrasound in women with and without urinary stress incontinence. Int Urogynecol J Pelvic Floor Dysfunct. 2001;12:308–311.
43. Wijma J, Potters AE, de Wolf BT, et al. Anatomical and functional changes in the lower urinary tract following spontaneous vaginal delivery. BJOG. 2003;110:658–663.
44. Dietz HP, Jarvis SK, Vancaillie TG. The assessment of levator muscle strength: a validation of three ultrasound techniques. Int Urogynecol J Pelvic Floor Dysfunct 2002;13:156–9.
45. Peschers U, Schaer G, Anthuber C, et al. Changes in vesical neck mobility following vaginal delivery. Obstet Gynecol. 1996;88:1001–1006.
46. Morkved S, Salvesen K. Pelvic floor muscle strength and thickness of the pelvic floor muscles in continent and incontinent nulliparous women. Neurourol Urodyn. 2001;21:358–359.
47. Schaer GN, Siegwart R, Perucchini D, et al. Examination of voiding in seated women using a remote-controlled ultrasound probe. Obstet Gynecol. 1998;91:297–301.

48. Dietz HP, Wilson PD, Clarke B. The use of perineal ultrasound to quantify levator activity and teach pelvic floor muscle exercises. Int Urogynecol J Pelvic Floor Dysfunct. 2001;12:166–168.
49. Peschers UM, Gingelmaier A, Jundt K, et al. Evaluation of pelvic floor muscle strength using four different techniques. Int Urogynecol J Pelvic Floor Dysfunct. 2001;12:27–30.
50. Sapsford R. Rehabilitation of pelvic floor muscles utilizing trunk stabilization. Man Ther. 2004;9:3–12.
51. Solomon MJ, Pager CK, Rex J, et al. Randomized, controlled trial of biofeedback with anal manometry, transanal ultrasound, or pelvic floor retraining with digital guidance alone in the treatment of mild to moderate fecal incontinence. Dis Colon Rectum. 2003;46:703–710.

# 2.7
Magnetic Resonance Imaging

Thomas Treumann, Ralf Tunn, and Bernhard Schüssler

Key Message

The aim of this chapter is to briefly introduce the technical background of MRI, to provide insight into the anatomy of the pelvis and pelvic floor, to outline the possible damage to the pelvic floor (part 1), and to depict dynamic changes in the pelvis and pelvic floor during contraction, straining, and defecation (part 2).

Introduction and Technique

Stress urinary incontinence and genital prolapse are clinical entities that originate from changes of soft tissue structures, i.e., pelvic floor muscles (PFMs), nerves, and connective tissues, surrounding the pelvic organs. Clinical examination of the perineal region, the PFMs, the vagina, and rectum at rest, at Valsalva maneuver, and during PFM contraction (see Chapter 2.1) allow for a basic assessment of various parameters, such as hypermobility of the urethra, muscle function, and pelvic organ prolapse. However, when meticulous assessment of the underlying changes in anatomy is needed, imaging techniques may offer better insight.

Imaging techniques, when applied for this purpose, should meet the following criteria:

- Large field-of-view imaging. This gives an overview over the pelvis and the pelvic floor with the position, size, and structure of the pelvic organs.
- High spatial resolution and high soft tissue contrast. This allows for the separate visualization of structural damage to pelvic organs and pelvic floor structures, e.g., as a result of childbirth.
- Cross-sectional images in various planes. This allows for a three-dimensional understanding of organ position and movement.
- Generation of dynamic sequences. This offers the ability to depict changes in the pelvic floor and pelvic organ prolapse during contraction, coughing, and straining.

The cross-sectional imaging techniques currently being used are ultrasound (US), computed tomography (CT), and magnetic resonance imaging (MRI). Apart from the fact, that MRI is the most expensive of the three techniques, it is the only technique that matches all four requirements; therefore, it is superior to US and CT in pelvic floor imaging.

When conservative treatment is planned, MRI is not a necessary precondition. However, it serves well for two reasons:

1. To understand functional anatomy of PFMs and sphincters under dynamic conditions and, thus, to enhance the results of clinical inspection and palpation in dysfunction of the pelvic floor.
2. To visualize soft tissue damage.

MRI Technique

Magnetic resonance imaging is based on the property of magnetism of the atomic nucleus of hydrogen. Hydrogen is present in nearly every molecule of the human body. When exposed to an external magnetic field, the nucleus of the hydrogen atom begins to rotate around the longitudinal axis of the magnetic field. In this state, the nucleus can be excited by radio frequency electromagnetic waves. After some time, the nucleus is emitting the absorbed energy in the form of similar radiofrequency waves. This is called the relaxation process. The answering waves can be registered by an antenna outside the patient and are used for the calculation of cross-sectional images. The images have a high soft-tissue contrast. By applying different forms of radio frequency excitations, the tissue contrast can be varied. Thus, tissues can be characterized by MRI. The main image-contrast types are called T1-weighted and T2-weighted images. Images with intermediate contrast are called proton density (PD) images.[1]

A single MR image represents one slice of the human body. A slice is characterized by its imaging plane (transverse, coronal, or sagittal) (Fig. 2.7.1), by its slice thickness (i.e., 1–10 mm or more), and by its position in relation to the center of the magnetic field. Images are depicted in two ways:

1. Multiple parallel slices with the same technical characteristics (plane, resolution, and contrast), acquired together as a block or stack of images over a certain period of time.

2. As a dynamic sequence of the same image at different points of time, repeated at the same position, while the patient is exercising (e.g., contraction or straining). A sequence of such images is a dynamic sequence and can be arranged as a movie (see DVD).

Static MRI

Anatomy, Normal Variation, and Pathology

The best results for static imaging are obtained in high-field MR systems with a magnetic field strength of 1.5 Tesla units. Such systems are closed ring units with a tube-like gantry in horizontal orientation, where the patient is brought into the magnet in the supine or prone position. For pelvic organs, T2-weighted images are superior to T1-weighted images. For PFMs, PD images are superior to T1- and T2-weighted images. The spatial resolution should be below 1 mm. The slice thickness should be 3 mm or 4 mm. Acquisition time for a series is usually between 3 and 5 minutes, depending on the characteristics of the MRI system and on the spatial resolution of the image resolution.

Static MRI of the pelvic floor in the supine position holds great promise because of its potential for identifying injuries to muscles and fascia of the pelvic floor. It allows muscle thickness to be measured. Anatomical variation between normal women, as well as anatomic changes in pelvic floor dysfunction, can be studied.

Although static MRI is not yet ready for clinical use, visualization of the anatomy and defects may contribute to a better understanding and, hence, better conservative treatment of patients with pelvic floor disorders.

MRI: Anatomy in Healthy Nulliparous Women

At the level of the middle and proximal urethra, the levator ani muscle sling (the puborectalis and pubococcygeus portion) is seen to the left and the

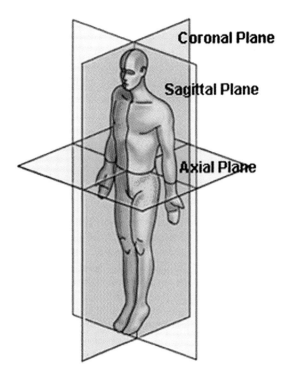

FIGURE 2.7.1. Magnetic resonance imaging imaging planes.

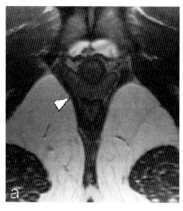
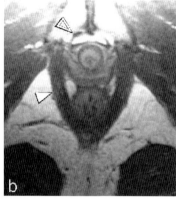

FIGURE 2.7.2. Axial section at the level of middle urethra showing difference in levator ani muscle thickness and configuration. In this and subsequent illustrations, two individuals are compared with one displayed in the left column and the other in the right column. **a** represents a thin muscle (31-year-old nullipara) and **b** a thicker muscle (36-year-old nullipara). Note also that the muscle is shaped more like a V in **a** and more like a U in **b**. The closed arrowhead marks the right levator ani muscle and the open arrowhead marks the insertion of the arcus tendineus of the fascia pelvis into the pubic bone in **b**.

right, surrounding the levator hiatus and rising from the inside of the pubic bone. Length, area, and volume measures of the levator ani muscle vary considerably in continent nulliparous women, with two- to threefold differences between the minimum and maximum measure (Fig. 2.7.2). In 10 % of the women investigated, a visible attachment of the levator ani muscle to the inside of the pubic bone is absent on both the left and the right (Fig. 2.7.3), and in another 10% it absent on only one side. This finding is associated with a significantly greater mean urogenital hiatus area, measured on an axial plane at the level of the proximal urethra, compared to those with an intact levator ani muscle attachment.

At axial planes at the level of midurethral attachment to the arcus tendineous fasciae pelvis (DeLancey level III), the intact morphology of the endopelvic fascia is shown with a typical vaginal configuration because of the musculofascial connection between the lateral vaginal wall and levator ani muscle.[2] The anterior vaginal wall has a symphysis-oriented, concave (hammock-like or "butterfly-shaped") configuration at the level of the middle/proximal urethra in all of the nulliparous women investigated (Fig. 2.7.2). Because of technical difficulties, such as slice thickness, of the MRI images, however, direct attachment of the vaginal wall to levator ani muscle at the level of attachment of the urethra and the bladder base

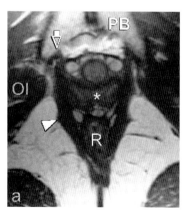
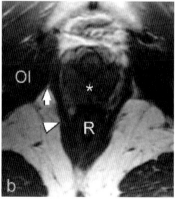

FIGURE 2.7.3. Levator ani muscle thickness (closed arrowhead) shown on coronal images at the level of the urethra (**a and b**) and the level of the rectovaginal space (**c and d**) comparing thin muscle (**a and c**; 31-year-old nullipara) with a thicker muscle (**b and d**; 28-year-old nullipara). Note that the levator ani muscle ends at the perineal membrane (open arrowhead). The puborectal muscle is selected in *d* (*). B = bladder, OI = obturator internus muscle, R = rectum.

(DeLancey level II and III) (Fig. 2.7.4), as well as the arcus tendenious of the pelvic fascia, is not clearly visible in all women.

The three layers of the urethra (mucosa, submucosa/smooth muscle, and striated muscle layer) are reproducible with MRI. The striated muscle layer is composed of the circular shaped external urethral sphincter muscle, the omega shaped compressor urethral muscle (proximal urethra; the muscle is covering the anterior part of the urethra and vaginal wall, respectively), and the urethrovaginal sphincter muscle (middle urethra). This composition of the striated muscle layer of the urethra causes the dorsal part to appear smaller compared to the lateral and anterior part (Fig. 2.7.4). There is also a great variation in diameter area or volume measures of these muscles, which parallels findings in the levator ani muscle.

Reversible and Permanent Birth-Related Changes of the Levator Ani Muscle

Having understood variations of normal found by MRI, investigating women after childbirth is an exciting technique to elucidate birth-related changes within muscle and connective tissue. As expected from clinical understanding, a complete loss of levator ani tissue was found in a few women only.[3] In all women, however, by 1 day postpartum, the T2-weighted signal intensity of the levator ani muscle was higher compared with the obturator internus muscle. By 6 months, these differences had disappeared in most women.

The change in MR signal intensity suggests a change in the chemical composition of the muscle during the first 6 months after delivery. There are several possible explanations for these observed changes. An increase in extracellular fluid,

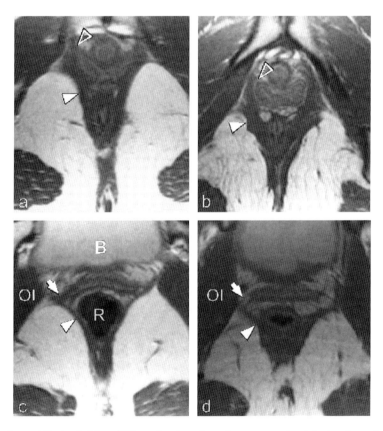

FIGURE 2.7.4. Axial planes at the level of the middle urethra (**a and b**) and bladder base (**c and d**) showing the differences in the relationship between the vagina and the levator ani muscle (closed arrowhead); both a direct connection in a 31-year-old nullipara (**a**, open arrowhead; **b**, closed arrow) and an indirect connection in a 29-year-old nullipara (**b**, open arrowhead; **d**, closed arrow) are seen. B = bladder, OI = obturator internus muscle, R = rectum.

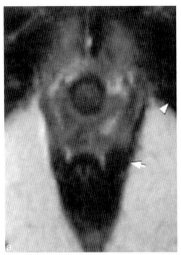 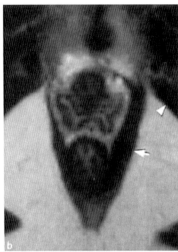

FIGURE 2.7.5. Changes in T2-signal intensity of the LA (arrow) on transverse sections following spontaneous vaginal delivery. 32-year-old woman, para 3, (a, b) with higher signal intensity 1 day postpartum (a) and normal muscle signal intensity 6 months postpartum compared to the obturator internus muscle (arrowhead).

glycogenolysis, and accumulation of lactate are known causes of changes in muscle signal intensity. Such changes are reflected by an increase in signal intensity on T2-weighted sequences (lighter muscle) and are reversible (Fig. 2.7.5, a and b).[3,4] Changes in signal intensity caused by a decrease of striated muscle and increase of connective tissue or fat content in the levator ani muscle are permanent (Fig. 2.7.5, c and d, and Fig. 2.7.6).[5]

Vaginal birth may lower the perineum and stretch the levator hiatus. However, significant changes at 2 weeks postpartum are suggestive of a return of normal levator ani geometry.[3] Elevation in perineal body position, as well as a decrease in the area of the urogenital hiatus by 27% and of the levator hiatus by 22%. Although the elongation of the muscle, which is reflected by the increased size of the urogenital hiatus and the descent of the perineal body, returns to normal in the resting supine position relatively rapidly, the internal chemical changes are not completely back to normal as late as 6 months after delivery. This may reflect different types of parturition-induced changes. In multiparous women the signal intensity of the levator ani muscle needs a longer time (6 months) to return to normal compared to primiparous women (2 months).

As one with clinical experience might expect, remarkable variations in levator ani structure changes are found among different individuals. One woman's delivery may leave the pelvic floor almost unchanged, whereas others may have profound changes in muscle geometry and signal intensity. Larger studies are necessary to

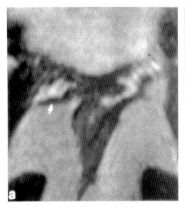 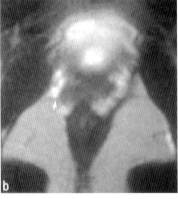

FIGURE 2.7.6. T2-weighted transverse sections at the level of the bladder neck of a 20-year-old woman, para 1, after spontaneous vaginal delivery showing disappearance of LA signal at different follow-up times. (a) Disrupted LA signal in the paravaginal region on the right (arrow) is seen 2 weeks postpartum. (b) Lack of visible signal in the area of the right LA (arrow) 6 months postpartum suggests atrophic changes.

Women with Stress Urinary Incontinence

Comparing the anatomy of the urinary continence mechanism in nulliparous continent women and women with stress urinary incontinence shows a wide range of overlapping findings.[2,6] Changes such as the loss of the symphyseal concavity ("butterfly" shape) of the anterior vaginal wall are significantly more common in women with stress urinary incontinence, but have also been reported in women without symptoms and, hence, could be both constitutional and delivery related.[6–8]

However, recent data in a large series of women with stress incontinence have revealed relevant pathomorphologic changes of the levator ani muscle, endopelvic fascia, and urethra.[9] In the levator ani muscle, unilateral loss of substance in approximately one third of women, a higher signal intensity in approximately one quarter, and altered origin in approximately one fifth (Fig. 2.7.7). The discovery of an absence of the levator ani origin at the pubic bone in some continent nulliparous women, which was associated with a significantly larger urogenital hiatus, makes one suspect that these women have a higher constitutional risk for the development of pelvic floor dysfunction.

The association of changes in signal intensity associated with morphologic and biochemical muscle changes have been confirmed by studies comparing MRI and histologic findings. MRI findings also correlate with the results of electromyography studies.[10] The increased signal intensity can, thus, be explained by histologically demonstrated myogenic changes of the PFMs.

The urethral sphincter muscle shows a reduced thickness in its posterior portion (37%), an omega shape (13%), or higher signal intensity (50%); configurations that are considered abnormal were associated with an increased signal intensity in 70% of women with stress urinary incontinence. Central defects of the endopelvic fascia were present in 39% of women and lateral defects were present in 46%. There was a significant association between loss of the symphyseal concavity of the anterior vaginal wall and lateral fascial defects and levator ani changes. These findings confirm theories on the pathogenesis of stress

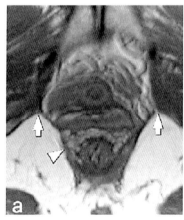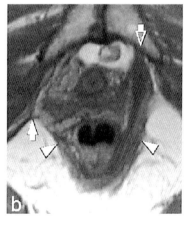

FIGURE 2.7.7. Axial PD-weighted MR images at the level of the proximal urethra showing different defects of the levator ani muscle (solid arrowheads) in women with stress urinary incontinence. (a) Bilateral nonvisualization of the origin of the levator ani at the pubic bone while origin at the arcus tendineus levator ani (ATLA) is seen (solid arrows) in a 34-year-old woman with stress urinary incontinence. (b) Intact levator ani on the left with depiction of its origin at the inner surface of the pubic bone (open arrow) and atrophy of anterior muscle portions on the right where only its origin at the ATLA is depicted in a 48-year-old woman with stress urinary incontinence. Increased signal intensity of the levator ani relative to the obturator internus muscle in a 51-year-old woman with stress urinary incontinence.

urinary incontinence, such as the hammock hypothesis by DeLancey[11] and the integral theory by Petros and Ulmsten.[12] In both theories the intact attachment of the vaginal wall to the levator ani muscle are considered crucial for maintaining continence.

Dynamic MRI of the Pelvis

For this technique, the best results are obtained in high-field systems, with the patient in a supine position.[13] Although it may be difficult for the patient to contract, strain, and defecate in the supine position, dynamic MRI in closed magnets is preferred to MRI in open upright devices such as the Fonar® system because closed high-field units offer much better image contrast, spatial resolution, and temporal resolution, and therefore allow for more precise evaluation of the pelvis and pelvic floor. Procedural restrictions seem not to play a major role. Comparison of the two methods has shown that no significant pathology was missed in the supine position compared with the upright position during dynamic MRI.[14]

As the urethra, the urethral sphincter, the vagina and uterus, the anal sphincter, the anal canal, the rectum, and the coccyx are located in the midline of the pelvis, this plane offers optimal insight into the organs, as far as their reaction to PFM contraction and straining is concerned. It has to be kept in mind, however, that the puborectalis muscle sling is mainly located lateral to the midline and is not seen in midline MRIs, except for the small part going around the anus posteriorly. The function of the puborectalis muscle is assessed indirectly by observation of the shortening and widening of the anogenital hiatus (Fig. 2.7.8, A–C).

Standard Reference Lines[14]

For the interpretation and grading of perineal descent and organ prolapse, three main reference lines are used (Fig. 2.7.9):

1. The pubococcygeal line (PCL)
2. The hiatus line (H-Line)
3. Descent of the pelvic floor (M-Line)

The pubococcygeal line represents a fixed static line from the posterior point of the pubis symphysis to the last visible coccygeal or sacrococcygeal intervertebral disk. The H-Line represents the level of the puborectalis muscle sling and leads from the symphysis, to the muscle sling posterior, to the rectoanal junction. Its length represents the width of the anogenital hiatus.

As a measure for the position of the muscular pelvic floor in relation to the bony pelvis, a third line is introduced. The M-line measures the distance of the puborectalis muscle sling to the PCL.[15]

In asymptomatic nulliparous women, the position of the pelvic floor does not exceed 2 cm below the pubococcygeal line at rest and 5 cm during a Valsalva maneuver. The width of the hiatus (H-Line) is between 4 and 6 cm at rest, not exceeding 8.5 cm during Valsalva. An H-line of more than 9 cm is considered to be pathologic.

The depth of a rectocele is measured as the distance from the center of the anal canal to the anterior wall of the rectocele. Anything less than 3 cm long is not called a rectocele. The depth of a cystocele is measured as the distance from the internal urethral orifice to the lowest point of the bladder wall. A cystocele starts at 1 cm in length.

Visualization of PFM Function

Dynamic MRI sequences (see DVD) visualize PFM function and changes in the pelvic organs under various conditions. For physiotherapy treatment, the understanding of the following standard situations of the pelvic floor is important (Fig. 2.7.8, A–C):

1. Pelvic floor at rest
2. Maximum contraction of the PFMs
3. Maximum relaxation of the PFMs and descent of the pelvic floor and organs during Valsalva maneuver and defecation.

Pelvic Floor at Rest

Dynamic MRI investigation of the pelvic floor starts with the pelvic floor in a resting position (Fig. 2.7.8 A) at normal intraabdominal pressure. Anteriorly, the abdominal wall is seen. The pelvic organs are supported by the pelvic floor. Within

2.7. Magnetic Resonance Imaging

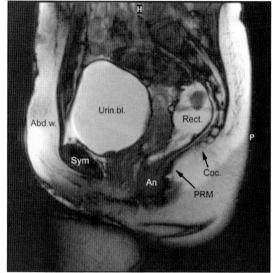

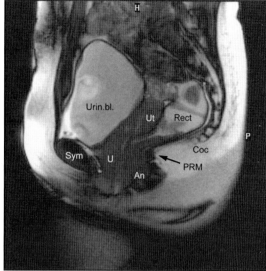

FIGURE 2.7.8. **A.** T_2 – weighted images of a 26 year old nulliparous woman (midline section) Position of pelvic organs at rest. Abd.w. = abdominal wall, Sym. = Symphysis pubis, Urin. Bl. = Urinary bladder, U. = urethra, Ut. = uterus, Rect. = rectum, Coc. = coccyx, PRM = puborectalis muscle (midline part), An. = anal sphincter and canal. **B.** Position of pelvic organs during contraction of the pelvic floor muscles. **C.** Position of pelvic organs Valsalva maneuver.

the pelvis, from anterior to posterior, the urinary bladder, the vagina, and the rectum can be identified. The urethral sphincter muscle is seen below the urinary bladder. The bladder contains urine, and the rectum and vagina are filled with contrast medium.

The puborectalis muscle sling can be identified posterior to the rectoanal junction. Posterior to the anorectum, the anococcygeal ligament leads to the coccyx. The sacrum, coccyx, and anal ligament form the posterior border of the pelvis. In coronal images, the ileococcygeus muscle is seen as a curved structure leading from the perineum to the inner surfaces of the ischium and os pubis.

Anteriorly, the puborectalis muscles are seen in cross-section lateral to the vagina and urethra.

Pelvic Floor Contraction

During contraction, the ileococcygeus muscle elevates the pelvic floor, and the puborectalis muscle sling shortens the anogenital hiatus. Pelvic floor contraction elevates the pelvic organs upwards and moves the urethra towards the back of the pubic symphysis. (Fig. 2.7.8 B) Because the coccyx has remained mobile, it is pulled anteriorly by tension on the anococcygeal ligament.

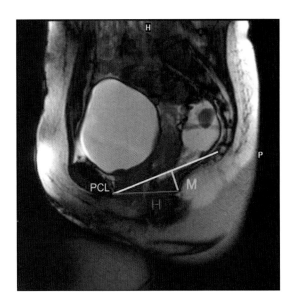

FIGURE 2.7.9. For quantification of prolapse on MRI, the following reference lines have gained acceptance:
1. The PCL level (PCL)
2. H: The width of the genital hiatus
3. M: The level of the PFM in relation to the pubococcygeal level

Quantification of pelvic muscle contraction is completed by measuring the anogenital hiatus width and the muscular pelvic floor position in relation to the PCL, and then comparing these findings to the corresponding values at rest. An elevation of the pelvic floor and a shortening of the hiatus of 1 cm are normal values for pelvic floor contraction (Fig. 2.7.9).

Pelvic Floor During Valsalva Maneuver

Valsalva maneuver is performed by increasing the intraabdominal pressure while relaxing the PFMs. This leads to a descent of the pelvic floor, widening of the anogenital hiatus, elongation of the anococcygeal ligament, and the posterior movement of the pelvic organs. (Fig. 2.7.8 C)

In the anterior compartment, the urethra and the bladder rotate around the posterior aspect of the pubic symphysis.

In the middle compartment, the vagina and the uterus normally move posteriorly. In genital descent, the vagina shortens and descends, and the uterus goes into an upright position into the longitudinal axis of the vagina and may descend within the vagina down to (descent) or below the pelvic floor (prolapse).

In the posterior compartment, the anterior rectal wall may bulge the posterior vaginal wall. The posterior vaginal wall may be bulged inside the vagina or everted to the outside.

During defecation, the anterior wall of a rectocele can move towards the anal canal, leading to a low intussusception. If the puborectalis muscle sling does not relax during Valsalva, the rectum can be compressed between anococcygeal ligament and the peritoneal sac or the uterus, leading to stool-outlet obstruction or fragmented defecation. In addition to that, the rectum can develop an internal intussusception at the point where the rectum is pressed over the hypomochlion of the ileococcygeus muscle (see DVD).

The pouch of Douglas usually ends at the vaginal vault. As an anatomic variation, the pouch of Douglas can extend between the rectum and the vagina. Investigations in nulliparous women have shown that the pouch of Douglas can extend as far down as to the level of the pelvic floor, thus, totally separating the posterior vaginal wall from the anterior rectal wall. This is suggested to be the origin of a posterior enterocele later in life (Fig. 2.7.10).[16] Recurrent pressure upon the pelvic floor, relaxation of the PFM, or ventral dislocation of the anterior compartment (e.g., Burch Colposuspension) may lead to a gradual widening of the pouch of Douglas and to

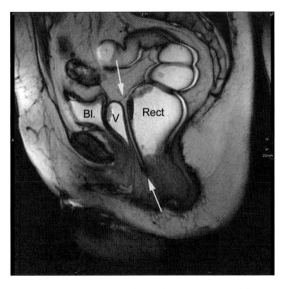

FIGURE 2.7.10. Open pouch of Douglas in a 56-year-old female. Arrows indicate entrance and end of this pocket.

2.7. Magnetic Resonance Imaging

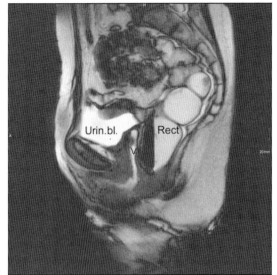

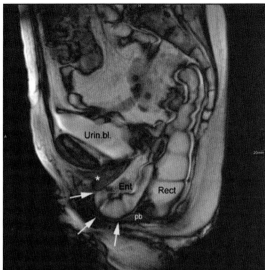

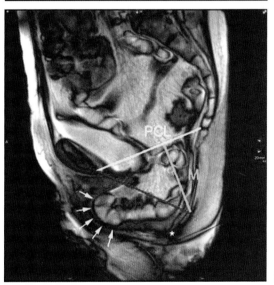

FIGURE 2.7.11. **A.** Images of a 72-year-old female with perineal descent and an enterocele twenty years after hysterectomy. At rest, there is normal position of the pelvic organs. Urinary bladder (Urin.bl.) is filled with urine. Vagina (V) and rectum (Rect) are filled with ultrasound gel. **B.** During Valsalva manoeuvre, the levator ani is totally effaced and elongated. An enterocele containing sigmoid colon (Ent) protrudes into the space between the vagina (V) and the rectum (Rect). Part of the posterior vaginal wall is everted (arrows due to an enormous descent of the perineum relative to the plane between the interior border border of the tuber ishiadicum. Furthermore, the anterior rectal wall as well as the posterior vaginal wall are pushed outwards.). * = urethral orifice. pb = perineal body. **C.** MRI of the same patient after defecation: After examination of the rectum this space is filled with the peritoneal sac, which is descending down to the perineum and everting parts of the posterior vaginal wall (arrows). Hiatus width (H) is abnormal with 10 cm. Position of muscular pelvic floor (M) in relation to PCL is abnormal with 7 cm. (* = anus) (pb = perineal body)

the development of an enterocele. An enterocele is the descent of peritoneal fat, sigmoid colon, or small bowel within the perineal sac down to the perineum, between the rectum and vagina. Such an enterocele can bulge or evert the posterior vaginal wall, similar to a rectocele. It can also evert the anterior rectal wall into or out of the anal canal. MRI is the method of choice to differentiate if prolapse of the posterior vaginal wall is caused by a rectocele, an enterocele, or both (rectocele first; enterocele after defecation).

Laxity of the pelvic floor can cause a ballooning of the perineum, thus, effacing the gluteal crease. MRI has shown that these changes in outer shape are the result of a tremendous elongation of the PFMs and the anococcygeal ligament (Fig. 2.7.11, A and B).

Significance of Dynamic MRI of the Pelvic Floor

Dynamic MRI is a method to quantify the laxity and descent of the pelvic floor and to depict abnormal behavior of the pelvic organs during strain and contraction. MRI is of importance for pelvic organ prolapse and dysfunctional defecation.

As far as the anterior compartment is concerned, MRI and ultrasound findings are similar. In the posterior compartment, MRI is superior to clinical evaluation and ultrasound in the detection and differentiation of rectoceles and enteroceles, causing a bulge of the posterior vaginal wall. It has to be kept in mind, however, that the hernial sac can contain structures that are not located in the midline. To get these structures into the images, additional planes (transverse, coronal) are required. However, this is not part of the standard investigation protocol. Sometimes, the final answer is given only during surgery. For rectal intussusception, MRI is the only imaging method for detection. For rectal prolapse, MRI can be used to differentiate rectal mucosal prolapse from rectal wall prolapse. Mobility of the coccyx can be assessed by MRI, but the role of the coccyx function and dysfunction of the pelvic floor is not quite clear. The coccyx probably does not play a major role.

References

1. Tress BM, Brant-Zawadski M. Nuclear magnetic resonance imaging. Basic principles. Med J Aust. 1985;142:21–24.
2. Tunn R, DeLancey JO, Howard D, et al. Anatomic variations in the levator ani muscle, endopelvic fascia, and urethra in nulliparas evaluated by magnetic resonance imaging. Am J Obstet Gynecol. 2003;188:116–121.
3. Tunn R, DeLancey JO, Howard D, et al. MR imaging of levator ani muscle recovery following vaginal delivery. Int Urogynecol J. 1999;10: 300–307.
4. Fleckenstein JL, Watumull D, Conner KE, et al. Denervated human skeletal muscle: MR imaging evaluation. Radiol.1993;87:213–218.
5. De Smet AA. Magnetic resonance findings in skeletal muscle tears. Skeletal Radiol. 1993;22:479–484.
6. Tunn R, Paris S, Fischer W, et al. Static magnetic resonance imaging of the pelvic floor muscle morphology in women with stress urinary incontinence and pelvic prolapse. Neurourol Urodyn. 1998;17:579–589.
7. Huddleston HT, Dunnihoo DR, Huddleston PM, et al. Magnetic resonance imaging of defects in DeLancey's vaginal support levels I, II, and III. Am J Obstet Gynecol. 1995;172:1778–1782.
8. Klutke C, Golomb J, Barbaric Z, et al. The anatomy of stress incontinence: magnetic resonance imaging of the female bladder neck and urethra. J Urol. 1990;143:563–566.
9. Tunn R, Schaer G, Peschers U, Bader W, Gauruder A, Hanzal E, Koelbl H, Koelle D, Perucchini D, Petri E, Riss P, Schuessler B, Viereck V. Updated recommendations on ultrasonography in urogynecology. Int Urogynecol J Pelvic Floor Dysfunct. 2005;16(3):236–241.
10. McDonald CM, Carter GT, Fritz RC, et al. Magnetic resonance imaging of denervated muscle: comparison to electromyography. Muscle Nerve. 2000;23:1431–1434.
11. DeLancey JO. Structural support of the urethra as it relates to stress urinary incontinence: the hammock hypothesis. Am J Obstet Gynecol. 1994; 170:1713–1720.
12. Petros PE, Ulmsten U. An integral theory of female urinary incontinence. Experimental and clinical considerations. Acta Obstet Gynecol Scand 1990; 153:7–31.
13. Lienemann A, Fischer T. Functional imaging of the pelvic floor. Eur J Radiol. 2003;47:117–122.
14. Bertschinger KM, Hetzer FH, Roos JE, et al. Dynamic MR imaging of the pelvic floor performed with the patient sitting in an open-magnet unit versus with patient supine in a closed-magnet unit. Radiol. 2002;223:501–508.
15. Stoker J, Halligan S, Bartram C. Pelvic floor imaging. A review. Radiol. 2001;218:621–641.
16. Baessler K, Schüssler B. The depth of the pouch of Douglas in nulliparous and parous women without genital prolapse and in patients with genital prolapse. Am J Obstet Gynecol. 2000;182: 540–544.

2.8 Electrophysiology

Clare J. Fowler and David B. Vodušek

Key Message

Nerve and muscle cells produce and conduct electrical activity. Two main types of investigations are used – electromyograph, to detect signals for striated muscle, and conduction studies, which employ electrical, magnetic, and mechanical stimuli to detect sensory, motor, and reflex responses of the nervous system. These tests may be used to investigate patients with central and peripheral nerve lesions affecting the urogenital anal region and for research purposes.

Introduction

Movement is achieved by activation of the motor units in muscles. A motor unit is defined as all muscle fibers within an anatomical muscle that are innervated by the same motor axon. The essential components of the motor units are shown in Figure 2.8.1. Coordinated muscle activity, voluntary activation of a muscle, and the resting tone of the muscle are determined by the pattern of activations of motor neurons, which lie in the anterior horn of the spinal cord. Activation results from descending spinal input and segmental spinal reflexes.

Skilled voluntary movement, as needed to perform tasks which require manual dexterity need individual motor units to be activated in a highly refined pattern generated by the motor cortex. In contrast, patterns of activation of axial muscles necessary to maintain posture arise from the vestibular nuclei and reticular formation. More detailed information about general motor control can be found in a number of standard physiotherapy texts. Motor units of muscles that comprise the pelvic floor are tonically active. Unlike most other skeletal muscles, this activity cannot be voluntarily suppressed, except as part of the coordinated process of micturition or defecation, which requires relaxation of the pelvic floor.

As explained in preceding chapters, the muscles of the pelvic floor include those that form its base and provide general support to the pelvic organs, i.e., the pubococcygeus and levator ani and, as separate structures, the striated muscle of the urethral and anal sphincters. It seems likely, although is by no means proven, that whereas the muscles that provide pelvic organ support are activated in response to postural changes and changes in pressure within the abdominal cavity, control of the sphincters is more intimately related to the function of the bladder and rectum, respectively.

Striated Muscle Activity and Electromyography

Urethral Sphincter

This small circular muscle surrounding the urethra has a critical role in urinary continence. Studies in experimental animals and humans

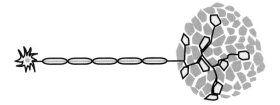

FIGURE 2.8.1. Schematic representation of a motor unit showing the motor cell body, its myelinated axon in the peripheral nerve, and muscle fibers, which are all innervated by that motor axon.

have shown that activity in this muscle increases with increasing bladder distension. This is part of the "guarding reflex," which is organized at a spinal level. Increases in motor unit activation occur with sudden rises in abdominal pressure, such as occur during coughing, sneezing, etc.

Voiding is achieved by the activation of a group of neurons that lie in the dorsal part of the pons that send inhibitory input to Onuf's nucleus in the sacral spinal cord, the nucleus formed by the anterior horn cells that innervate the urethral sphincter. Cessation of motor unit activity in the sphincter is the first recordable event of voluntary micturition, and this is followed some seconds later by a contraction of the detrusor smooth muscle. This coordinated sphincter relaxation and detrusor contraction depends on intact spinal cord connections between the pons and the sacral part of the cord (Fig. 2.8.2). If spinal disease disrupts these connections, sacral segmental reflexes emerge, which cause the detrusor muscle to contract in response to low bladder volumes, i.e., detrusor overactivity. A second feature of the spinal disruption is that the synergistic behavior of the sphincter and the bladder is lost and detrusor sphincter dyssynergia occurs, i.e., the striated urethral sphincter contracts when the detrusor muscle is contracting.

In addition to the striated urethral sphincter's role in micturition and bladder emptying, voluntary contraction of the sphincter has a powerful inhibitory effect on detrusor contraction. This effect has been studied little in animals because of the obvious difficulty of experimental design, but there is some neurophysiological data pointing to an inhibitory influence. Curiously, there have been few studies of this "procontinence maneuver" in patients and volunteers, but it is, of course, by this mechanism that incontinence may be avoided if urgency is experienced. The pathway involved requires connections between the motor cortex and sacral cord and is therefore often affected in patients who have spinal cord disease. It is, however, available to patients with idiopathic detrusor overactivity, i.e., an overactive bladder in the absence of neurological disease.

Striated Anal Sphincter

The neural processes involved in fecal continence have been studied even less than those of bladder control. However, the same pattern of tonic activation during the filling phase is thought to occur, with suppression of activity at the onset of defecation.

Increases in motor unit activity in the sphincter are, again, crucial in maintaining fecal or flatal continence, and it seems likely that activation of this muscle has an inhibitory effect on rectal contraction.

Methods of Recording EMG Activity from the Pelvic Floor

The method preferred by electromyographers for recording muscle motor unit activity is usually the concentric needle electrode (CNE). In previous years, some very specialized recordings were made using single-fiber needles, but these are no longer commonly being used. They have an expensive construction and therefore, non-disposable. Some approximation to the signal recorded by a CNE may be obtained by using surface electrodes of various sorts, but the precision of what is being recorded, and the details of the electrical signal are inevitably less refined using such devices.

Concentric Needle Electrode

A CNE consists of a central recording wire that forms the core of an outer cannula. As with all electrophysiological recordings, a measured change in voltage is recorded between an active and a reference electrode. In the CNE, the central wire forms the active electrode and the cannula

2.8. Electrophysiology

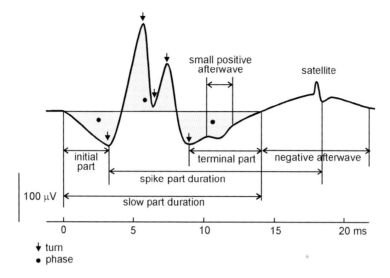

FIGURE 2.8.2. Component phases of motor unit potential (bottom) recorded with a CNE (top). The spike of the motor unit is produced by the muscle fibers closest to the electrode, whereas the late phases result from activity in more distant fibers.

forms the reference. The distal end of this combination is sharpened to a bevel tip to facilitate its insertion into tissue, and the recordings are made from a hemispherical recording volume with a diameter of 0.5 mm. The CNE will, therefore, be recording the extracellular currents of approximately 20–30 muscle fibers, and because each of these are part of separate motor units, when a CNE is used to record from the striated muscle of either sphincter approximately five or six motor units are found firing tonically while at rest. With voluntary recruitment this number increases, which also happens if the subject coughs, sneezes, etc. The ongoing activity in these muscles precludes the possibility of detecting spontaneous fibrillations, which is the hallmark of denervation in other skeletal muscles.

In the clinical neurophysiology laboratory, the electromyographer would inspect the configuration of individual motor units, a procedure easily performed using a CNE in conjunction with an EMG machine that has a "trigger and delay line." Captured motor units are stored, and their amplitude, duration, and number of phase reversals (polyphasicity) measured up in some detail (Fig. 2.8.2), either by setting cursors or, nowadays, by using automated analysis techniques. Individual motor unit analysis of the urethral, but more commonly of the anal, sphincter is carried out for diagnostic purposes in the investigation of suspected cauda equina lesions, and the innervation of those muscles, differential diagnosis of various forms of Parkinsonism. Used in these circumstances, it is the prolongation and enlargement of motor units that is being sought. These are, in fact, the consequence of reinnervation and provide no evidence about the extent of denervation that has occurred. Using a CNE, the only clue to the extent of a denervating injury is the reduction in the number of firing motor units that can be reflexively or voluntarily recruited. Unfortunately, this is inevitably a somewhat subjective assessment, although the sophisticated software that exists in some EMG machines attempts to provide a statistical and more objective approach. For the physiotherapist's assessment of the pelvic floor, the only advantage to be gained from using a concentric needle electrode would be the certainty of knowing precisely which muscle information is being recorded from. This information is lost when using surface electrodes, but the design of the CNE has been such as to optimize its "in–out" insertion for what the electromyographers call "sampling" a muscle, and not long-term recording situations.

Hook Wire Electrodes

A recording system which combines the precision for a selected site together with the

FIGURE 2.8.3. A kinesiological EMG recording with hook wire electrodes from the right and left pubococcygeus muscle in a female. The detection site in the right muscle detects continuous firing of motor unit potentials during relaxation ("tonic" activity). Also, on the right, a more prolonged recruitment of motor units has occurred. Such pelvic floor muscle activity pattern has been called "tonic" (on the right). There is no firing of motor units during relaxation at the particular detection site in the left pubococcygeus muscle, and recruitment of motor units is brisk and relatively short lasting. Such activity pattern has been called "phasic" (on the left). Such patterns can be recorded in a normal pubococcygeus muscle at different detection sites and, in the case presented, do not signify pathology. A kinesiological EMG recording of a pathological pattern of activity of pubococcygeus muscles can be observed in Figure 1.2.9.

possibility of making long-term recordings is the "hook wire electrode." The essential part of this recording system consists of a twisted pair of coated wires with exposed tips shaped into a hook. These are inserted into the particular muscle to be studied through a hollow core needle and the needle then withdrawn. The hook is retained in the muscle for long-term recordings and is removed at the end of the recordings by tugging gently. The difficulty of using this type of electrode is that, whereas with a CNE the position of the electrode can be adjusted to give the best EMG signal by listening over the audio output system of the recording machine, this is not possible with the hook wire electrodes because the quality of the signal can only be examined when the ensheathing needle has been removed, and at that point little adjustment of the wires is possible. Notwithstanding these difficulties, some groups have succeeded in making very useful "kinesiological" recordings from the pelvic floor and shown asymmetries in activity at rest and on coughing in women with stress incontinence (Fig. 2.8.3).

Surface Electrodes

The principle of recording EMG with surface electrodes is that the active electrode is placed over the main body of the muscle and the reference is placed over the tendon insertion of that muscle. A surface electrode has a large pick up area and furthermore, the soft tissues, i.e., connective tissue, fat, and skin, between the muscle and the electrode, may significantly attenuate the amplitude of the signal. Thus, with a CNE the amplitude of the EMG activity from the levator ani might be expected to be up to 500 mV, but an amplitude of one tenth of that would be considered "good" if recorded using a surface electrode.

The advantages of surface electrode recordings are that the attachment of the electrodes is pain free and that movement is less likely to dislodge the recording position. However, surface electrodes will pick up EMG activity from muscles distant from the one being studied, the amplitude of the signal may be low, and the contribution of interference considerable. These limitations notwithstanding, this is the type of recording electrode most commonly used in physiotherapy practice when recording from pelvic floor structures.

An important practical aspect of using these recording devices is that the raw recording should be carefully inspected for the veracity of its EMG content before any subsequent analysis is performed (Fig. 2.8.4). Ideally an audio output of the signal should be obtained so that any recognizable electrical interference can be minimized. When what is thought to be a good EMG signal is obtained, the subject should be asked to voluntarily contract, and an increase in ongoing EMG activity listened for. Because the EMG signal is

2.8. Electrophysiology

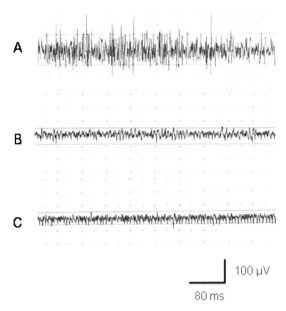

FIGURE 2.8.4. EMG recording with surface electrodes from rectus abdominis **(A)** well separated, artifact free recording, **(B)** recording electrodes poorly placed, **(C)** poor EMG signal with high mains interference content.

recorded as a waveform, continuously crossing from positive to negative, and the need in muscle building sessions is often only to look at the level of increase in the signal with voluntary activation, various online analysis facilities are available, which perform rectification, i.e., converting all signals to positive, and integration of the original EMG signal. Although this has the advantage that a single line may be watched and observed to go up when there is an increase in muscle activity, if the fundamental signal is of poor quality this may not be recognized. Poorly attached recording electrodes, which allow high amplitude, mains interference may also display a rise in the integrated signal if, during a contraction, the configuration of the electrodes moves and the interference component increases.

Types of Surface Electrodes

Sticky pads incorporating a small metal surface are easily applied to the skin, and many papers have been published on the use of such electrodes attached to either side of the anal sphincter or vagina, with a third electrode placed over a bony point to act as a ground electrode. In an effort to increase the proximity of the surface electrode to a structure deeper within the pelvis, such as the urethral or anal sphincter, especially devices on which the recording electrode is mounted have been made. In particular, there is a commercially available sponge electrode that can be used to record from the urethral sphincter, a range of device probes that are used to record activity from the pubococcygeus muscles in women, and a variety of anal probes that are inserted into the anal canal to record activity from the anal sphincter and puborectalis muscle in men and women.

Other Neurophysiological Investigations

There are various other neurophysiological tests that can be performed in the laboratory, which are sometimes referred to as "electrodiagnostic tests" or more inaccurately "EMG" (because that term should properly be reserved for electromyographic recordings). The other tests are reflex studies, nerve conduction studies, or evoked potentials.

Reflex Studies

Reflex studies look at the timing of a muscle response after an electrical stimulation of skin or a nerve. Specifically, in the pelvic region, where there are many physiological reflexes, two have been studied; the bulbocavernous and the anal reflex. Each requires a recording system to pick up the respective muscle response, and stimuli are given to the urogenital area. A reflex analogous to the bulbocavernosus in men is not easily recorded in women, but recordings from the anal sphincter are more robust.

Nerve Conduction Studies

The principle of nerve conduction studies is that an electrical pulse is given at a point along a nerve, and a response is recorded either from a muscle which that nerve innervates or from the nerve itself, from some distance away. The recordings that are taken are referred to as "compound muscle action potentials" or "sensor action potentials," respectively.

If a motor nerve is stimulated at two points, giving a proximal and distal motor latency, and the figures subtracted from one another (so that slowing of the nerve in its distal parts where it enters the muscle are removed from the sum) the figure left is known as the "conduction time." This, divided into the measured distance between the two stimulation sites, gives the measured conduction velocity of that nerve segment in meters per second. However, the conduction velocity of a nerve is not as valuable as might be supposed. This is because, unless any pathology affecting the nerve causes a loss of the fastest conducting axons, conduction velocity may be reduced by very little, possibly by as little as 10%, despite severe nerve fall out and muscle weakness. The real value of the conduction times of nerves is in detecting either generalized or focal regions of demyelination. Demyelination is the process whereby the axon of the nerve remains in tact, but its insulating sheath of concentric wraps of myelin is disrupted, leading to an impairment of the physiological salutary conduction of that nerve. If there is uniform demyelination along the whole nerve, as occurs in an inherited demyelinating neuropathy, the measured conduction velocity can be reduced from the normal range of 50–60 m/sec down to 20–10 m/sec. If, however, impulses still manage to get through, albeit slowly, the clinical deficit associated with this laboratory-measured, slow-conduction velocity can be quite minimal.

Regions of focal demyelination occur where a nerve is entrapped, most famously at the carpal tunnel of the wrist. In this condition, stimulation of the median nerve above the wrist, while recording from a muscle of the thumb, the abductor pollicis brevis, will demonstrate focal slowing of conduction velocity. It should be noted, however, that the degree of slowing is not likely to correlate with the weakness of the muscle.

Of greater possible clinical relevance, but unfortunately more subject to technical variability, is the amplitude of the compound muscle action potential because this envelope contains a summation of all the motor units that can be activated by electrical stimulation of the nerve.

Pudendal Motor Latency

Based on the established clinical value of conduction studies to detect the entrapment condition of carpal tunnel syndrome, it was proposed that stimulation of the pudendal nerve at the level of the ischial spine and recording from muscles it innervates, such as the pubococcygeus or the urethral or anal sphincter, would be of value in detecting nerve damage following childbirth. Theoretically, however, there have always been arguments against this test as a means of detecting significant pelvic floor weakness. Chief amongst these was the concern that a prolongation of latency does not give an indication of the extent of muscle denervation. The investigation that might have reflected axonal loss was the compound muscle action potential recorded either from the urethral sphincter or pelvic floor muscles. Initial reports did not include any data on amplitude of the recorded compound muscle action potentials and, unfortunately, more recent attempts to design recording electrodes that could pick up well from these muscles have not been successful, and the recorded amplitudes are often only a few microvolts. Although it seems likely that stretching of the pudendal nerve occurs with parturition, attempts to demonstrate this using pudendal motor latency have not proved useful, and measurement of this conduction figure is certainly not used in clinical laboratories to investigate women with suspected stress incontinence, nor is it now recommended as a research investigation.

Evoked Potentials

Evoked potentials are the electrical response recorded from neural tissue, either the spinal cord or cortical surface, in response to repeated peripheral stimulation. First introduced to test conduction in the optic nerves, the visual-evoked response is recorded with scalp electrodes over the visual cortex while the subject looks at a checked board on which the black and white squares alternate. Critical in detecting the cerebral response is the process of "averaging," whereby electroencephalographic activity time locked to the onset of the stimulus becomes more pronounced and background activity from the random ongoing waveforms is diminished.

This same principle of averaging the cortical signal in response to a repeated stimulus such as an electrical pulse to a peripheral nerve is the basis of the somatosensory evoked potentials (SSEPs). Although stimulation of the median nerve at the wrist or tibial nerve at the ankle are the SSEPs most commonly performed in clinical neurophysiological laboratories, it is also possible to record an SSEP after stimulation of the pudendal nerve. For this the subject holds the stimulator on the dorsal nerve of the clitoris or penis and electrical stimuli given at an intensity of two to three times that of threshold for sensation, i.e., a non-painful intensity.

Although when first introduced it was hoped that this recording would provide insight into the afferent innervation of the pelvis, this has not proved to be the case. Electrical stimulation depolarizes the largest myelinated nerve fibers of a sensory nerve, those which convey muscle afferent activity or the sensory modalities of light touch and vibration and does not activate those small unmyelinated nerve fibers involved in visceral sensation. Although the pudendal evoked potential is technically easy to record, it clinical value in detecting a spinal cord deficit can be equally well accomplished by recording the standard tibial evoked responses – it is rare to find an abnormality of the former without the latter also being abnormal.

Conclusion

Over the course of many years, various neurophysiological techniques have been developed that have quite quickly been transferred to the investigation of the pelvic floor. Only a few of these have been shown to be of enduring value, whereas some, such as single-fiber EMG and use of magnetic stimulation, are no longer recommended even as research techniques. For the investigation of neurological patients with pelvic organ complaints some of the more complex investigations may still be used but for physiotherapy practice – both clinical and research – surface EMG recordings are likely to continue to give the best results.

Acknowledgment. We would like to thanks Prof Grace Dorey PhD, FCSP for her advice about the text.

2.9
Outcome Measures in Pelvic Floor Rehabilitation

Kate H. Moore and Emmanuel Karantanis

Introduction

Throughout this textbook, many different measures are used to evaluate the pelvic floor and urinary or fecal incontinence. In this chapter, those measures that are also suitable for evaluating posttreatment response, or "outcome," are considered. In the past 3–4 decades, numerous tests, scoring systems, and quality-of-life instruments have been created. Unfortunately many of these tests have not been formally validated as outcome measures. Therefore they may not give an accurate picture of the "quantity" of a patient's response to treatment. Also, when many different outcome measures are used to gauge response to any treatment (by different authors), it is almost impossible to compare results.

This chapter gives a brief overview of some of the early nonvalidated outcome measures that are still in use, and then describes more recent tests that have been fully validated. The system of categorization used is that of the Standardization Committee of the International Continence Society (ICS),[1] which recommends that there should be five main groups, or "domains" of outcome measures:

1. Patient's observations (symptoms)
2. Quantification of symptoms (e.g. urine loss on diary or pad test)
3. Physician's observations (anatomical and functional e.g. perineometry, and compliance with treatment)
4. Quality of life measures
5. Socioeconomic evaluations.

How to "Validate" an Outcome Measure

First, it is useful to briefly define what is meant by "validation." The design of symptom score or questionnaire is not a simple process, and the development of a physical test (e.g. perineometry) for use as an outcome measure also requires considerable work.

The process of validation for symptom scores and quality of life tests requires one to demonstrate validity, reliability, and responsiveness to change after treatment. These items must also be demonstrated for tests of physical function, but some differences in methods occur between psychometric tests and physical tests.

Validity refers to whether the test measures what it is supposed to measure. There are three main aspects:

1. Content/face validity is the assessment of whether a questionnaire makes sense to the patients and clinicians, i.e. that it is understandable, unambiguous, and clinically sensible. Construct validity involves assessing how the questionnaire performs in a range of setting and patient groups. One example is differentiating between patient groups, such as hospital patients and individuals in the community. Criterion validity describes how well a questionnaire correlates with a "gold standard" measure that already exists. This "criterion validity" also is important for tests of physical function, e.g. leakage, on ultrasound of the bladder neck

may need to be validated against leakage on videourodynamics.

2. The reliability of a test refers to its ability to measure quantities reproducibly. There are two aspects: (a) internal consistency and (b) reproducibility. Internal consistency refers to the extent to which items within a questionnaire are consistent with each other. It can be assessed by statistical techniques such as item–total correlation or Cronbach's alpha. Reproducibility is an important concept for both questionnaires and physical tests that measure outcome, as one must demonstrate that responses are stable over a short period of time in a pretreatment sample (i.e. by giving the same test to respondents 2–6 weeks apart, before treatment starts). Reproducibility is assessed by using analysis of variance (to find the standard deviation for repeated measurements), or by determining the intraclass correlation to determine variability between and within subjects, or by using the Bland-Altman method to derive the limits of agreement.[2,3] Many studies of test–retest reliability quote the correlation coefficient between the first and second test, but this is not a correct measure of reliability because two tests of the same measure will generally correlate (or "co-relate") to a large degree. Correlation between two tests does not measure the degree of variation between the two tests.

3. The responsiveness of a test describes its response to change after treatment. This is usually done either by measuring effect sizes in randomized trials, or by analyses of covariance. Therefore in order for a questionnaire to be fully validated and "robust," multiple experiments assessing all of the aforementioned components must be undertaken.

Patient Symptom Scores as Outcome Measures

In the 1970s, before the advent of validated continence scores, grading systems were developed to measure the severity of incontinence. These were "doctor controlled," with the doctor making the final impression as to what the response should be. Popular formats included the Stamey grading system[4] and the Ingelmann-Sundberg Scale.[4] These 3-point scales (Table 2.9.1) were

TABLE 2.9.1. Three point scales

Stamey Grading	
I	Leak with stressful activities eg. cough. sneeze etc
II	Leak with minimal activities eg. walk/standing up
III	Leak at all times, all activities irrespective of position
Ingelmann – Sundberg Scale	
I	Incontinence only when coughing, sneezing, or lifting heavy objects
II	Incontinence during daily activities eg. rising from a chair, or fast walking
III	Dripping incontinence in the upright position

never validated, yet they are still commonly used today. For example, Stamey scores are commonly used in studies assessing urethral bulking agents[6,7] because the score continues to be stipulated as an outcome measure for research by the US Food and Drug Administration, whereas several European groups continue to use the Ingelmann-Sundberg Scale.[8]

The Lagro-Janssen 12-point score comprises 4 categories which each have a score of 1–3. These comprise frequency of urine loss (weekly, more than weekly, daily), amount of urine loss (drops, a little, a lot), use of pads (none, occasionally, most of the time) and effect upon lifestyle (none, some, a lot).[9] The score is used to define mild (4–6), moderate (7–9), and severe incontinence (10–12). Although studies of repeatability are not available, it is interesting that the terms used in this 1991 score are very similar to the more recent ISI and ICIQ-SF, which are descibed in the following sections.

The Incontinence Severity Index was developed by Sandvik et al.[10] It is a very short measurement tool, having only two items. The responses for the first and second question are multiplied together to give the total score a range of 0 to 8 (Table 2.9.2).

TABLE 2.9.2. The incontinence severity index

How often is urine leakage experienced?
Never = 0
1 to several times a month = 2
1 to several times per week = 3
Every day and/or night = 4
How much urine is lost each time?
A few drops = 1
A little = 1
More = 2

		ICIQ-SF			
Initial number		CONFIDENTIAL	DAY	MONTH	YEAR
					Today's date

Many people leak urine some of the time. We are trying to find out how many people leak urine, and how much this bothers them. We would be grateful if you could answer the following questions, thinking about how you have been, on average, over the PAST FOUR WEEKS.

1 Please write in your date of birth: DAY   MONTH   YEAR

2 Are you *(tick one)*:   Female ☐   Male ☐

---

3 **How often do you leak urine?** *(Tick one box)*

never ☐	0
about once a week or less often ☐	1
two or three times a week ☐	2
about once a day ☐	3
several times a day ☐	4
all the time ☐	5

---

4 We would like to know how much urine <u>you think</u> leaks.
How much urine do you <u>usually</u> leak (whether you wear protection or not)?
(Tick one box)

none ☐	0
a small amount ☐	2
a moderate amount ☐	4
a large amount ☐	6

---

5 **Overall, how much does leaking urine interfere with your everyday life?**
Please ring a number between 0 (not at all) and 10 (a great deal)

0   1   2   3   4   5   6   7   8   9   10
not at all   a great deal

ICIQ score: sum scores 3+4+5 ☐☐

---

6 **When does urine leak?** *(Please tick all that apply to you)*

never – urine does not leak	☐
leaks before you can get to the toilet	☐
leaks when you cough or sneeze	☐
leaks when you are asleep	☐
leaks when you are physically active/exercising	☐
leaks when you have finished urinating and are dressed	☐
leaks for no obvious reason	☐
leaks all the time	☐

Thank you very much for answering these questions.

Copyright © "ICIQ Group"

FIGURE 2.9.1. The ICIQ-SF Questionnaire.

Severity on a 48-hour pad test and severity of the score correlates strongly (r = 0.59; P < 0.001).[11]

The St. George Score for urinary incontinence was developed to parallel the Wexner scale for fecal incontinence (see below). It asks whether patients leak with stress or with urge, whether they have damp or soaked pads, and how much the leakage affects quality of life. Each of the four questions is then graduated by the patient as to how often the problem occurs (weekly, daily, more than daily, etc). The 5 subscales, thus, each have 4 maximum points (total maximum 20). The construct validity, reliability, and responsiveness to change have been published,[12] and its suitability for telephone administration of long-term follow-up has been shown.[13]

The Groutz Score is not just a symptom score, but combines results of a patient questionnaire (0, cured; 1, improved; 2, failed) along with point scores from the 24 hour voiding diary (0, no leaks; 1, 1–2 leaks; 2, ≥3 leaks) and a 24-hour pad test (0, <8 g; 1, 9–20 g; 2, >20 g).[14] This is a very attractive scoring system because it combines the first 3 domains of the ICS recommendations. Because patients grade their degree of cure in the questionnaire, it cannot be used for baseline assessment, which is a minor shortcoming in statistical analysis.

The ICIQ-SF is the most recently validated questionnaire. It is a symptom severity and quality of life instrument, developed under the auspices of the International Consultation on Incontinence (ICI) and has undergone full psychometric testing.[15,16]

The final ICIQ comprises three scored items (Fig. 2.9.1) and an unscored self-diagnostic item. It allows the assessment of the prevalence, frequency, and perceived cause of urinary incontinence, and its impact on everyday life using just 3 questions on one page, producing an overall maximum score of 21. Women can complete the questionnaire quickly and completely with low levels of missing data (mean 1.6%).

It discriminates well across different groups of individuals, indicating good construct validity. Convergent validity was acceptable, with most items demonstrating "moderate" to "strong" agreement with other questionnaires. Reliability testing revealed "moderate" to "very good" stability in test–retest analysis and a Cronbach's alpha of 0.95. The ICIQ correlates strongly with the 24-hour pad test (r = 0.458; P = 0.000).[17] The test is also responsive to treatment. Because it is very simple, yet very "robust" and is recommended by the WHO/ICI, it will probably become the gold standard in this area.

The Wexner Score for fecal incontinence was originally designed as a 20-point score concerning 3 types of incontinence and one category for impact upon lifestyle, but was later modified to include a score for the need to wear a pad, the need to take constipating medication, or fecal urgency (Table 2.9.3).

The internal consistency, test–retest reliability, construct validity, criterion validity, and sensitivity to change of the Wexner score appear to be adequate.[18] The score is widely used as an outcome measure in North America, the United Kingdom, Australia, and several European countries.

TABLE 2.9.3. Modified faecal incontinence score (Vaizey et al 1999)

	Never	Rarely	Sometimes	Weekly	Daily
Incontinent Solid Stool	0	1	2	3	4
Incontinent Liquid Stool	0	1	2	3	4
Incontinent to Gas	0	1	2	3	4
Alters lifestyle	0	1	2	3	4
				No	Yes
Need to wear pad/plug				0	2
Take constipating meds				0	2
Unable to defer 15 min				0	2

Quantification of Symptoms

Frequency of Leakage on Bladder Diary/Frequency Volume Chart

Moving away from scores and questionnaires, we now consider tests that actually measure leakage severity. The most commonly used is the bladder chart. This is a generic term used to indicate several types of records: the micturition chart, which only requires the patient to document times of voiding and incontinence episodes; the frequency volume chart (FVC), which also requires patients to record their fluid intake and their voided volumes, and the changing of pads; and the urinary diary, which includes the details of the FVC but also includes symptoms and activities at leakage episodes, including urgency, and assessment of the severity of each leakage episode. Although the urinary diary provides more detail about the episodes of leakage, it results in poorer compliance because of the detail required.[19]

A FVC or urinary diary is a useful outcome measure, with the simplest parameter being the number of leakage episodes per 24 hours. One of the main debates about diaries is the degree to which a clinician can trust what a patient annotates. The quality of the information depends upon patient compliance, which needs to be assessed with the patient face-to-face.

Another controversy about diaries is the duration over which they should be undertaken. Seven-day diaries have been shown by several authors to have excellent test-retest reliability[20–22], are the most sensitive and accurate, but also produce the least compliance[22]. Recently, a test-retest reliability study of the seven-day voiding diary was performed ($n = 138$)[23] The test-retest reliability for the number of weekly incontinence episodes was high ($r = 0.83$). The correlation between the first 3 days and the last 4 days was very high ($r = 0.9$). They concluded that a 3-day diary is an appropriate outcome measure in stress incontinence.

Although longer durations may be most reliable, the ICI committee for research methodology suggested that, in most cases, a single 24-hour diary is sufficient.[24]

Pad Tests

The Pad Test

The paper towel test was first published by Miller et al.[25] This study provided the protocol and formula for calculating the degree of leakage made on a paper towel folded in 3 layers (placed against the perineum), based on the diameter of the patch formed by the leak, when certain provocative maneuvers are performed with a comfortably full bladder. The same authors[26] later found that the test failed to correlate with other measures of severity, and concluded that the test was inaccurate.

1-hour Pad Tests

Sutherst[27] and Walsh[28] originally described the use of perineal pads in the objective assessment of urinary loss. The test procedures were then standardized,[29] and after being endorsed by the International Continence Society in 1988[30] they became a widely used outcome measure. Patients were asked to attend with a comfortably full bladder, then drink 500 ml of water over 15 min, then perform a preset group of activities to provoke leakage (walking and stair climbing for 30 min; standing up 10 times, coughing 10 times, running 1 min, bending 5 times, and washing hands for 1 minute). The voided volume was then measured.

The repeatability of the 1-hour pad test has been heavily criticized. In the mid-1980s, four groups of authors studied repeatability in small sample sizes (18–23 patients per study). Klarskov and Hald[31] found that the physical tasks dictated by the test were quite demanding, and many patients completed them to a variable degree. Lose et al.[32] found wide divergence of results in approximately 50% of patients, which they attributed to differing degrees of diurese during the two tests, in accordance with Jorgensen et al.[33] and Christensen et al.[34]

Lose et al.[35] later assessed repeatability using the Bland–Altman method[3] in 25 incontinent women at a standardized bladder volume, but found differences of up to 24 g between two test results. Because a large bladder volume may increase leakage,[36] they improved repeatability by catheterization and retrograde filling of a large standard volume (which appears unnecessarily invasive). A larger ($n = 56$) repeatability study was undertaken by Simons et al.[37] Bladder volume was standardized by repeated ultrasound during natural fill, so that the test volume at the second occasion was close to that of the first test. Nevertheless, substantial variations between the two tests were found. The test–retest reliability of the 1-hour pad test was found to be clinically inadequate, making it unsuitable for use as an outcome measure.

The 1-hour test also appears to be less sensitive than a longer test for detecting incontinence, particularly in women with mild incontinence.[38,39] Another problem is that patients must be sufficiently robust to climb stairs, lift shopping bags, and jump 10 times. Consequently, many authors have adapted and modified the test, making any interpretation between studies impossible.[40] The 1-hour pad test also fails to correlate with other measures of severity. Theofrastos et al.[41] and Fantl et al.[42] et al. showed that the UDI/IIQ did not correlate with the 1-hour pad test measures of severity, nor did urodynamic assessment of severity correlate.[43]

Until recently, the 1-hour pad test was the only objective method that could be used to differentiate mild, moderate, and severe incontinence. The volume leaked was divided by the total volume of the bladder at the start of the test, and the percentage of total volume leaked was divided by the total volume of the bladder at the start of the test. The percentage of total volume leaked was used to calculate 33% centiles for the patient sample. This yielded definitions of "mild" (1–10 g), "moderate" (11–50 g), and "severe" (>50 g).[31]

The 2-hour Pad Test

The 2-hour pad test was designed by Sutherst et al.[27] to reduce the problem of variable and often inadequate diuresis rates that are noted within 45 min of drinking 500 ml in the 1-hour test.[44] In the 2-hour pad test, after emptying the bladder (thus, starting at a known zero volume), the patient drinks 1 L, and then waits for 1 hour before commencing the same provocations given in the 1-hour test. This was found to be more sensitive,[45] but repeatability studies are lacking, mainly because the test requires 2 hours of careful nursing attention in the laboratory. The test does not correlate with a visual analogue score of severity.[46] A home 2-hour pad test was investigated by Wilson et al,[47] but has not been further validated.

The 24-hour Pad Test

In the mid-1980s, several authors investigated 24-hour and 48-hour home pad tests.[48-50] Women were given a set of pre-weighed pads in sealed bags, to be worn for a 24-hour or 48-hour period. Such tests are not embarrassing, and reflect everyday function/ provocation of incontinence. Several authors have demonstrated that there is no loss of accuracy by evaporation from the sealed plastic bag for durations of 72 hours,[47] one week,[38] and two weeks.[50] Thus, wet pads can be returned via the mail from any location.

The 24-hour pad test has been found to be more sensitive than the 1-hour test.[38] In 31 incontinent women who underwent both tests, 13 patients were dry after the 1-hour test, compared to 3 on the 24-hour test, giving a false-negative rate of 39% for the 1-hour test. The 1-hour and 24-hour pad test results did not correlate significantly.

Normal ranges for the 24-hour pad test in continent women have been controversial. Versi et al,[50] and Mouritsen et al.[51] tested 24 "young nursing and physiotherapy students" and 25 nursing and physiotherapy staff (median age 41) respectively. Such young patients are not representative of the typical age group presenting with urinary incontinence. Lose et al.[38] recruited 23 continent women with a more appropriate age group (34–69 years), and Ryhammer et al.[52] tested 78 continent postmenopausal women, (age range 45–57 years). In all these studies, home kitchen scales were used to weigh the pads, with accuracy of only ± 1 gm. These studies gave normal values of between 3 and 8 grams, which seems a large amount of fluid to be tolerated by an asymptomatic woman. Recently, the normal values were defined in 120 women of median age 48 (IQR 32–60); using scales accurate to 0.1 g, a median value of 0.3 g (IQR 0.2–0.6; 95th centile 1.3 g) was obtained.

Characterization of Mild, Moderate, and Severe

Because PFM exercises are more likely to be curative in patients with mild incontinence, and surgery is often offered to patients with severe leakage, a pad test should be able to differentiate severity groups and help to prognosticate the outcome of treatment. Recently, mild, moderate, and severe leakage were characterized as 1.3–20 g, 21–74 g, and >75 g per 24 hours,[53] respectively.

Repeatability

The repeatability of the 24-hour pad test has been controversial. Three studies have concluded that repeatability is adequate,[19,54,50] whereas one study has disagreed.[38] Few studies have assessed repeatability according to the type of urodynamic diagnosis, and no studies have assessed the contribution of activity to the repeatability of the test. This is particularly relevant when considering a woman who leaks once per week during sporting activities. A study of 108 women who performed 7 consecutive 24-hour pad tests showed that, on repeated measures analysis of variance, there was no statistically significant difference in mean pad weight during the 7 days of pad testing. Over the 7 days, 51% of patients did not change their severity grade at all and 18% changed by 1 grade on 1 day.[55]

Criterion Validity

The 24-hour pad test has seldom been compared to other measures of severity, but as mentioned, the correlation with the ICIQ is highly significant.

Anatomical and Functional Observations

The most common anatomical and functional parameters that have been used as an outcome measure comprise the following: (a) PFM strength assessment, (b) urodynamic testing, (c) urethral pressure profile, and (d) leak-point pressures. These tests are all described in detail in Chapter 2.1 and 2.3, but their suitability as an outcome measure will be summarized here.

The measurement of PFM strength is fully described in Chapter 2.1. Measuring PFM strength (or power) is only one component of the assessment, which includes testing endurance, repetitions, and fast contractions.[56] The assessment of PFM strength is usually performed using digital assessment, or by use of a perineometer. The Oxford scoring system has become the predominant grading method for digital assessment of PFM strength.[57] During digital examination, the clinician asks the woman to tighten her pelvic floor, and then assigns a score out of 5 (Table 2.1.6). The examiner must coach the patient first, to ensure that the patient is performing the test correctly, as Bump et al.[58] has shown that patients who are simply given verbal instruction (without digital examination) are unable to contract the muscles correctly in 51% of cases. Without this method, subsequent outcome testing is not possible.

Although the Oxford score has been in use since 1989, the test–retest reliability has only recently been demonstrated.[59] More recently, the perineometer has been developed as an alternative measure of PFM strength. This involves the vaginal insertion of a pressure-measuring device that measures vaginal pressures both at rest and during pelvic floor tightening.

Bo et al. compared the findings of the Oxford score to that of a perineometer and found no significant correlation, but the study was limited to 25 physiotherapy students.[60] In contrast, Isherwood et al.[59] found good agreement between the Oxford scale and perineometry in 263 women with a wide age range. Similarly, Kerschan-Schindl et al.[61] showed better repeatability and reliability with the digital examination rather than the perineometer. An advantage of the Oxford score over the perineometer is that an examining finger can detect any counterproductive Valsalva or strain that patients may perform.

Urodynamic Assessment

Urodynamics describes the laboratory investigation of incontinence. The main components are filling cystometry, the cough test at maximum bladder capacity, uroflowmetry, and determination of the postvoid residual. The use of X-ray (fluoroscopy) at the time of urodynamics is the gold standard, although more recently, translabial ultrasound has been adopted to look for the degree of funneling or the degree of bladder neck mobility associated with leakage on coughing. Further description is given in Chapter 2.3.

Two studies have shown that cystometric capacity measurement repeatability is poor.[62,63] In a study of 50 women with stress incontinence symptoms who were filled to a minimum of 300 ml during cystometry, the cough test has been shown to be repeatably positive and repeatably negative.[64]

During urodynamic tests, many clinicians classify severity of USI in women as mild, moderate, or severe. However, there is no standardized approach to grading severity. Thus, clinicians' assessment of the degree of leakage during cough is highly subjective, varies from unit to unit, and, as such, is a poor outcome measure. For example, fluid in the proximal urethra can only be seen on fluoroscopy, but is often considered evidence of urethral sphincter incompetence. Yet, when urodynamic testing is used to judge "cure," the authors do not always state whether fluoroscopy was used. They may employ twin channel cystometry and define "cure" as no leakage of urine from the external urethral meatus.

The classic type I, II, and III categorization of stress incontinence on videourodynamic testing with urethral pressure measurement developed by McGuire[65] does not appear to have been subjected to repeatability assessment and is not generally regarded as an outcome measure.

Only one study has correlated urodynamic severity grading with severity on the 1-hour pad test. In this study, women at maximum capacity in an erect oblique position were asked to cough as hard as possible. Severity of leakage was determined arbitrarily: severe, leakage occurred on the first cough; moderate, on the second or third cough; slight, if leak only after multiple coughs and encouragement. The study showed no significant correlation between the clinician's impression and 1-hour pad test.[43]

Urethral Pressure Profilometry

Urethral pressure profilometry (UPP) may involve the use of water-perfused catheters the Brown and Wickham method,[66] or microtransducers[67] (see Chapter 2.3).

A major problem with UPP measurement is the lack of standardization. Recently, the ICS published a standardization report guiding future authors as to the basic requirements to be documented in UPP studies.[68] Clear documentation on patient position, bladder volume, maneuvers, and probe types is required by the ICS, but the use of any standardized methods or techniques over others was not recommended.

The UPP technique has been shown to be repeatable. Using the Brown and Wickham fluid perfusion method, Martin and Griffiths[69] found agreement between 2 observations to be ±2 cm H_2O pressure. Meyhoff et al.[70] examined the reproducibility of various parameters of the UPP in both consecutive observations on the same day, and 2 observations separated by 1 month. The variability in same-day observations was adequate, but they noted significant variation in observations repeated after a 1-month interval. Hilton and Stanton[67] analyzed the short- and long-term variance of the microtip transducer UPP method and found good reliability.

Only one study has attempted to correlate UPP results with severity on pad tests and validated questionnaires. The study showed no correlation between maximal urethral closure pressure and the 1-hour modified pad test, or between the UDI and IIQ.[41]

Valsalva Leak Point Pressures

The VLPP is defined as the minimal pressure within the bladder to cause urinary leakage in the absence of a bladder contraction.[71] VLPP research regarding reproducibility is scant. A study by Song et al.[72] found VLPP is reproducible; however, the correlation coefficient was used.

Recently, Fleischmann et al.[73] studied VLPP pressures in 65 women with stress incontinence. The 24-hour pad test results, the number of leakage episodes on FVC, and the degree of urethral mobility did not correlate with the VLPP value. This study disputes McGuire's earlier finding that low VLPP readings correlated with worse leakage. A study by Nitti and Combs[74] found a strong correlation between the subjective degree (SEAPI-QMN) of urinary incontinence and VLPP in 51 women with stress incontinence.

Leak-point pressure measurement has been hindered by the same lack of standardization seen with the UPP. The VLPP is thought to be most accurately measured at bladder volumes of 250–300 ml,[75] but this varies from study to study. The VLPP remains to be standardized regarding catheter size, bladder volume, and the presence of cystocoele.[76] In addition, the pressure recording at the moment of leakage is difficult to standardize.

The definition of high or low VLPP is not standardized. Generally, a VLPP of 60 cm H_2O or less is thought to represent severe leakage ("intrinsic sphincteric deficiency"),[77] although this definition varies, e.g. some authors use a VLPP cut-off for ISD of <100 cm H_2O.[78] Furthermore, there is a poor correlation between VLPP and UPP.[71]

Quality of Life Scores

In the last decade, attempts have been made to standardize and validate self-completed questionnaires as instruments assessing subjective severity and quality of life associated with incontinence.

The Urogenital Distress Inventory (UDI) and the Incontinence Impact Questionnaire (IIQ) were developed by Shumaker et al.[79] to assess the impact of urinary incontinence symptoms on quality of life for women. The original forms of the IIQ and UDI had 30 and 19 items, respectively. Later work by Uebersax et al.[80] created a 7-item version of the IIQ and a 6-item version of the UDI. These are now widely used in both clinical and research applications.

The Kings Health Questionnaire, was first developed at Kings College Hospital London by Kelleher and Cardozo and published in 1997.[81] It has 21 items and was designed to measure symptoms and quality of life for men and women with urinary incontinence. The 21 items are distributed to form 8 domains that are scored between 0 and 100, with 100 indicating a greater impact on quality of life. Domains include social limitations and sexual function. It was developed in the United Kingdom, making it most appropriate for that population, but has been translated to many languages. Its disadvantages are its length and complicated scoring system.

Many other quality of life tests have been validated for use as outcome measures in male incontinence, or for patients with the overactive bladder. These are not considered here, as they do not directly relate to rehabilitation of the pelvic floor; for further details see Donovan et al.[82]

Socioeconomic Evaluation

The ICS Standardization Committee recommends that the costs of incontinence should be an important outcome measure. If a treatment reduces leakage, it should also reduce the costs of managing that leakage. A balance occurs when the cost of the treatment is less than the pretreatment management costs, but the time frame over which all these costs are calculated is important. For further details of economic evaluation see Hu et al.[83]

The ICS recommends that incontinence costs be gathered in seven aspects:

1. The costs of clinicians' time and services.
2. The costs of laboratory tests and imaging studies.
3. Expenses for procedures and medications.
4. The costs of disposable and reusable pads, undergarments, bed protectors, etc.
5. The costs of managing side effects and adverse events arising from treatment (e.g., de novo detrusor overactivity after bladder neck surgery).
6. The costs of travel to obtain treatment.
7. The loss of wages from receiving care or surgery.

These items have been assembled into a nurse-administered cost index, for which the construct validity, test–retest reliability, and responsiveness to treatment change have been published (the Dowell-Bryant Incontinence Cost Index[84,85]). The ICS recommends that such measurements of cost be built into the design of any new therapeutic trial.

Full economic analysis of the efficacy of treatment usually involves use of a specific type of quality of life test that allows one to calculate the benefit in terms of Quality of Life Years (or QALY). For more details see Hu et al.[83]

Conclusion

When reporting the success of any new treatment, fully validated objective measures should be employed. Clinicians should not "create" a new outcome test unless they are prepared to validate it thoroughly. Our common research interests would be best served by choosing validated outcome measure at the start of any intervention trial.

References

1. Lose G, Fantl JA, Victor A, et al. Outcome measures for research in adult women with symptoms of lower urinary tract dysfunction. Neurourol Urodyn. 1998;17:255–262.
2. Karantanis E, O'Sullivan R, Moore KH. The 24-hour pad test in continent women and men: normal values and cyclical alterations. Br J Obstet Gynaecol. 2003;110:567–571.
3. Bland JM, Altman DG. Statistical methods for assessing agreement between two methods of clinical measurement. Lancet. 1986;1:307–310.
4. Stamey T. Urinary incontinence in the female. Campbells Urology, 4th ed. Philadelphia: WB Saunders Co.; 1979:2272–2293.
5. Ingelmann-Sundberg A. Urininkontinens hos kvinnan. Nord Med. 1953;50:1149–1152.
6. Bent A, Dmochowski RR, Herschorn S, et al. Evaluation of Uryx versus contigen as periurethral bulking agents in female stress incontinence: a mulitcenter randomized controlled study. Int Urogynecol J Pelvic Floor Dysfunct. 2002;13:S24.
7. Sand PK, Dmochowski RR. Clinical experience with coaptite urological bulking agent. Int Urogynecol J Pelvic Floor Dysfunct. 2000;13:S20.
8. Soulie R. Bladder suspension by retropubic endoscopy. Techniques and preliminary results (24 cases). [in French] Prog Urol. 1996;6:60–69.
9. Lagro-Janssen TLM, Debruyne FMJ, Smits AJA, et al. Controlled trial of pelvic floor exercises in the treatment of urinary stress incontinence in general practice. Br J Gen Prac. 1991;41:445–449.
10. Sandvik H, Hunskaar S, Seim A, et al. Validation of a severity index in female urinary incontinence and its implementation in an epidemiological survey. J Epidemiol Community Health. 1993;47:497–499.
11. Sandvik H, Seim, A, Vanvik A, et al. A severity index for epidemiological surveys of female urinary incontinence: comparison with 48-hour pad-weighing tests. Neurourol Urodyn. 2000;19:137–145.
12. Blackwell AL, Yoong W, Moore KH. Criterion validity, test retest reliability and sensitivity to change of the St. George Urinary Incontinence Score. Br J Urol Int. 2004;93:331–335.
13. Moore KH, O'Sullivan RJ, Simons A, et al. Randomised controlled trial of nurse continence advisor therapy compared with standard urogynaecology regimen for conservative incontinence treatment: efficacy, costs and two year follow up. BJOG. 2003;110:649–657.
14. Groutz A, Blaivas JG, Rosenthal JE. A simplified urinary incontinence score for the evaluation of treatment outcomes. Neurourol Urodyn. 2000;19:127–135.
15. Donovan JL, Badia X, Corcos J, et al. Symptom and quality of life assessment. In: Abrams P, Cardozo L, Khoury S, Wein A, editors. Incontinence, volume 1, second edition. Bristol: Health Publication, Ltd.; 2002:269–316.
16. Avery K, Donovan J, Peters TJ, et al. ICIQ: A brief and robust measure for evaluating the symptoms and impact of urinary incontinence. Neurourol Urodyn. 2004;23:322–330.
17. Karantanis E, Fynes M, Moore KH, et al. Comparison of the ICIQ-SF and 24-hour pad test with other measures for evaluating the severity of urodynamic stress incontinence. Int Urogynecol J. 2004;15:111–116.
18. Vaizey C, Garapeti E, Cahill J, et al. Prospective comparison of faecal incontinence grading systems. Gut. 1999;44:77–80.
19. Groutz A, Blaivas JG, Chaikin DC, et al. Noninvasive outcome measures of urinary incontinence and lower urinary tract symptoms: a multicenter study of micturition diary and pad tests. J Urol. 2000;164:698–701.
20. Locher JL, Goode PS, Roth DL, et al. Reliability assessment of the bladder diary for urinary incontinence in older women. J Gerontol A Biol Sci Med Sci. 2001;56:M32–M35.
21. Robinson D, McClish DK, Wyman JF, et al. Comparison between urinary diaries completed with and without intensive patient instructions. Neurourol Urodyn. 1996;15:143–1488.
22. Wyman JF, Choi SC, Harkins SW. The urinary diary in evaluation of incontinent women: a test-retest analysis. Obstet Gynecol. 1988;71:812–817.
23. Nygaard I, Holcomb R. Reproducibility of the seven-day voiding diary in women with stress urinary incontinence. Int Urogynecol J Pelvic Floor Dysfunct. 2000;11:15–17.
24. Payne C, Van Kerrebroeck P. Research methodology in urinary incontinence. In: Abrams P, Cardozo L, Khoury S, Wein A, eds. Incontinence.

1 vol. Bristol: Health Publication, Ltd.; 2002: 1047–1077.
25. Miller JM, Ashton-Miller JA, Delancey JO. Quantification of cough-related urine loss using the paper towel test. Obstet Gynecol. 1998;91:705–709.
26. Miller JM, Ashton-Miller JA, Carchidi LT, et al. On the lack of correlation between self-report and urine loss measured with standing provocation test in older stress-incontinent women. J Womens Health. 1999;8:157–162.
27. Sutherst J, Brown M, Shawer M. Assessing the severity of urinary incontinence in women by weighing perineal pads. Lancet. 1981;1:1128–1130.
28. Walsh JB, Mills GL. Measurement of urinary loss in elderly incontinent patients. A simple and accurate method. Lancet. 1981;1:1130–1131.
29. Bates P, Bradley W, Glen E, et al. Fifth report on the standardisation of terminology of lower urinary tract function. Bristol: International Continenence Society; 1983.
30. Abrams P, Blaivas JG, Stanton SL, et al. The standardisation of terminology of lower urinary tract function. The International Continence Society Committee on Standardisation of Terminology. Scand J Urol Nephrol Suppl. 1988;114:5–19.
31. Klarskov P, Hald T. Reproducibility and reliability of urinary incontinence assessment with a 60 min test. Scand J Urol Nephrol. 1984;18:293–298.
32. Lose G, Gammelgaard J, Jorgensen T. The one-hour pad-weighing test: Reproducibility and the correlation between the test result, the start volume in the bladder and the diuresis. Neurourol Urodyn. 1986;5:17–21.
33. Jorgenson L, Lose G, Andersen JT. One-hour pad-weighing test for objective assessment of female urinary incontinence. Obstet Gynecol. 1987;69:39–42.
34. Christensen SJ, Colstrup H, Hertz JB, et al. Inter- and intra-departmental variations of the perineal pad weighing test. Neurourol Uroydyn. 1986;5:23–28.
35. Lose G, Rosenkilde P, Gammelgaard J, et al. Pad-weighing test performed with standardized bladder volume. Urology. 1988;32:78–80.
36. Jakobsen H, Kromann-Andersen B, Nielsen KK, et al. Pad weighing tests with 50% or 75% bladder filling. Does it matter? Acta Obstet Gynecol Scand. 1993;72:377–381.
37. Simons AM, Yoong WC, Buckland S, et al. Inadequate repeatability of the one hour pad test: the need for a new incontinence outcome measure. BJOG. 2001;108:315–319.
38. Lose G, Jorgensen L, Thunedborg P. 24-hour home pad weighing test versus 1-hour ward test in the assessment of mild stress incontinence. Acta Obstet Gynecol Scand. 1989;68:211–215.
39. Thind P, Gerstenberg TC. One-hour ward test vs 24-hour home pad weighing test in the diagnosis of incontinence. Neurourol Urodyn. 1991;10:241–245.
40. Soroka D, Drutz HP, Glazener CM, et al. Perineal pad test in evaluating outcome of treatments for female incontinence: a systematic review. Int Urogynecol J Pelvic Floor Dysfunct. 2002;13:165–75.
41. Theofrastous JP, Bump RC, Elser DM, et al. Correlation of urodynamic measures of urethral resistance with clinical measures of incontinence severity in women with pure genuine stress incontinence. The Continence Program for Women Research Group. Am J Obstet Gynecol. 1995;173:407–412.
42. Fantl JA, Harkins SW, Wyman JF, et al. Fluid loss quantitation test in women with urinary incontinence: a test-retest analysis. Obstet Gynecol. 1987;70:739–743.
43. Versi E, Cardozo LD. Perineal pad weighing versus videographic analysis in genuine stress incontinence. Br J Obstet Gynaecol. 1986;93:364–366.
44. Haylen BT, Frazer MI, Sutherst JR. Diuretic response to fluid load in women with urinary incontinence: optimum duration of pad test. Br J Urol. 1998;62:331–333.
45. Richmond DH, Sutherst JR, Brown MC. Quantification of urine loss by weighing perineal pads. Observation on the exercise regimen. Br J Urol. 1987;59:224–227.
46. Frazer MI, Haylen BT, Sutherst JR. The severity of urinary incontinence in women. Comparison of subjective and objective tests. Br J Urol. 1989;63:14–15.
47. Wilson PD, Mason MV, Herbison GP, et al. Evaluation of the home pad test for quantifying incontinence. Br J Urol. 1989;64:155–157.
48. Pierson CA. Assessment and quantification of urine loss in incontinent women. Nurse Pract. 1984;9:18–19.
49. Ekelund P, Bergstrom H, Milsom I, et al. Quantification of urinary incontinence in elderly women with the 48-hour pad test. Arch Gerontol Geriatr. 1988;7:281–287.
50. Versi E, Orrego G, Hardy E, et al. Evaluation of the home pad test in the investigation of female urinary incontinence. Br J Obstet Gynaecol. 1996;103:162–167.
51. Mouritsen L, Berild G, Hertz J. Comparison of different methods for quantification of urinary

leakage in incontinent women. Neurourol Urodyn. 1989;8:579–587.
52. Ryhammer AM, Laurberg S, Djurhuus JC, Hermann AP. No relationship between subjective assessment of urinary incontinence and pad test weight gain in a random population sample of menopausal women. J Urol. 1998;159:800–803.
53. O'Sullivan R, Karantanis E, Stevermuer TL, et al. Definition of mild, moderate and severe incontinence on the 24-hour pad test. Br J Obstet Gynaecol. 2004;111:859–862.
54. Rasmussen A, Mouritsen L, Dalgaard A, et al. Twenty-four hour pad weighing test: reproducibility and dependency of activity level and fluid intake. Neurourol Urodyn. 1994;13:261–265.
55. Karantanis E, Allen W, Stevermuer TL, et al. The repeatability of the 24-hour pad test. Int Urogynecol J Pelvic Floor Dysfunct. 2005;16:63–68
56. Laycock J. Clinical evaluation of the pelvic floor. In Schussler J, Laycock J, Norton P, Stanton S (eds), *Pelvic Floor Re-education*, Vol 1. London: Springer–Verlag; 1994:42–48.
57. Sampselle C, Brink C, Wells T. Digital measurement of pelvic floor muscle strength in childbearing women. Nurs Res. 1989;38:134–138.
58. Bump RC, Hurt WG, Fantl JA, et al.. Assessment of Kegel pelvic muscle exercise performance after brief verbal instruction. Am J Obstet Gynecol. 1991;165:322–327.
59. Isherwood PJ, Rane A. Comparative assessment of pelvic floor strength using a perineometer and digital examination. BJOG. 2000;107:1007–1011.
60. Bo K, Finckenhagen HB. Vaginal palpation of pelvic floor muscle strength: inter-test reproducibility and comparison between palpation and vaginal squeeze pressure. Acta Obstet Gynecol Scand. 2001;80:883–887.
61. Kerschan-Schindl K, Uher E, Wiesinger G, et al. Reliability of pelvic floor muscle strength measurement in elderly incontinent women. Neurourol Urodyn. 2002;21:42–47.
62. Mortensen S, Lose G, Thyssen H. Repeatability of cystometry and pressure-flow parameters in female patients. Int Urogynecol J Pelvic Floor Dysfunct. 2002;13:72–75.
63. Brostrom S, Jennum P, Lose G, et al. Short-term reproducibility of cystometry and pressure-flow micturition studies in healthy women. Neurourol Urodyn. 2002;21:457–460.
64. Swift SE, Yoon EA. Test-retest reliability of the cough stress test in the evaluation of urinary incontinence. Obstet Gynecol. 1999;94:99–102.
65. McGuire EJ. Urodynamic findings in patients after failure of stress incontinence operations. In Zinner NR, Sterling AM (eds). *Female Incontinence*. New York: Alan R Liss; 1980.
66. Brown MC. The urethral pressure profile. Proc R Soc Med. 1970;63:701.
67. Hilton P, Stanton SL. Urethral pressure measurement by microtransducer: the results in symptom-free women and in those with genuine stress incontinence. Br J Obstet Gynaecol. 1983;90:919–933.
68. Lose G, Griffiths D, Hosker G, et al. Standardisation of urethral pressure measurement: report from the Standardisation Sub-Committee of the International Continence Society. Neurourol Urodyn. 2002;21:258–260.
69. Martin S and Griffiths DJ. Model of the female urethra: Part 1-Static measurements of pressure and distensibility. Med Biol Eng. 1976;14:512–518.
70. Meyhoff HH, Nordling J, Walter S. Short and long term reproducibility of urethral closure pressure profile parameters. Urol Res. 1979;7:269–271.
71. McGuire EJ, Fitzpatrick CC, Wan J, et al. Clinical assessment of urethral sphincter function. J Urol. 1993;150:1452–1454.
72. Song JT, Rozanski TA, Belville WD. Stress leak point pressure: a simple and reproducible method utilizing a fiberoptic microtransducer. Urology. 1995;46:81–84.
73. Fleischmann N, Flisser AJ, Blaivas JG, et al. Sphincteric urinary incontinence: relationship of vesical leak point pressure, urethral mobility and severity of incontinence. J Urol. 2003;169:999–1002.
74. Nitti VW, Combs AJ. Correlation of Valsalva leak point pressure with subjective degree of stress urinary incontinence in women. J Urol. 1996;155:281–285.
75. Faerber GJ, Vashi AR. Variations in Valsalva leak point pressure with increasing vesical volume. J Urol. 1998;159:1909–1911.
76. Haab F, Zimmern PE, Leach GE. Female stress urinary incontinence due to intrinsic sphincteric deficiency: recognition and management. J Urol. 1996;156:3–17.
77. McGuire EJ, Cespedes RD, O'Connell HE. Leak-point pressures. Urol Clin North Am. 1996;23:253–262.
78. Hsu TH, Rackley RR, Appell RA. The supine stress test: a simple method to detect urethral sphincter dysfunction. J Urol. 1999;162:460–463.
79. Shumaker SA, Wyman JF, Uebersax JS, et al. Health-related quality of life measures for women with urinary incontinence: the Incontinence Impact Questionnaire and the Urogenital Distress Inventory. Continence Program in

Women (CPW) Research Group. Qual Life Res. 1994:3:291–306.
80. Uebersax JS, Wyman JF, Shumaker SA, et al. Short forms to assess life quality and symptom distress for urinary incontinence in women: the Incontinence Impact Questionnaire and the Urogenital Distress Inventory. Continence Program for Women Research Group. Neurourol Urodyn. 1995; 14:131–139.
81. Kelleher CJ, Cardozo LD, Khullar V, et al. A new questionnaire to assess the quality of life of urinary incontinent women. Br J Obstet Gynaecol. 1997;104:1374–1379
82. Donovan JL, Badia X, Corcos J, et al. Symptom and Quality of Life Assessment. Report of World Health Organisation Consensus Conference: "Incontinence", Eds: Abrams P, Cardozo L, Koury S, Wein A. Health Publications Ltd., Plymouth; 2001:965–983.
83. Hu TW, Moore KH, Subak L, et al. "The economics of incontinence". Report of World Health Organisation, Editors: Abrams P, Cardozo L, Koury S, Wein A. Plymouth, UK: Health Publications Ltd., Plymouth; 2001;14:965–983.
84. Dowell CJ, Bryant CM, Moore KH et al. Calculation of the direct costs of urinary incontinence: the DBICI, a new test instrument. Br J Urol. 1999; 83:596–606.
85. Simons AM, Dowell CJ, Bryant CM, et al. Use of the Dowell Bryant incontinence cost index as a post-treatment outcome measure. Neurourol Urodyn. 2001;20:85–93.

Part III
Techniques of Pelvic Floor Rehabilitation and Muscle Training

3.1
Concepts of Neuromuscular Rehabilitation and Pelvic Floor Muscle Training

Jo Laycock

Key Messages

- The principles of muscle training are overload, specificity, and reversibility.
- Pelvic floor muscles work synergistically with the deep abdominal muscles, the lumbar multifidi muscles, and the respiratory diaphragm, forming a cylinder to support the lumbar spine.
- With contraction of the transverse abdominis muscle, the pelvic floor muscle is coactivated, and this pattern can be used in pelvic floor re-education.
- Reduced endurance of pelvic floor contraction is mainly a slow-twitch fiber disorder and requires endurance training by increasing the length of contractions and the number of repetitions.
- Reduced pelvic floor strength is predominantly a fast-twitch fiber disorder and requires strength training with maximum voluntary contractions until the muscles fatigue.
- In patients with stress urinary incontinence, even moderate pelvic floor contractions before increases in intraabdominal pressure can reduce stress incontinence.
- In patients with OAB, during an episode of urgency, several maximum contractions may help by inhibiting the overactive detrusor muscle.
- Adherence to a pelvic floor exercise program is influenced by the patient's understanding of the causes, the consequences of the condition, and the perceived sequelae of noncompliance.

Introduction

Muscle training involves the improvement in the tone of muscles or muscle groups, providing a readiness for action, and includes improvement in stiffness, strength, endurance, coordination, and function. Much of the evidence of the efficacy of muscle training and the principles involved has come from sports medicine literature.[1,2] In addition to increasing the cross-sectional area and neuromuscular function, strength training has been shown to increase the connective tissue within and around the muscles,[22] and, in theory, should improve the support of the pelvic organs afforded by the pelvic floor muscles. However, there are a number of differences between pelvic floor muscle rehabilitation and the training of muscles of sports women and men, which should be considered alongside the general principles described below. The main differences are the difficulty of teaching and assessing correct pelvic floor muscle contractions in functional positions and the motivation for people to carry out a program of exercises. Teaching and assessing a correct pelvic floor muscle contraction is addressed in Chapter 2.1 and also in the section on biofeedback (see Chapter 3.2). Motivation and adherence to training programs is discussed in the following sections.

Another fact has to be considered; athletes in training usually have normal and strong muscles, and training aims to hypertrophy these muscles and improve performance. However, patients with pelvic floor muscle dysfunction may have damaged muscles, damaged muscles and fascial attachments (see Chapter 1.1), partially denervated/re-innervated muscles (see Chapter 1.2), and lack of pelvic floor awareness. The aim here is to rehabilitate, re-educate, and so restore normal or near-normal function, and to introduce a compensational muscle action to counterbalance other insufficiencies, such as connective tissue weakness and irreparable changes. Yet another problem in pelvic floor rehabilitation is the often difficult diagnosis of the underlying causes, the damage, and the functional defects.

Principles of General Muscle Training

A muscle-training program should satisfy the following principles: overload, specificity, and reversibility.[1-3]

Overload

This term implies the need to make the particular muscles work more than they have been used to; to work harder, to work longer, to push their performance to the limit. Overload can be applied to strength, endurance, and function/technique. Exercise regimens progress by demanding more effort over a longer period of time to push the performance to a higher limit.

Specificity

Specificity describes "training for a purpose," and examines the specific objective of the exercise. Overload and specificity can generally be considered together. For example, a sprinter may train specifically for speed, recruiting mainly fast-twitch muscle fibers, whereas a marathon runner requires endurance, which is when predominately slow-twitch muscle fibers are recruited; thus, these runners would apply specificity to their training programs. By running faster than previously (sprinter) or longer than previously (marathon runner), they are also applying overload.

Reversibility: Loss of Hypertrophy

With athletes, strength training aims to hypertrophy the appropriate muscles; this is an on-going process, which takes 15–20 weeks; top athletes are continuously improving. Reversibility implies that the training effect is reversed; hypertrophy is lost and performance is reduced if the training stops; this is a gradual process.[1,2]

Principles in Pelvic Floor Muscle Training

Overload and Specificity

Pelvic floor muscle assessment of a patient with incontinence will establish whether the muscles are weak (reduced strength – predominately a fast-twitch fiber disorder), or easily fatigued (reduced endurance – mainly a slow-twitch fiber disorder), or dysfunctional. The latter case is seen when a woman either cannot perform a voluntary pelvic floor muscle contraction, or can perform a strong pelvic floor muscle contraction while lying supine, but cannot manage this while standing or during coughing. Specificity implies that if the muscles are weak, then strength training is required. This involves maximum voluntary contractions until the muscles fatigue (overload). If on assessment the muscles are found to tire easily (producing only short contractions and/or few repetitions), endurance training is needed. This will involve the patient aiming to increase the length of contraction (hold time) and increase the number of repetitions. This is done by repeating longer, submaximal contractions.

Many patients present with a deficiency in both strength and endurance. Continuous reassessment during a course of treatment is necessary to ensure that overload is satisfied. This highlights a further difference between training athletes and training women with incontinence; the exercise science literature recommends 8–12 maximum contractions in three sessions, and three or four training sessions per week, with rest days for recovery.[1-3] Many patients with weak pelvic floor muscles who are in a rehabilitation program cannot perform eight maximum contractions of sufficient strength and endurance to have a training effect; thus, they need to practice several times each day, at least 6 days a week.

Improving function is the most important training parameter, and this should be assessed and addressed accordingly. In a study by Miller et al., older women with stress urinary incontinence on coughing were taught "The Knack." This is an anticipatory pelvic floor muscle contraction before and during a cough.[4] This study is a fine example of specificity, and demonstrated that it is not necessarily the strength of the contraction, but the ability to use the pelvic floor muscles specifically when needed, to prevent incontinence, that is important. Another good example of specificity is "holding on," which is described by Norton, for patients with fecal urge incontinence.[5] A randomized controlled trial[6] demonstrated that patients who only practiced "holding on" to defer defecation improved, as well as those practicing "holding on" sphincter exercises and biofeedback.

Taking specificity a step further dictates the way pelvic floor muscle exercises are taught. Patients are instructed to "squeeze and lift the muscles between the legs, as if preventing urination/defecation"; this is thought to enhance cortical awareness of the specific function of the muscles. Simply instructing the patient to "squeeze and lift" may reduce this effect.

Rothstein et al. suggest that to integrate an activity or skill into normal, automatic, or unconscious function, many repetitions must be performed under diverse functional situations.[7] This supports the use of ambulatory surface EMG biofeedback to ensure appropriate muscle recruitment in different positions. Furthermore, Grimby and Hannerz directly relate proprioceptive dysfunction to dysfunction of tonic or slow motor unit recruitment,[8] and so submaximal pelvic floor muscle contractions with extended hold times should be practiced.

As with athletes, all patients will start at different levels, and so will require patient-specific exercise programs that are applicable to their needs. As fitness improves, the program should become more challenging.

Reversibility: Loss of Function with Deterioration of Cofactors

In pelvic floor rehabilitation, therapy continues until agreed goals are reached; e.g., this may be continence, or reduced incontinence. However, because of the ageing process[19] and other factors, e.g. constipation,[20] when muscles weaken because of reduction of motor units it is necessary to continue with pelvic floor muscle exercises not only to keep the training effect but also to compensate for these additional confounders so as to prevent incontinence from recurring. This is not true reversibility, but implies that incontinence symptoms may worsen if pelvic floor muscle exercises are discontinued or reduced. However, Bo suggests that telling patients that they should continue with pelvic floor muscle exercises for life may demotivate them.[11] The key is, when strength training has been carried out successfully, to integrate the pelvic floor muscles into daily routine and functional tasks. They should start to act automatically whenever there is an increase in intraabdominal pressure, thus, maintaining a training effect.

Pelvic Floor Muscle Exercises: Coactivation of Synergists

During low levels of activity, the smooth muscle fibers and fatigue-resistant slow-twitch fibers in the periurethral part of the levator ani are predominately activated. When activity increases, fast-twitch fibers are also recruited.[9,10] The pelvic floor muscles act along with the abdominal muscles (especially transversus abdominis [TrA]), the lumbar multifidi muscles, and the respiratory diaphragm, forming a cylinder of support for the spinal column (Fig. 3.1.1).[11]

The muscles of this cylinder all respond to changes in intraabdominal pressure; this includes changes in posture, coughing, walking, talking, deep breathing, and using the upper limbs. The timing of this response may be critical, and studies have shown that a delay in recruitment or inhibition of the multifidis muscles may be caused by pain[12] or muscle weakness.[13] Constantinou and Govan demonstrated that recruitment of the periurethral muscles on coughing occurred 250 msec before an increase in intraabdominal pressure.[14] This suggests that a delay in periurethral muscle recruitment may be a cause of incontinence. However, Verelst and Leivseth[15] have shown no difference in this parameter between continent and incontinent patients.

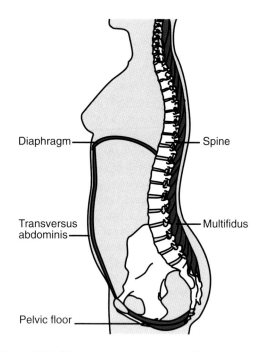

FIGURE 3.1.1. Diagrammatic representation of the muscular cylinder (capsule) of support for the lumbar spine. It includes the diaphragm, multifidi, transversus abdominis and pelvic floor muscles.

Clearly, in health, all of these cylinder muscles orchestrate their recruitment as required, and so rehabilitation of the pelvic floor muscles should include attention to these other muscle groups. There is evidence of functional activation of the pelvic floor muscles – in particular pubococcygeus – with the TrA muscle,[16] and so it has been suggested that exercises for the TrA can be used to indirectly activate the pelvic floor muscles, and should be part of the rehabilitation program. The TrA muscle is activated by instructing the patient to "gently draw in" the lower abdominal wall, minimizing any movement of the upper abdominal wall and the rib cage. This contraction should be around 25% of a maximum voluntary TrA contraction. As the aim is to enable the patient to use the pelvic floor muscles and TrA functionally, a normal breathing pattern should be used. Once both TrA and the pelvic floor muscles have been identified, they should be exercised together, aiming for 10-s sustained contractions repeated 10 times.[10] However, there are no randomized controlled trials (RCTs) to support the training of TrA alone, and Dumoulin et al.[22] showed no difference between pelvic floor muscle strength training alone and combined with TrA training. Clearly, more research is needed to clarify the optimum way the pelvic floor muscles are trained.

A good posture requires that the muscles of the cylinder of support for the spinal column (pelvic floor, multifidi, and transverse abdominis) work synergistically together. An improvement in posture initially produces increased activity in the pelvic floor muscles, and so good posture should be encouraged in sitting and standing. Exercising on an air cushion or balance ball while maintaining an upright posture is thought to recruit the pelvic floor muscles,[10] but more research is needed in this area.

Another coactivation concerns the glutei muscles, which contribute to movement of the coccyx (see Fig. 1.1.8, A and B). A mobile coccyx probably supports the ventro-cranial movement of the puborectalis muscle and elevation of the bladder neck.

Concept of Pelvic Floor Muscle Recruitment in Stress Incontinence

Involuntary loss of urine caused by an increase in intraabdominal pressure, e.g., coughing, assumes that bladder pressure has exceeded urethral pressure; consequently, therapies aimed at increasing urethral pressure should reverse this situation and ameliorate the problem (Fig. 3.1.2). Urethral pressure is caused by a number of factors including smooth and striated muscles, and elastic and vascular tissues. In addition, muscular and fascial support of the urethra and bladder maintains the organs in position (see Chapter 1.1). Of the factors listed above, striated muscles are the target of physiotherapy, and these include the striated urethral sphincter muscles and the levator ani muscle. The urethral sphincter is responsible for increasing intraurethral pressure, and the pelvic floor muscles, particularly the pubococcygeus and puborectalis, are responsible for elevating the proximal urethra and maintaining this position during increased intraabdominal pressure. Clearly, both of these muscle groups should be targeted in an exercise program. In

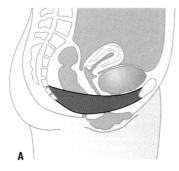 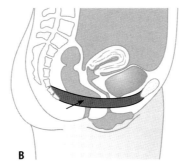 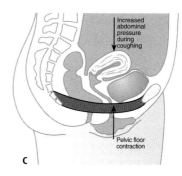

FIGURE 3.1.2. **A.** Pelvic floor at rest. **B.** A pelvic floor contraction results in cranio-ventral displacement of the bladder neck (arrow shows the vector). **C.** This contraction can act as an opposition during coughing e.g. Physiologically this contraction is generated via a reflex and co-contraction. In incontinent patients it is trained as a pre-contraction ("Knack"). (*Source*: modified from Schüssler B. Aims of pelvic floor evaluation. In: Schüssler B, Laycock J, Norton P, Stanton S, editors. Pelvic floor re-education: principles and practice. London:Springer-Verlag; 1994:39.)

continent women, co-contraction of the pelvic floor and sphincter muscles is demonstrated before and during coughing; consequently, this should be taught to women with SUI on coughing. This has been described as "The Knack".[4] This technique can be applied to any activity causing incontinence, e.g., lifting. In many cases, a moderate contraction is sufficient to prevent urine loss. Using ambulatory EMG biofeedback or perineal ultrasound (see CD and Chapter 2.6), patients can monitor their performance.

Concept of Pelvic Floor Muscle Recruitment in Urge Incontinence

The perineodetrusor-inhibiting reflex described by Mahoney et al. clearly describes detrusor inhibition as a response to pelvic floor muscle tone.[17] Consequently, one can assume that the greater a pelvic floor muscle contraction, the greater the inhibition. As most patients with an overactive bladder cannot maintain a strong contraction during an episode of urgency, they may need to perform several maximum contractions of shorter duration, with a minimal rest period (e.g. 1s) between contractions. As the detrusor is made of smooth muscle, it is very slow to respond, and so a short pause in the pelvic floor muscle contraction should not diminish the inhibitory effect. This is a further example of the principle of specificity in a training program.

Adherence to an Exercise Protocol

The evidence supporting factors that influence motivation/adherence to an exercise program for incontinence is scant. The literature on adherence relating to heart disease and asthma highlights a number of interesting options described by Chiarelli.[18] Those, as well as other principles and strategies to enhance recall, are shown in Boxes 1–3.[18] The aims of pelvic floor rehabilitation are to teach an appropriate exercise program, which, along with advice on types and quantities of fluid intake, along with medication, will address the patient's symptoms. Clearly, such a program requires patient cooperation and will not be suitable for everyone.

Box #1
Adherence to an exercise regimen depends on:

- The patient's understanding of the diagnosis and cause of symptoms
- The patient's comprehension and recall of the exercise program
- The patient's beliefs and understanding of the long-term consequences of the condition
- Any barriers to treatment
- The perceived benefits and sequelae of non-compliance
- Ways of increasing the opportunities to practice
- Providing cues or reminders to practice

> **Box #2**
> **Factors to encourage adherence**
>
> - Perception of self-efficacy
> - Improving communication with the patient
> - Goal setting
> - Program with minimal lifestyle changes; least disruptive to daily routine
> - Simple (less complex) program
> - Check adherence

> **Box #3**
> **Strategies to enhance recall**
>
> - Primacy – presenting the most important information first
> - Stressing and repeating important facts
> - Using specific advice
> - Check on comprehension by asking patient to repeat instructions in their own words
> - Use written instructions, and, wherever possible, have the patient write these
> - The use of reminders e.g. small colored stickers around the house

References

1. Astrand PO, Rodahl K. Textbook of work physiology: physiological basis for exercise. McGraw Hill Company; 1988.
2. American College of Sports Medicine Position Stand. The recommended quantity and quality of exercises for developing and maintaining cardiovascular and muscular fitness in healthy adults. Med Sci Sports Exerc. 1990;22:265–274.
3. Bo K. Pelvic floor muscle exercise for the treatment of stress urinary incontinence: an exercise physiology perspective. Int Urogyecol J. 1995;6:282–291.
4. Miller J, Ashton-Miller J, DeLancey JOL. A pelvic muscle precontraction can reduce cough related urine loss in selected women with mild SUI. J Am Geriatr Soc. 1989;46:870–874.
5. Norton C, Chelvanayagam S. Conservative management of faecal incontinence in adults. In: C Norton and S Chelvanayagam, editors. Bowel continence nursing. Beaconsfield, UK: Beaconsfield Publishers; 2004:118–119.
6. Norton C, Kamm M. Randomised controlled trial of biofeedback for faecal incontinence. Neurourol and Urodyn. 2002;21:295–296.
7. Rothstein J, Roy S, Wolf S. The rehabilitation specialist's handbook. Philadelphia: FA Davis Co.; 1991.
8. Grimby L and Hannerz J. 1976. Disturbance in voluntary recruitment order of low and high frequency motor units on blocades of proprioceptive afferent activity. Acta Physiol Scand. 1976;96:207–216.
9. Petros PP, Skilling PM. Pelvic floor rehabilitation in the female according to the integral theory of female urinary incontinence. First report. Eur J Obstet Gynecol Reprod Biol. 2001;94:264–269.
10. Jones R, Comerford M, Sapsford R. Pelvic floor stability and trunk muscle co-activation. In: Laycock J, Haslam J, editors. Therapeutic management of incontinence and pelvic pain. London: Springer-Verlag; 2002:66–71.
11. Bo K. Pelvic floor muscle training is effective in treatment of female stress urinary incontinence, but how does it work? Int Urogynecol J. 2004;15:76–84.
12. Hodges P, Richardson C. Altered trunk muscle recruitment in people with low back pain with upper limb movement at different speeds. Arch Phys Med Rehabil. 1999;80:1005–1012.
13. Hides JA, Richardson CA, Jull G. Multifidus muscle recovery is not automatic after resolution of acute, first-episode low back pain. Spine. 1996;21:2763–2769.
14. Constantinou CE and Govan DE. Spatial distribution and timing of transmitted and reflexly generated urethral pressures in healthy women. J Urol. 1982;127:964–969.
15. Verelst M, Leivseth G. Are fatigue and disturbances in pre-programmed activity of pelvic floor muscles associated with female stress incontinence? Neurourol and Urodyn. 2004;23:143–147.
16. Sapsford R, Hodges PW. Contraction of the pelvic floor muscles during abdominal manoeuvres. Arch Phys Med Rehabil. 2001;82(8):1081–1088.
17. Mahoney DT, Laferte RO, Blais DJ. Integral storage and voiding reflexes. Neurophysiologic concept of continence and micturition. Urology. 1977;9:95–106.
18. Chiarelli PE. Improving patient's adherence. In: Laycock J, Haslam J, editors. Therapeutic management of incontinence and pelvic pain. London: Springer-Verlag; 2002:73–74.
19. Wagg A. The ageing lower urinary tract. In: Laycock J, Haslam J, editors. Therapeutic management of incontinence and pelvic pain. London: Springer-Verlag; 2002:39–44.

20. Snooks SJ, Barnes PRH, Swash M and Henry MM. Damage to the innervation of the pelvic floor musculature in chronic constipation. Gastroenterol. 1985;89:977–981.
21. Dumoulin C, Lemieux M, Bourbonnais D, et al. Conservative management of stress urinary incontinence: a single-blind, randomized, controlled trial of pelvic floor rehabilitation with or without abdominal muscle rehabilitation compared to the absence of treatment. Neurourol Urodyn. 2003;22:543–544.
22. Stone M. Implications for connective tissue and bone alterations resulting from resistance exercise training. Med Sci Sports Exerc. 2003;20:162–168.

3.2
Exercise, Feedback, and Biofeedback

Pauline E. Chiarelli and Kate H. Moore

Key Messages

- Feedback is essential in motor learning.
- Intrinsic feedback: the subject self-assesses or senses the contraction.
- Augmented feedback: involves the help of an assessor or device.
- Biofeedback techniques include digital (tactile), verbal, visual, electromyographic (visual or auditory), and manometric (simple or computerized) methods; perineal and abdominal ultrasound; and vaginal retention sensation, as provided by weighted vaginal cones or catheters and visual feedback during cystourethroscopy.
- Proprioceptive, kinesthetic, and tactile perception of a pelvic floor muscle (PFM) contraction are valuable to develop "contraction awareness."
- Biofeedback can be used for autonomic functions, such as detrusor overactivity and bladder sphincter dyssynergia, during cystometry.

Motor Learning and Biofeedback

Motor learning is a complicated process resulting in the acquisition of new and, after practice, relatively permanent motor skills. There are three stages to motor learning. The first stage is called the cognitive stage and involves a person learning exactly what is required to perform a particular task and exactly how to perform that task correctly.[1] In this stage of learning, feedback is essential if the motor task is to be reproduced precisely. The second stage involves "fine tuning" of the new skill. Mistakes are made less and less frequently until, finally, the task becomes automatic and does not require much attention during its execution, which is the third stage of skill acquisition.[2] Although practice of the specific motor learning task is of prime importance, feedback is considered the next most important variable.[3]

When learning how to correctly perform a motor task, e.g. an effective PFM contraction, sensory information or feedback is received by the person either during or after each contraction attempt. This feedback can be internal (intrinsic: the subject self-assesses or senses the contraction) or external (augmented: feedback with the help of an assessor or device).[2] In the case of a PFM contraction, intrinsic feedback means "feeling" the muscle contraction, which involves proprioceptive, kinesthetic, or tactile cues.[2] During initial attempts at PFM contraction, many women cannot perform or perceive a contraction and therefore receive little intrinsic feedback. For this reason, digital assessment of PFM contractions with verbal feedback to the patient/subject is very important in the early stages of PFM rehabilitation.[4] Subsequently, other forms of augmented feedback might be helpful.

Definition

Biofeedback in pelvic floor re-education constitutes an operant, cognitive, and respondent process by which the subject learns to reliably improve pelvic floor function. It entails repetitive responses and repeated teaching. It can involve pelvic floor functions that are not usually under voluntary control (such as detrusor muscle contractions), as well as responses that are usually under voluntary control, but are no longer easily regulated because of trauma or disease (e.g. PFMs following traumatic vaginal delivery). Biofeedback can be used to teach women to sustain, strengthen, direct, or eliminate a body action or reaction. The ultimate clinical goal of biofeedback is "self"-control without the need to use external biofeedback.

Biofeedback in the Rehabilitation of the PFMs

In PFM rehabilitation, commonly used biofeedback techniques include digital (tactile), verbal, visual, electromyographic (visual or auditory), and manometric (simple or computerized) methods; perineal and abdominal ultrasound; and vaginal retention sensation, e.g. as provided by weighted vaginal cones. Biofeedback can also be given during a cystourethroscopy showing the conscious woman an open bladder neck, e.g. and how it can be closed.

Although augmented biofeedback initially consisted of a simple pressure biofeedback device introduced by Arnold Kegel in 1948[4a] (Chapter 3.5), biofeedback has been improved by computerization and joined by a number of other biofeedback methods, including electromyography (EMG), which records electrical activity within specific muscles, sensory biofeedback (cones), PFM contraction indicator devices,[5] and dynamic ultrasound.[6,7]

The PFMs are a "hidden" muscle group. Women with PFM dysfunction are often unaware of the presence of this muscle group and are unable to initiate any voluntary contraction. In trying to perform a PFM contraction, many woman use primarily other muscle groups such as gluteals and hip adductors[8] or perform a Valsalva maneuver.[9]

Enhancement of Pelvic Floor Awareness

In the early stages of PFM retraining, the woman's muscle response might be minimal, and verbal feedback from the clinician during vaginal palpation can be used to describe a muscular response that would otherwise be unnoticed by the woman. This positive feedback should motivate the woman to contract the right muscles and continue with the exercise program. Extrinsic feedback seems an appropriate strategy during early intervention.[4]

To be able to contract, a muscle needs to have an intact nerve supply. In cases where nerve supply is at least partially intact, but where women are unable to initiate a PFM contraction, verbal instructions describing how a woman might correctly perform a PFM contraction have been shown to be quite unhelpful.[9] Forty-seven women with stress or urge incontinence or pelvic organ prolapse were first given brief verbal instruction about how to perform PFM contraction. The women were then asked to perform a correct PFM contraction while a urethral pressure profile was undertaken to ascertain the effectiveness of their contraction attempt. Only 49% were deemed to have performed an effective PFM contraction.[9] This study illustrates the fact that simply telling women how to contract their PFM is not sufficient. To maximize their PFM contraction performance, contraction attempts need to be enhanced by biofeedback.

Tactile and Verbal Feedback

The utilization of tactile sensation is an inexpensive and readily available form of biofeedback in pelvic floor re-education. Proprioceptive, kinesthetic, and tactile[2] perception of a PFM contraction are valuable to develop "contraction awareness." Proprioception, to encourage the "squeeze" component of a PFM contraction, can be provided by applying a gentle stretch to the PFM by parting the inserted fingers laterally or in the ventral and dorsocaudal direction (Fig. 3.2.1). Because the levator animuscle is well endowed with stretch receptors, in most cases, the woman reports feeling the muscles being stretched and can be encouraged to "pull the examining fingers in an upward direction." This method can also be used to provide the resistance necessary within a muscle-strengthening program. The woman should be taught the different actions of an effective PFM contraction, which not only includes a

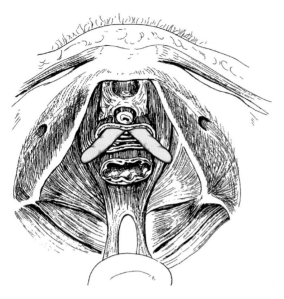

FIGURE 3.2.1. Stretching of the levator ani muscle (modified from Laycock J, Clinical Evaluation of the Pelvic Floor. In: Schüssler B, Laycock J, Norton P, Stanton S, editors. Pelvic Floor Re-education: Principles and Practice. Springer-Verlag London Ltd, 1994:43).

"squeeze" action to pull the urethra forward against the pubic bone, but also a "lifting" action.[10]

Another inexpensive biofeedback method involves the use of a large-gauge urethral catheter that is inserted vaginally and then inflated. This provides beneficial visualization of the catheter elevation during pelvic floor contraction and can also be used in standing (Fig. 3.2.2). Voluntary retention of the inflated vaginal catheter during coughing while standing offers excellent biofeedback and is also a powerful teaching tool.

Although digital assessment and feedback of a PFM contraction by the clinician will support the exercise program, digital self-assessment by the woman herself might be considered. However, digital self-assessment is limited because of the likely lack of knowledge of a normal pelvic floor. It has been shown that it is not carried out regularly following professional instruction[11] because it can be physically difficult to perform.

Visual Biofeedback

Simple visual biofeedback can be provided using a hand mirror to watch the PFMs as they contract, elevate the perineum, and narrow the genital hiatus. This has been used successfully among postpartum women.[12]

Devices that provide visual biofeedback of PFM during contraction include the "Pelvic Floor Educator" (Neen Healthcare) (Fig. 3.2.3).[5] It employs the use of a vaginal probe with a thin, but long, plastic rod attached to it. During a correct pelvic floor contraction – and only then – the rod will tilt downward, whereas during pushing or coughing the rod will move upwards.[13]

Kinesthetic Biofeedback

Weighted vaginal cones were initially designed and described by Plevnik in 1988.[14] Varying weights are inserted intravaginally, above the level of the PFM, while the woman performs upright, sedate activities of daily living. Feedback occurs when the cone begins to slip downwards in the vagina (kinesthetic perception), causing the woman to contract her PFM in order to prevent the cone from escaping the vagina.

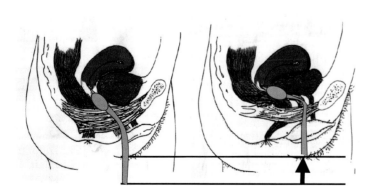

FIGURE 3.2.2. An inflated indwelling catheter in the vagina provides biofeedback when it is lifted during pelvic floor contraction. Additional traction on the catheter by the subject or clinician will enhance the retention feeling and the pelvic floor contraction can be trained to prevent slipping of the catheter.

3.2. Exercise, Feedback, and Biofeedback

FIGURE 3.2.3. The plastic rod is attached to a vaginal probe. With a sufficient pelvic floor contraction the rod rotates dorsally as the vagina is displaced ventrally.

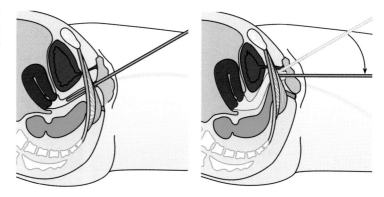

(Fig. 3.2.4). Anecdotally, older women tend to be less accepting of the use of weighted vaginal cones within PFM exercise programs.

Electromyographic Biofeedback

The electrical activity of motor units is generated by depolarization, registered as motor unit action potentials, and can be recorded in EMG studies (Chapter 1.2). Electromyographic signals are picked up using fine wire electrodes implanted within a muscle; alternatively, surface electrodes can be used, but they are considered less specific than fine wire electrodes. However, fine wire EMG is not widely used in PFM retraining because of the invasive nature of the procedure. The EMG biofeedback can be given as visual or auditory feedback. Endovaginal and endoanal "surface" electrodes have been developed and used to provide EMG feedback during PFM rehabilitation. (Fig. 3.2.5). Feedback can be given in numbers, bars, or curves on a screen showing how strong a contraction is performed over time.

Manometric Visual Biofeedback

Kegel designed the first manometric device to give women an incentive to increase the contractile power of their PFM by providing biofeedback (Fig. 3.5.1). The perineometer records changes in vaginal pressure during PFM contraction. Depending on the design and position of the probe, it will also detect increased abdominal pressure during straining or coughing.

Vaginal and anal pressure biofeedback units can directly record the changes in pressure caused by the contraction of the circumvaginal

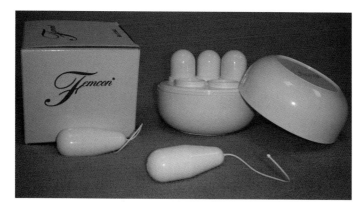

FIGURE 3.2.4. Set of vaginal cones, weighing 20–70 g. The vaginal weight is inserted into the vagina above the level of the levator ani. The sensation that the cone might slip generates a pelvic floor contraction.

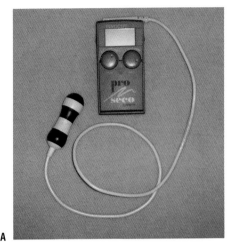 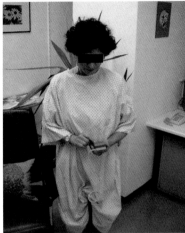

Figure 3.2.5. (A and B) Vaginal EMG-electrode and hand-held device used to provide biofeedback. Strength, endurance and timing of pelvic floor contractions can be practiced in everyday situations.

musculature (e.g. in mm Hg) and display them on a handheld gauge. More sophisticated means of amplifying the muscle response have emerged, including using computers to display the muscle response onscreen. This technique also allows the examiner to assess the pressure–time relationship of contractions, endurance, and coordination to monitor training progress, and to compare results intra- and interindividually.

Sonography: Dynamic Ultrasound

Real-time perineal and abdominal ultrasound offers the opportunity to specifically observe the elevation of the bladder neck during a PFM contraction and bladder neck descent during a Valsalva maneuver or coughing (see Chapter 2.5). Perineal ultrasound has been shown to selectively depict PFM activity better than intravaginal EMG and perineometry.[17]

Transperineal ultrasound has been used as a reliable method of demonstrating to women the contraction of their PFM and the elevation of their bladder neck. This was used successfully in women initially unable to voluntarily contract their PFM.[7] More recently, transabdominal ultrasound has been shown to be an alternative, although the visualization of the bladder neck is limited (see Chapter 2.5).[18] In response to the needs of physiotherapists, a small, mobile unit required only for biofeedback in muscle training is now available.

Biofeedback to Control Detrusor Overactivity

Biofeedback can be used in patients with unstable detrusor contractions and bladder sphincter dyssynergia. During cystometric investigations, the intravesical pressure is recorded and presented to the patient as an auditory and visual signal. Patients watch the urodynamic trace throughout the test while they try to inhibit detrusor contractions with PFM contractions. The treatment has had some measure of success, but patients have to be highly motivated, and the intervention is time consuming in terms of operator time.[19]

Tactile and Visual Biofeedback During Cystourethroscopy

During cystourethroscopy, the bladder neck movement during coughing and pelvic floor contraction is visible. The patient can be encouraged to close the bladder neck with a pelvic floor contraction and keep it closed during coughing (see video on DVD). This method of biofeedback can only be applied if there is an indication to perform an office cystoscopy.

Summary

To date, there is little evidence that the use of biofeedback improves the outcomes of regular pelvic floor treatment programs. Although various forms of biofeedback are readily available for use and can be very helpful in the individual patient, the value of biofeedback has also to be measured in terms of improved outcomes versus cost.

However, the most important judgment has to come from the subjects, and future studies should address the woman's perspective on the usefulness of biofeedback. Understanding pelvic floor function, teaching correct muscle coordination, and assessing progress are features beneficially provided by biofeedback.

References

1. Gentile A. Skill aquisition: action, movement and neuromotor processes. In: Carr J, Shepherd R, Gordon J, editors. Movement science: Foundations for physical therapy in rehabilitation. Rockville, MD: Aspen Systems; 1987.
2. Kisner C, Colby L. Therapeutic exercise: Foundational concepts. In: Kisner C, Colby L, editors. Therapeutic exercise: Foundations and techniques. Philadelphia: FA Davis Co; 2002:3–33.
3. Nicholson D. Teaching psychomotor skills. In: Shepard KF, Jensen GM, editors. Handbook of teaching for physical therapists. Boston: Butterworth-Heinemann; 1997:271.
4. McNevin N, Gabriele W, Carlson C. Effects of attentional focus, self control and dyad training on motor learning: Implications for physical rehabilitation. Phys Ther. 2000;80(4):373–385.
4a. Kegel AH. Progressive resistance exercise in the functional restoration of the perineal muscles. Am J Obstet Gynecol 1948;56:238–248.
5. Haslam J. Biofeedback for the assessment and reeducation of the pelvic floor musculature. In: Laycock J, Haslam J, editors. Therapeutic management of incontinence and pelvic pain. London: Springer-Verlag; 2002:75–81.
6. Dietz H, Clarke B. The urethral pressure profile and ultrasound imaging of the lower urinary tract. Int Urogynecol J. 2001(B);12(1):38–41.
7. Dietz H, Wilson P, Clarke B. The use of perineal ultrasound to quantify levator activity and teach pelvic floor muscle exercise. Int Urogynecol J Pelvic Floor Dysfunct. 2001;12:166–169.
8. Bo K, Larsen S, Oseid S, et al. Knowledge about and the ability to perform correct pelvic floor muscle exercises in women with urinary stress incontinence. Neurourol Urodyn. 1988;7(7):261–262.
9. Bump RC, Hurst G, Fantl JA, et al. Assessment of Kegel pelvic muscle exercise performance after brief verbal instruction. Am J Obstet Gynecol. 1991;165(2):322–329.
10. Schüssler B. Radiological evaluation of the pelvic floor and viscera. In: Schüssler B, Laycock J, Norton P, Stanton S, editors. Pelvic floor re-education: Principles and practice. London: Springer-Verlag; 1994:75–82.
11. Chiarelli P. Female urinary incontinence in Australia: prevalence and prevention in postpartum women [doctoral thesis]. Newcastle (Australia): University of Newcastle; 2001.
12. Chiarelli P, Cockburn J. Promoting urinary continence in women following delivery: randomised controlled trial. Br Med J. 2002;324(25):1241–1247.
13. Chiarelli P, Brown W. Perineal elevation – the reliability testing of a new measure of pelvic floor muscle function. In: Continence Foundation of Australia 8th National Conference on Incontinence. Sydney: Continence Foundation of Australia; 1999.
14. Plevnik S. New method for testing and strengthening of pelvic floor muscles. In: 15th annual meeting of the International Continence Society. London: ICS; 1988:95–1049.
15. Basmajian. Muscles alive. Their functions revealed by electromyography. 3rd edition. Baltimore: Williams and Wilkins; 1974.
16. LeCraw D, Wolf S. Electromyographic biofeedback (EMGBF) for neuromuscular relaxation and reeducation. In: Gersh M, editor. Electrotherapy in rehabilitation. Philadelphia: FA Davis Company; 1993:291–327.
17. Peschers U, Gingelmaier A, Jundt K, et al. Evaluation of pelvic floor muscle strength using four different techniques. Int Urogynecol J. Pelvic Floor Dysfunct. 2001;12:27–30.
18. Thompson J, Briffa K, Court S. The comparison between transperineal and transabdominal ultrasound in the assessment of women performing pelvic floor exercises. In: Continence Foundation of Australia (CFA) 12th National Conference. Sydney: CFA; 2003:39.
19. Cardozo L. Detrusor instability. In: Stanton S, editor. Clinical gynaecologic urology. St. Louis: C V Mosby; 1984:193–203.

3.3 Electrical Stimulation

Wendy F. Bower

Key Messages

- Large-diameter nerve fibers supplying the pelvic floor and urethral striated muscles are stimulated directly by electrical stimulation.
- 35–50 Hz creates a strong pelvic floor contraction.
- 5–10 Hz, with a pulse width of 0.5–1.0 ms, facilitates inhibition of an overactive detrusor via inhibitory motor-neurons of the pudendal nerve.
- Typical sessions last approximately 20–40 min daily to once per week for 12 weeks.
- Anatomical vaginal and anal probes allow close stimulation of the pelvic floor.
- Surface electrodes can be placed on the perineum and anus if indicated, but the application time should be longer because of the high impedance of the skin.
- Electrical stimulation at 35–50 Hz enhances awareness of the pelvic floor, is better than no treatment, and is particularly useful in women who are unable to contract the pelvic floor muscles (PFMs).

Introduction

Electrotherapy is the application of electrical current to provoke nerve activity and to stimulate function in different organ systems. The aim of electrical stimulation within the pelvic floor region is to induce skeletal muscle training, remodel smooth muscle and or connective tissues and to modulate bladder, bowel, or sexual dysfunction.[1] Table 3.3.1 outlines the applications of electrical stimulation for pelvic dysfunction.

Mechanisms of Action of Pelvic Electrotherapy

Activation of the PFMs by electrical stimulation has been shown in animal experiments to be primarily caused by the direct activation of pudendal nerve efferents. Single-fiber electromyographic (EMG) studies in humans have shown that pelvic floor stimulation elicits pelvic striated muscle responses. Electrical stimulation depolarizes the peripheral somatic motor fibers of the pudendal nerve within the PFM, which contract the urethral striated sphincter and draw in the pelvic floor skeletal musculature. This artificial recruitment of PFM induces a training effect, allowing the patient to develop sensory awareness of muscle activity. It also provides an opportunity for motor planning and practice of pelvic floor contractions while the muscle is weak and response is imperceptible. Axonal budding after denervation, and an increase in vascularization and hypertrophy have all been reported after stimulation of pelvic skeletal muscles.[2] Although regular recruitment of any muscle results in hypertrophy and enhanced fiber efficiency, voluntary contraction gives superior muscle remodeling compared to passive electrical recruitment.[3–6]

Depolarization of sensory fibers of the pudendal nerve also generates a reflex response (electric neuromodulation) with inhibition of an

TABLE 3.3.1. Application of electrical therapy for pelvic dysfunction (ICS Standardization Terminology of LUT Rehabilitation)

AIMS	Mode of action
Facilitate voiding	– Stimulate afferent fibers to convey bladder filling sensation
	– Stimulate efferent fibers / detrusor muscle to induce bladder contraction
Restore continence	– Improve urethral closure
	– Re-educate and condition PFM
	– Abolish/reduce detrusor overactivity
Control pelvic pain	– Relax PFM overactivity
	– Increase local blood flow
Improve sexual function	– Enhance vascular and skeletal muscle competence
Improve anal continence/defecation/constipation	– Increase gut motility
	– Enhance puborectalis function and external anal sphincter closure
	– Induce appropriate relaxation of overactive PFM

overactive sphincter. Besides these so-called peripheral effects, there are also central effects, which consist essentially of reorganization, reoordination, and awareness of the lower urinary tract, especially with respect to pelvic floor function.

The overactive bladder is frequently thought to be caused by deficient central inhibition. Electrical stimulation at optimal parameters serves to rebalance the descending excitatory information by artificially activating bladder-inhibiting reflexes. Peripheral effects, such as altered neurotransmitter concentration, are also seen. It is known, for example, that beta-adrenergic activity in the detrusor increases after repeat electrical stimulation, whereas cholinergic activity is decreased.[7]

Methods of Peripheral Electrical Stimulation

The effect of electrical stimulation depends on the distance between the nerve and the electrode, the excitability of the nerve fibers, the quality of central pulse transmission, and the properties of the effector organ.[8] Conduction speed of nerve fibers increases with fiber diameter and degree of myelination. Group A fibers are the largest and are involved in motor function and sensory perception, group B fibers serve the viscera in a sensory and motor capacity and are moderate in diameter, and group C fibers are unmyelinated and primarily involved in somatic pain responses.[9]

The intensity of the stimulation required depends on the size of the afferent nerve fiber and its distance from the electrode. Because of pain perception, all clinical applications of electrical stimulation are well below the threshold for maximal reflex inhibition. The large-diameter nerve fibers serving striated urethral and paraurethral muscles can be stimulated directly, utilizing a frequency close to the natural firing rate of the motorneurons. Although 50 Hz creates a strong, forceful contraction with minimal fast-twitch muscle relaxation between pulses, it can produce muscle fatigue.[10] Thus, a frequency of 35–40 Hz is commonly used, and attention is paid to lengthening the rest period between cycles of wavefronts (known as the "duty cycle").

The inhibitory motor neurons facilitate maximal inhibition of the overactive detrusor at 5–10 Hz.[11] If stimulation is applied close to non-muscular pudendal nerve afferents, weaker impulses will affect bladder inhibition. Pulse width is optimally 0.5–1.0 ms, with stimulation being either continuous or involving a very short duty cycle.

Historically, there are two main methods of electrical stimulation; chronic low intensity stimulation over a long period or maximum vaginal/anal stimulation over a short period. Chronic stimulation is based on daily stimulation over

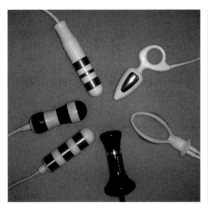 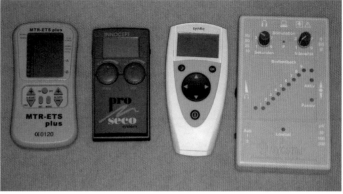

FIGURE 3.3.1. A. Different available vaginal and anal probes. **B.** Different hand-held displays.

1–12 hours via vaginal or anal plugs for a duration of up to 12 months.[12] The impulses are biphasic and weak, with a maximum of 12 V. The duty cycle is set so that the rest phase is at least double the working time. This method of stimulation favors PFM weakness. The time commitment required by the patient reduces the acceptance of this method.

Most clinicians prefer to use maximum electrical stimulation for overactive bladder symptoms or mixed urinary incontinence. Individual treatment sessions last around 20 minutes and are applied anywhere from daily to once per week for a period of at least 12 weeks. Monophasic currents are delivered by vaginal and/or anal plugs at an intensity close to the patient's limit of tolerance. Because the vagina offers a route of low impedance (low resistance of the vaginal mucosa), current delivery via probe electrodes has been favored. The currently available probes are anatomically shaped to allow good contact to the pelvic floor muscle and/or anal sphincter (Fig. 3.3.1).

However, stimulating the branches of the pudendal nerve that are mainly or exclusively sensory nerves using surface electrodes (dorsal penile/clitoridal nerve) is also common practice. Electrodes can be placed at the anus or perineum and just inferior to the pubic symphysis or coccyx. Treatment of the overactive bladder variously utilizes anogenital electrodes, transcutaneous delivery (transcutaneous electrical nerve stimulation [TENS]), sacral nerve electrode stimulation, percutaneous posterior tibial nerve needle electrodes, or an electromagnetic field induced in the pelvic region. Surface electrodes, such as those used in TENS, offer the possibility of longer-term current application, and perhaps enhance the effect on neural plasticity. Stimulation can be provided by a Mains-powered or portable battery–operated machine, making daily home treatment realistic.

Electrical Stimulation in Practice

Electrotherapy aims at developing PFM awareness, increasing PFM bulk, strengthening the perineal closure reflex, or inhibiting detrusor overactivity. As part of an overall treatment approach, adjunctive electrical stimulation has a role for patients of both genders who present with PFM weakness (with or without urinary incontinence), fecal incontinence, urge or mixed type incontinence, PFM overactivity, or pelvic/perineal pain syndromes.

It is important to place the role of electrical stimulation in context; can the patient expect a cure or appreciable improvement, is it merely "worth a try" until scheduled surgery, or is it adjunctive to other intervention methods? Similarly, the motivation and beliefs of a patient affect compliance in attending appointments or using the treatment at home. Previous negative sexual experiences may also limit the use of genital electrical stimulation. Rarely is electrotherapy a stand-alone treatment, but more often occurs with associated pelvic floor exercise regimes or

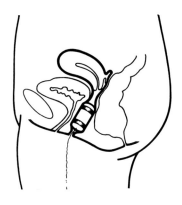

FIGURE 3.3.2. Position of a vaginal probe between the levator ani crura and an anal probe (*Source*: Laycock, J. Plevnik S, Senn E. Electrical Stimulation In: Schüssler B, Laycock J, Norton P, Stanton S, editors. Pelvic floor re-education. Principles and practice. London: Springer-Verlag; 1994144.)

bladder-training maneuvers. Some patients are unwilling to focus so acutely on self-help, preferring medication or surgical correction.

Contraindications to the use of electrical stimulation include: poor patient cognition, pregnancy, puerperium, atrophic vaginitis or recurring infection, recent or recurrent hemorrhage, and adverse skin reaction under surface electrodes. Some initial irritative symptoms may be reported[13] when commencing stimulation with body cavity electrodes, which can be managed effectively with shorter progressive treatment sessions.[14] The presence or history of a local malignancy is generally considered a contraindication to electrotherapy because current may stimulate rapidly dividing cells. However, in practice, this concern has not been proven. For similar reasons, the use of electrotherapy in pregnancy has never been advocated.

Evidence for Use of Electrical Neurostimulation

In general, the results of electrical stimulation in the management of lower urinary tract (LUT) symptoms and PFM dysfunction are difficult to interpret because electrical stimulation offers numerous combinations of current types, waveforms, frequencies, intensities, and electrode placements. Electrotherapy description should ideally specify the mode of application, stimulation site, electrical parameters, protocol, and equipment characteristics (International Continence Society standardization LUT rehabilitation).

Stress Urinary Incontinence

Electrical stimulation is adjunctive to voluntary exercise in increasing the strength and recruitment of a weak PFM and should be the introduction to specific functional recruitment of the PFM. Clinicians usually are not in favor of electrical stimulation as a stand-alone modality once a patient has achieved palpable voluntary recruitment of the PFM.

Electrical stimulation is known to be better than no treatment at all.[6] In regard to electrical stimulation versus placebo electrical stimulation, "in women with SUI the findings of two good quality trials using similar stimulation protocols are contradictory."[15,16] Most of the other trials favored electrical stimulation over placebo stimulation.[10,17,18] However, in some cases, the placebo devices generated stimulation of a limited output, which the investigators considered would not have a therapeutic effect. There is no trial comparing vaginal versus anal electrical stimulation for the indication of stress incontinence.

Pooled data for electrical stimulation versus PFM training alone found self-reported cure and improvement to be greater in women not using electrotherapy. Not all studies measured urine leakage or utilized valid quality of life measurement tools. It could be argued that once a patient has sufficient muscle awareness and motor control to recruit the PFM that electrical stimulation is no longer appropriate and, therefore, studies should only include subjects with imperceptible voluntary muscle activity. No evidence is currently available to indicate that adding electrical stimulation to biofeedback-assisted PFM training is beneficial.[19]

Urge and Mixed Incontinence

Symptoms of mixed stress and urge incontinence should theoretically respond to a tailored program that includes stimulation at both 50 Hz and <20 Hz. This addresses deficiencies in both

PFM strength and central inhibition. A 40-minute program divided evenly between the two protocols has been suggested; however, if surface electrodes are used to deliver the inhibitory current, longer duration treatment is probably warranted. Evidence from two studies[20,21] did not find added self-report or objective benefit from electrical stimulation given in addition to active PFM exercises. Again, it is possible that the subjects were already competent in PFM recruitment and not initially a suitable subject for electrical stimulation.

Fecal Incontinence

Anal continence requires sensory feedback from the anorectum and the ability to discern rectal distension.[22] The internal anal sphincter will relax in response to fecal fullness, leaving continence dependent on closure pressure exerted by the external anal sphincter.[23] It has been postulated that in addition to actual skeletal muscle hypertrophy, electrical stimulation may induce plastic changes centrally and increase the representational area of the anorectum.[23] A Cochrane review of trials using electrical stimulation for fecal incontinence highlighted the difficulties in comparing treatment approaches that used different protocols, parameters, and clinical applications. Nonetheless, it was noted that electrical stimulation "may have a therapeutic effect" in patients with fecal incontinence and that efficacy could only be determined after appropriately sized randomized controlled trials.[24]

Overactive Pelvic Floor and Voiding Dysfunction

Voluntary relaxation of the external urethral sphincter and the PFM normally precedes activation of the micturition reflex and subsequent voiding. Either detrusor weakness (hypocontractility or acontractility) or incomplete relaxation of the PFM may precipitate urinary retention and voiding difficulties. Fowler et al.[25] described a series of young women with urinary retention and abnormal electrical activity in the urethral sphincter EMG whose sphincter overactivity was associated with impaired relaxation of the pelvic floor muscles.

To date, none of the pharmacological therapeutic options to treat pelvic floor spasticity have proved efficacious. Although alpha-blockers improve obstructive voiding symptoms in men, this is not the case in patients with impaired capability of relaxation of the pelvic floor muscles. A study of intrasphincteric injection of botulinum toxin in six women with nonneurogenic pelvic floor dysfunction did not show any benefit.[26,27] Electrotherapy applied across the sacral region has application for both urinary urge incontinence and urinary retention.[28] It is not known whether the mode of action is enhanced PFM proprioception facilitating greater awareness of PFM and sphincter relaxation or deactivation of the urethral guarding reflex.[29] Shaker and Hassouna reported the treatment of 20 patients with nonobstructive retention. Voiding function improved significantly during stimulation, but most of the patients were unable to void without having the sensation of the stimulation. This may support the proposition that the stimulation has a modulatory, but not stimulatory, effect on the afferent nerve pathways.[30] Additional clinical and experimental studies are required to elucidate the true efficacy of electrotherapy for the overactive PF.

References

1. Abrams P, Cardozo L, Fall M, Griffiths D, et al. The standardisation of terminology of lower urinary tract function: report from the sub-committee of the International Continence Society. Neurourol Urodyn. 2002;21:167–178.
2. Brubaker L. Electrical stimulation in overactive bladder. Urology. 2000;55(Suppl 5A):17–23.
3. Dudley GA. Harris, RT. Strength and power in sport. Komi PV, editor. Use of electrical stimulation in strength and power training. Oxford: Blackwell Scientific Publications; 1992:329–337.
4. Bouchard C, Shephard RJ, Stephens T, editors. Physical activity, fitness, and health: status and determinants. Adjuvants to physical activity. Physical activity, fitness and health. Consensus statement. Champaign: Human Kinetics Publishers; 1993:33–40.
5. Bø K, Talseth T. Change in urethral pressure during voluntary pelvic floor muscle contraction and vaginal electrical stimulation. Int Urogynecol J. 1997;8:3–7.

6. Henalla S, Hutchins C, Robinson P, et al. Non-operative methods in the treatment of female genuine stress incontinence of urine. J Obstet Gynaecol. 1989;92:22–25.
7. Bower WF, Yeung CK. A review of non-invasive electro neuromodulation as an intervention for non-neurogenic bladder dysfunction in children. Neurourol Urodyn. 2004;23(1):63–67.
8. Ahlstrom K, Eriksen B, Fall M. Electrostimulation. Female urinary incontinence. Sjoberg N-O, Holmdahl TH, Crafoord K, editors. Parthenon Publishing Group; 2000.
9. Romanzi LJ. Pelvic floor exercise, biofeedback, electyrical stimulation, and behaviour modification. In: Blaivas JG, editor. Evaluation and treatment of urinary incontinence. Tokyo: Igaku-Shoin Med Publishers; 1996.
10. Laycock J, Jerwood D. Does pre-modulated interferential therapy cure genuine stress incontinence? Physiother. 1993;79:553.
11. Lindström S, Fall M, Carlsson CA, Erlandson BE. The neurophysiological basis of bladder inhibition in response to intravaginal electrical stimulation. J Urol. 1983;129(2):405–410.
12. Yamanishi T, Yasunda K. Electrical stimulation for stress incontinence. Int Urogynecol. 1998;9:281–290.
13. Indrekvam S, Sandvik H, Hunskaar S. A Norwegian national cohort of 3198 women treated with home-managed electrical stimulation for urinary incontinence – effectiveness and treatment results. Scand J Urol Nephrol. 2001;35:32.
14. Laycock J, Vodusek DB. Electrical stimulation. In: Laycock J, Haslam J, editors. Therapeutic management of incontinence and pelvic pain. London: Springer-Verlag; 2002.
15. Luber KM, Wolde-Tsadik G. Efficacy of functional electrical stimulation in treating genuine stress incontinence: a randomized clinical trial. Neurourol Urodyn. 1997;16:543.
16. Sand PK, Richardson DA, Staskin DR, et al. Pelvic floor electrical stimulation in the treatment of genuine stress incontinence: a multicenter, placebo-controlled trial. Am J Obstet Gynecol. 1995;173:72.
17. Yamanishi T, Yasuda K, Sakakibara R, et al. Pelvic floor electrical stimulation in the treatment of stress incontinence: an investigational study and a placebo controlled double-blind trial. J Urol. 1997;158:2127.
18. Blowman C, Pickles C, Emery S, et al. Prospective double blind controlled trial of intensive physiotherapy with and without stimulation of the pelvic floor in the treatment of genuine stress incontinence. Physiotherapy. 1991;77:661.
19. Knight S, Laycock J, Naylor D. Evaluation of neuromuscular electrical stimulation in the treatment of genuine stress incontinence. Physiotherapy. 1998;84:61.
20. Wang AC, Wang YY, Chen MC. Single-blind, randomized trial of pelvic floor muscle training, biofeedback-assisted pelvic floor muscle training, and electrical stimulation in the management of overactive bladder. Urology. 2004;63(1):61–66.
21. Berghmans B, van Waalwijk van Doorn E, Nieman F, et al. Efficacy of physical therapeutic modalities in women with proven bladder overactivity. Eur Urol. 2002 Jun;41(6):581–587.
22. Hobday DI, Aziz Q, Thacker N, et al. A study of the cortical processing of ano-rectal sensation using functional MRI. Brain. 2001;124(Pt 2):361–368.
23. Salvioli B, Bharucha AE, Rath-Harvey D, et al. Rectal compliance, capacity, and rectoanal sensation in fecal incontinence. Am J Gastroenterol. 2001;96(7):2158–2168.
24. Hosker G, Norton C, Brazzelli M. Electrical stimulation for faecal incontinence in adults. Cochrane Database Syst Rev. 2000;(2):CD001310.
25. Fowler CJ, Kirby RS. Electromyography of urethral sphincter in women with urinary retention. Lancet. 1986;1:1455–1457.
26. Phelan M, Franks M, Somogyi G, et al. Botulinum toxin urethral sphincter injection to restore bladder emptying in men and women with voiding dysfunction. J Urol. 2001;165:1107–1110.
27. Fowler CJ, Betts CD, Christmas TJ, et al. Botulinum toxin in the treatment of chronic urinary retention in women. Br J Urol. 1992;70:387–389.
28. Tanagho EA, Schmidt RA. Electrical stimulation in the clinical management of the neurogenic bladder. J Urol. 1988;140:1331–1339.
29. Shaker HS, Hassouna M. Sacral root neuromodulation in idiopathic nonobstructive chronic urinary retention. J Urol. 1998;159:1476–1478.
30. DasGupta R, Fowler CJ. The management of female voiding dysfunction: Fowler's syndrome – a contemporary update. Curr Opin Urol. 2003;13(4):293–299.

3.4
Extracorporeal Magnetic Stimulation

Alastair R. Morris and Kate H. Moore

Key Messages

- Magnetic stimulation is an electric current that is induced by a time-varying magnetic field.
- Deep structures can be stimulated without pain; stimulation is generated externally.
- Sessions with 10 Hz to reduce overactive bladder symptoms and with 50 Hz to treat stress incontinence are commonly used over 6–8 weeks, 1–2 times per week.
- Success rates vary and decline over time, with a maximum of 33% of patients being dry.

Introduction

Until recently, stimulation of the pelvic tissues was only possible by the direct application of electrical current. However, new devices utilizing a pulsed electromagnetic field have been developed, which may achieve a similar effect. This chapter will describe the theory upon which magnetic stimulation is based, indications for its use, and summaries of current clinical experiences.

Principles of Electromagnetic Induction Therapy

Magnetic stimulation therapy is based on the principle that an electric current can be induced by a time-varying magnetic field. Modern devices contain a coil through which a large alternating current is passed, thereby generating a constantly changing magnetic field. The changing magnetic field induces local eddy currents in tissues lying within it. The eddy currents in turn evoke depolarization of adjacent peripheral nerves, with stimulation of the pelvic floor muscles (PFMs).

The characteristics of the magnetic field depend on several factors. Strength is primarily determined by the current applied to the stimulating coil, and the larger the current the stronger the field. However, field strength reduces inversely with the square of the distance from the stimulating coil. Shape is determined principally by that of the coil itself. Single coils produce a doughnut-like field with centrally reduced stimulation, whereas double coils generate a field that is saddle shaped.

Unlike electrical current, a magnetic field is not significantly attenuated by skin tissue. Consequently, deep structures can be stimulated without causing the cutaneous pain that often limits conventional electrostimulation.[1] Additionally, probes do not need to be inserted into the vaginal or anal canal, allowing the patient to remain clothed throughout treatment. Some devices have been developed to allow treatment at home, but the majority still require hospital attendance.[2]

Reponse of Neuromuscular Tissue to Electrical and Magnetic Stimulation: How Does It Work?

Depolarization of neural tissue following either electrical or magnetic stimulation appears indistinguishable.[3,4] Thus, presumably, they affect the pelvic floor musculature in a similar fashion. Magnetic stimulation may lead to hypertrophy of the PFM, alter the proportions of type 1 and 2 fibers, and allow greater recruitment if a neurotrophic effect occurs. However, no studies have assessed histological changes in striated or detrusor muscle fibers after magnetic stimulation. Optimal electrical stimulation of striated muscle is achieved using a 50-Hz current,[5] whereas inhibition of detrusor contraction is maximal using 10 Hz.[6]

The acute mechanism of action of magnetic stimulation on detrusor function is unknown. Early studies suggest that electrical stimulation of the pudendal nerve evokes reflex inhibition of the detrusor muscle and contraction of the striated pelvic musculature.[6] In both normal patients[7] and those with idiopathic detrusor overactivity (DO),[8] there is an immediate fall in intravesical pressure after magnetic stimulation is applied over the sacral foramina, suggesting immediate detrusor relaxation. McFarlane suggested that this reflex cannot be mediated entirely by contraction of the external sphincter, and postulated that inhibition of parasympathetic efferent output by blockage of sensory afferent and sympathetic input is the most likely effect. Hyperreflexic bladder contractions in spinally injured patients can also be decreased by stimulating the sacral nerve roots with a handheld magnetic unit.[9]

In a group of patients with neuropathic DO, bladder capacity at initial desire to void increased from 175 ml (±67) to 225 ml (±65; P = 0.017) when magnetic stimulation using a chair device was applied during cystometry. The maximum cystometric capacity (MCC) also rose significantly from 290 ml (±62) to 348 ml (±69; P = 0.003), as did the maximum urethral closure pressure (P < 0.0001). Similarly, among 10 women with idiopathic DO, a significant rise in MCC (P = 0.02) was observed, in addition to a significant reduction in maximum detrusor pressure (P = 0.004).[11] These findings have also been noted during ambulatory monitoring of 8 patients with idiopathic DO following therapy.[12]

Hence, acute extracorporeal magnetic innervation (ExMI) appears to contract the urethra, reduce contraction of the detrusor, and allow a larger bladder capacity.

Indications

Magnetic stimulation is useful to treat stress urinary incontinence (SUI) and may have a role in treating the overactive bladder. It is contraindicated in patients with either a cardiac pacemaker or metallic hip implant.

Treatment Protocols

No standardized treatment regime has been accepted. Most published protocols are modeled on those traditionally used for electrical stimulation. These often require 2 to 3 treatments per week over a 6 to 8 week period. Individual treatment sessions generally comprise two 10-minutes cycles of maximal-tolerated therapy separated by a short rest period (often 2 minutes). The first period of stimulation generally uses a low frequency (5–15 Hz) and the second a higher one (50 Hz). No randomized controlled trials (RCTs) have reported data comparing the effects that this may have on pelvic floor or detrusor muscle.

Two categories of magnetic stimulation device are currently commercially available. These are a portable hand device that is placed over the sacral area to stimulate the sacral nerve roots or a chair-like device containing the magnetic coil in its base, such as the Neotonus© Chair (Fig. 3.4.1). Stimulation using the latter device is commonly called ExMI and has received Federal Drug

FIGURE 3.4.1. The Neotonus® chair.

Administration approval in the United States for the treatment of SUI. Although handheld devices require the constant presence of an operator, the chair-like device does not. No RCTs have been undertaken to compare the efficacy of these two types of device.

Studies Relating to Pure Urodynamic Stress Incontinence

Two RCTs in patients with proven urodynamic stress incontinence (USI) have been reported. In the first, 70 patients were randomized equally to receive active ExMI using a Neotonus Chair or sham therapy on an identical, but nonfunctioning, device.[13] Concurrent symptoms of OAB were not exclusion criteria. All patients also underwent a "low-intensity" PFM exercise training program supervised by a physiotherapist.[14] Unfortunately, the sham group was significantly more wet at the baseline of the 24-hour pad test. At 8 weeks, 17% of active treatment patients were dry versus 9% of sham patients, presumably the latter as a result of the PFM training program (Table 3.4.1). Both had similar significant reductions in pad test leakage.

TABLE 3.4.1. Randomized sham controlled trial of neotonus ExMI in USI[13]

	Pretreatment (Mean, SD)	Posttreatment (Mean, SD)
Active neotonus		
24-hour pad test	24.0 (4.7)	10.1 (3.1)*
20-min pad test	39.5 (5.1)	19.4 (4.6)**
PFX strength	1.6 (0.3)	2.7 (0.4)*
CMR score	5.0 (0.4)	5.3 (0.4)
Pads/day on FVC	6.9 (0.1)	0.6 (0.1)*
IQOL	63.7 (2.8)	71.2 (3.3)**
Kings HQ	9.6 (0.8)	6.9 (0.7)***
Sham chair group		
24-hour pad test	37.2 (7.2)	22.0 (5.2)**
20-min pad test	39.9 (7.4)	32.4 (6.7)
PFX strength	1.7 (0.3)	1.9 (0.4)
CMR score	4.4 (0.4)	4.6 (0.4)
Pads/day on FVC	1.2 (0.2)	1.0 (0.1)
IQOL	62.6 (4.0)	67.3 (4.4)*
Kings HQ	9.7 (0.9)	8.6 (1.0)

* $P < 0.05$; ** $P < 0.01$; *** $P < 0.001$.

The protocol of the second RCT ($n = 62$) utilized a hand held device placed over the sacrum.[15] However, active therapy comprised only a single treatment at 15 Hz, rather than the usual 50 Hz for SUI, and outcome was determined after only 1 week. Though intergroup differences in leaks per day and pad loss were significant, so were the reductions in the sham group alone. Why such a short treatment should result in an immediate and identifiable effect is unclear, as muscle hypertrophy normally takes 8 weeks to become clinically evident.

Studies Relating to Patients with USI and Urgency/Urge Incontinence

The reported success of treatment has varied. After twice weekly treatment over 6 weeks ($n = 64$), 36 patients were reviewed at 6 months with 10 (28%) being dry and 8 (22%) using less than 1 continence pad per day.[16] At 1 year, 27 patients were seen, of whom 11 (38%) remained cured with 12 (41%) improved.[17]

In contrast, in 91 patients treated twice weekly over 8 weeks, 34 (37%) were dry immediately after treatment, although by 3,6, and 12 months, 47%, 61.7%, and 94%, respectively, had become wet again.[18]

In another trial, 17 patients with USI and 20 with urge incontinence received combined 10-Hz and 50-Hz therapy.[19] Of the urge patients, 25% were dry on bladder diary, whereas 53% of the USI women were dry on 1-hour pad test and bladder diary. Urodynamic data was not fully reported, but only 8 of the 20 urge patients had DO, so the study is difficult to interpret.

Studies Relating to the Overactive Bladder

Two randomized trials involving sham therapy are currently available. The largest recruited 44 consecutive patients with idiopathic DO and administered 20 treatments at 10 Hz over 6 weeks, using the Neotonus Chair.[20] A sham chair containing a deflector plate was available, which was indistinguishable in sight and sound from the active chair. Only 29 patients (65%) completed the protocol, 15 (51%) receiving active therapy. At 6 weeks after treatment, active therapy only significantly decreased episodes of urgency P = 0.003 over sham therapy.

In the other trial, 37 women with OAB symptoms were randomized to a single treatment at 15 Hz using a handheld magnetic device placed over the sacrum, or sham therapy with assessment one week later.[19] Intergroup comparison showed leaks per day and volume per void decreased significantly with active treatment (P = 0.04 and P = 0.04, respectively). Again, it is unclear how a single treatment could achieve this significant benefit.

In a small open trial of 18 patients with idiopathic DO, only 9 completed the protocol of twice weekly treatments for 6 weeks; the rest found it too time consuming.[12] Of these 9, urge symptoms improved in 8, but urge incontinence improved in only 2 cases.

Fecal Incontinence

A small study of 16 women who were treated for fecal incontinence at both 5 and 50 Hz on the ExMI chair revealed a significant reduction of the modified Wexner score from 10/24 to 7/24 (P < 0.05) after 8 weeks.[22] The best response was seen in women with an intact anal sphincter and weak levator muscles; those with fecal urgency or a diarrhea component did not respond.

Place of ExMI in a Pelvic Floor Rehabilitation Program

The place of magnetic stimulation in a pelvic floor re-education program remains uncertain. It is generally well tolerated and avoids the placement of internal probes. However, the equipment is expensive and the chair device is not portable. Consequently, these techniques may be difficult to offer outside of specialist clinics, although small devices for home use have been trialed.[2] Current protocols are based on those developed for traditional electrical stimulation, so multiple weekly attendances are required. Women find this time consuming, so they must be highly motivated before commencing treatment. Many fail to complete their treatment regime because of the onerous time commitment needed.[12,20]

The optimal ExMI regime has yet to be defined, and this should be addressed by well-constructed RCTs. Adequate sham limbs in such studies are technically difficult to achieve, and few have been published so far. The high rate of attrition amongst women in such trials suggests that multicenter studies will be required to achieve adequate power. Short-term benefits in SUI are frequently reported, but consistent medium/long-term data to confirm ongoing benefit is lacking. Units offering ExMI participate in well-constructed clinical trials to help answer these questions.

In women with USI who are unable to isolate or adequately contract their PFM, magnetic stimulation has proven benefit over sham therapy.[13] It has also been pointed out that ExMI using a chair device was particularly well accepted by elderly women.[23] However, once magnetic stimulation therapy ends, its benefits often rapidly disappear and a continued program of treatment appears necessary to maintain any improvement.

References

1. Hallet M, Cohen LG. Magnetism. A new method for stimulation of nerve and brain. JAMA. 1989; 262:538–541.
2. But I. Conservative treatment of female urinary incontinence with functional magnetic stimulation Urol. 2003;61:558–561.
3. Olney RK, So YT, Goodin DS, et al. A comparison of magnetic and electrical stimulation of peripheral nerves. Muscle Nerve. 1990;13:957–963.
4. Brodack PP, Bidair M, Joseph A, et al. Magnetic stimulation of the sacral roots. Neurourol Urodyn. 1993;152:533–540.
5. Brubaker L. Electrical stimulation in overactive bladder. Urology. 2000;55 (Suppl 5A):17–23.
6. Fall M, Lindstrom S. Electrical stimulation. A physiological approach to the treatment of urinary incontinence. Urol Clin N Am. 1991;18: 393–407.
7. Craggs MD, McFarlane JP, Knight SL, et al. Detrusor relaxation of the normal and pathological bladder. Br J Urol. 1997;79(Suppl 4):58–59.
8. McFarlane JP, Foley SJ, De Winter P, et al. Acute suppression of idiopathic detrusor instability with

magnetic stimulation of the sacral nerve roots. Br J Urol. 1997;80:734–741.
9. Sherriff MK, Shah PJ, Fowler C, et al. Neuromodulation of detrusor hyperreflexia by functional magnetic stimulation of the sacral nerve roots. Br J Urol. 1996;78:39–46.
10. Yamanishi T, Yasuda K, Suda S, et al. Effect of functional continuous magnetic stimulation for urinary incontinence. J Urol. 2000;163;456–464.
11. Morris AR, Dunkley P, O'Sullivan R, et al. Idiopathic detrusor instability – a double blind, randomised trial of electromagnetic stimulation therapy versus sham therapy. Proc Intl Cont Soc Heidelberg. 2002;142–143.
12. Bradshaw HD, Barker AT, Radley SC, et al. The acute effect of magnetic stimulation of the pelvic floor on involuntary detrusor activity during natural filling and overactive bladder symptoms. BJU Int. 2003;91:810–813.
13. Gilling P, Kennett I, Bell D, et al. A double blind randomized trial comparing magnetic stimulation of the pelvic floor to sham treatment for women with stress urinary incontinence. Proc Aust Urol Soc. 2000.
14. Bo K, Hagen R, Kvarstein B, et al Pelvic floor muscle exercise for the treatment of female stress urinary incontinence: III. Effects of two different degrees of pelvic floor muscle exercise. Neurourol Urodyn. 1990;9:489–502.
15. Fujishiro T, Enomoto H, Ugawa Y, et al. Magnetic stimulation of the sacral roots for the treatment of stress incontinence: an investigational study and placebo controlled trial. J Urol. 2000;164:1277–1279.
16. Galloway NT, El-Galley RES, Russell H, et al. Extracorporeal magnetic innervation (ExMI) therapy for stress urinary incontinence. J Urol. 1999;53(6):1108–1111.
17. Unsall A, Saglam R, Cimentepe E. Extracorporeal magnetic stimulation for the treatment of stress and urge incontinence in women. Scand J Urol Nephrol. 2004;37:424–428.
18. Almeida FG, Bruschini H, Srougi M. Urodynamic and clinical evaluation of 91 female patients with urinary incontinence treated with perineal magnetic stimulation; 1-year followup. J Urol. 2004; 171;1571–1575.
19. Yokoyama T, Fujita O, Nishiguchi J, et al. Extra corporeal magnetic innervation treatment for urinary incontinence. Int J Urol. 2004;11:602–606.
20. Morris AR, O'Sullivan RO, Dunkley P et al. Extracorporeal magnetic stimulation if female detrusor overactivity: simultaneous cystometry testing and a randomized sham controlled trial. in press.
21. Fujishiro T, Satoru T, Enomoto H et al. Magnetic stimulation of the sacral roots for the treatment of urinary frequency and urge incontinence: an investigational study and placebo controlled trial. J Urol. 2002;168:1036–1039.
22. Shoberi SA, Chesson RR, Echols KT, et al. Evaluation of extracorporeal magnetic innervation for the treatment of faecal incontinence. Proc Int Cont Soc Florence. 2003; Abstract # 324.
23. Madersbacher H, Pilloni. Efficacy of extracorporeal magnetic innervation therapy (ExMI) in comparison to standard therapy for stress, urge and mixed incontinence: a randomized prospective trial. Proc Intl Cont Soc Florence. 2003; Abstract # 367.

3.5
Devices

Ingrid Nygaard and Peggy A. Norton

Key Messages

- Intravaginal resistance device; weighted vaginal specula, and cones (weights) can be used to assess and train the pelvic floor.
- Pelvic floor muscle training with vaginal cones (weights) has been shown to significantly improve stress incontinence symptoms.
- "Bladder neck–supportive devices" like special continence tampons, the Conveen continence guard, and urethral pessaries brace the bladder neck during increased intraabdominal pressure.
- Intraurethral inserts and urethral suction caps can be used to prevent urinary leakage, although efficacy data is scarce.
- To control pelvic organ prolapse, supportive and space-occupying pessaries can be inserted vaginally.

Introduction

There are a variety of devices that can be used to train pelvic floor muscles and to treat pelvic floor disorders. Those used for pelvic floor training are an important source of biofeedback and objective outcome measurement for pelvic floor rehabilitation. Devices worn in the vagina are non-surgical options in the treatment of pelvic organ prolapse and stress urinary incontinence that can be used in many clinical settings. In this chapter, we will outline the indication and use of these devices.

Devices Used to Train Pelvic Floor Muscles

Devices can be used both to assess the strength of pelvic floor muscles and to train these muscle groups. Although biofeedback and electrical stimulation are covered in Chapters 3.2 and 3.3, this section will describe vaginal devices used for both assessment and training.

Intravaginal-resistance devices consist mostly of perineometers, which are modified from Arnold Kegel's original design (Fig. 3.5.1.); these devices are inserted into the vagina and provide a visual display of pelvic floor contraction strength. A perineometer is essentially a simple pressure gauge that measures vaginal pressure, but it cannot distinguish between pressure generated by vaginal muscles from pressure generated by abdominal pressure.[1] Some modifications include EMG electrodes at the base of the probe to isolate pelvic floor muscle activity. They are available directly to patients on the internet through a variety of sources. Although some patients will find the visual display helpful, a recent study found digital examination worked well with a perineometer device in assessing pelvic floor strength.[2] Several groups have reported a vaginal speculum designed to assess pelvic floor muscular strength.[3]

Vaginal weighted cones were designed to both assess PFMS and to further train the pelvic floor with a set of increasing weights.[4-6] When placed in the vagina, the sensation of losing the cones

FIGURE 3.5.1. Original Kegel perineometer as sold in 1949.

Devices used to Treat Pelvic Floor Disorders

Surgery for pelvic floor disorders such as stress urinary incontinence and pelvic organ prolapse is functional surgery; instead of removing a diseased organ, these procedures are aimed at restoring or improving the function of the pelvic organs. By its nature, such functional surgery cannot be guaranteed to restore continence and support to its original state. Given that a third of surgeries for pelvic floor disorders fail,[11] alternatives to surgery may offer less risk and expense to many women for the management of pelvic floor disorders. Devices are widely available, but require some professional intervention to determine the correct use and fit, similar to a contraceptive diaphragm. Little has been published on their use, possibly because there is no industry support for (or profit from) conducting properly controlled clinical trials.

Intravaginal Devices for Pelvic Organ Prolapse

Vaginal pessaries have been used for many centuries, but improvements in materials and design have increased the usefulness of these devices for prolapse.

Indications for a pessary in the management of pelvic organ prolapse (POP) include patients who desire nonsurgical management of the condition. Although a few women are unable to undergo surgical management because of medical problems, a larger number of women might be interested in a pessary because it manages the prolapse without the need to undergo surgery. In our practices, pessaries are used successfully in women who cannot take time off for surgery, such as mothers with small children at home and women with busy careers outside the home.

Willingness to use a vaginal device may be cultural, especially in areas where contraceptive diaphragms are used. An Australian study[12] found that only 21% of 104 women who presented to a community continence clinic stated that they felt very comfortable about inserting a device into the vagina and half felt uncomfortable. At the University of Iowa, two thirds of 190 women

prompts a pelvic floor contraction so as to retain the cone. No fitting is required, and patients insert the device for a prescribed period of time daily. They are available to professionals and through direct marketing to patients themselves. A recent Cochrane review of vaginal cones concluded that vaginal weighted cones are better than control treatments (level 1 evidence) for self-reported cure or efficacy in women with proven stress urinary incontinence.[7] However, several randomized trials[8,9] have found equal efficacy in women using vaginal cones and those doing pelvic muscle training without devices; in some trials, pelvic muscle training performed without cones was more effective.[10] There is no information available to determine which women are more likely to benefit from vaginal cones.

(mean age 57.4 years, range 15–89) who were offered a trial of pessary to manage stress or mixed incontinence were interested in trying one.[13]

The best clinical scenario for pessary use in prolapse is an anterior and/or apical defect (cystocele, uterine prolapse, vaginal vault prolapse) in a woman with a narrow pubic arch and good pelvic floor strength. Obstetricians are familiar with the assessment of the angle at which the pubic rami meet at the symphysis; a wide arch in which three or more fingers can be placed is less likely to hold the ventral/caudal edge of a pessary. The pelvic floor braces the dorsal edge of many pessaries, and in the absence of intact pelvic floor muscles one must consider the use of pessaries that utilize suction or inflation ("space-occupying pessaries"). If the vaginal capacity is reduced after surgery, a narrower pessary may be needed (oval, Hodge, cube.) Reported risk factors for pessary failure include a shortened vagina and a wide levator hiatus.[14]

Supportive pessaries (which depend on some levator muscle support to stay in place) include the Gehrung, Hodge, Shaatz, and rings and ovals with support (Fig. 3.5.2). The pessary we use the most is the ring with support, in sizes 3 and 4 (refers to diameter in centimeters.) In a survey of members of the American Urogynecologic Society, 22% of respondents used the same pessary, usually a ring pessary, for all support

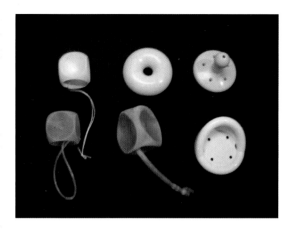

FIGURE 3.5.3. Space-occupying pessaries used to treat pelvic organ prolapse. Top row: left, cube; middle, doughnut; right, Gelhorn (all three by Milex). Bottom row: left, cube with drainage holes (Mentor); middle, inflatoball (Milex); right, Mar-land (Mentor).

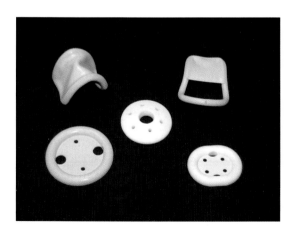

FIGURE 3.5.2. Support pessaries used to treat pelvic organ prolapse. Top left, Gehrung (Milex); top right, Hodge (Milex); middle, Shaatz (Mentor); bottom left, ring with support (Milex); bottom right, oval with support (Mentor).

defects.[15] In the remainder that tailored the pessary to the defect, a ring pessary was more common for anterior and apical defects, a Gelhorn was more common for complete procidentia, and a doughnut was more common for posterior defects. A similar management strategy was employed by Wu et al.,[16] who always used a flexible ring pessary as the first pessary tried. Seventy percent of women were successfully fitted with a size 3, 4, or 5 ring pessary. In one questionnaire survey,[17] physicians reported that ring and doughnut pessaries were the most common pessaries used. However, other centers use different strategies; Sulak et al[i] used a Gelhorn pessary in 96 out of 107 women with symptomatic pelvic organ prolapse.[18]

Supportive pessaries generally allow coitus while wearing the device. They are the easiest pessaries to use because they fold to a smaller dimension for insertion, but may not be sufficient to support large prolapses. Pessaries are easiest to insert lying down, easiest to remove standing up, and may require digital bracing per vaginam during bowel movements. Some women have difficulty removing the pessary; in such cases, we recommend tying a strand of dental floss around the ring so that the pessary can be pulled out by the floss.

Space-occupying pessaries include the cube, doughnut, Gelhorn, and inflatoball (Fig. 3.5.3).

These pessaries are more difficult to insert and remove, but work when the device would otherwise be extruded, such as with larger prolapse, poor pelvic floor strength, or wider pubic arch. Of these, we use the doughnut and the Gelhorn with the most frequency. The doughnut is simply pushed into the vagina (a difficult task if the introitus is scarred or atrophic), whereas the disk of the Gelhorn is fitted behind the symphysis, similar to a ring pessary, but the knob is aligned parallel to the axis of the vagina and facilitates placement and removal. The cube has a relative suction effect and may be effective in the case of lax vaginal walls, but generates significant discharge and, in our experience, is more prone to excoriation and ulceration than other pessaries. The inflatoball is pumped up with a small bulb and is similarly prone to excoriation, unless care is taken.

We recommend that women remove the pessary at least weekly, leave it out overnight, and then reinsert in the morning. In our experience, women rarely encounter excessive or malodorous vaginal discharge using this approach and, thus, have little use for creams other than estrogen. Space-occupying pessaries are sometimes difficult for a woman to remove on her own. In such a case, we try to estimate the appropriate interval between pessary removals in the following manner. We first examine women two weeks after initial pessary insertion. If discharge is minimal and no erosions are present, we examine next at four weeks. Similarly, if the examination is reassuring, we examine again after six weeks, and so on. The appropriate pessary interval is either a maximum of three months or the interval at which we see foul-smelling discharge or early erosions.

After an initial 2-week and 3-month check, we examine women who manage their own pessary without difficulty yearly. In women who retain the pessary for several months at a time, we believe that a visual inspection of the vagina should occur at least twice yearly. It is important to examine the anterior and posterior vaginal walls during the examination (by turning the speculum 90 degrees), as well as the obvious lateral walls that are visible when the speculum is placed in the usual fashion. We have seen several women with large recto-vaginal or vesicovaginal fistulae caused by pessaries; in all cases they were undergoing regular examinations by a physician. It is possible that unseen erosions under the speculum blades may have heralded the beginnings of such pressure ulcers.

Several series have demonstrated that pessaries are useful. In one study, 74% of 110 women were fitted successfully.[16] Of the 62 women who used a pessary for more than one month, 66% were still using it after 12 months. In another series, half of 107 women fitted with a pessary continued to use the device at the time of manuscript preparation (average length of use was 16 months).[18]

In a prospective study, 73 of 100 women with symptomatic POP were fitted successfully with a pessary.[19] Two months after fitting, only 3% of women reported a bulge, compared to 90% at baseline. Other symptoms that improved with pessary use included pressure, discharge, and splinting. One third of women had urge incontinence at baseline; this improved by 54%. Twenty-three percent had voiding difficulty at baseline, which improved by 50%. At 2 months, 92% were either very or somewhat satisfied with their pessary.

Intravaginal Devices for Stress Urinary Incontinence

Devices in this category are worn in the vagina and work as a "back-stop" to brace the bladder neck during increased intraabdominal pressure. The vaginal pessary has undergone some modifications for use in women with stress incontinence, but there are other devices for this indication that are not pessaries. Many women report use of their contraceptive diaphragm as being effective.[20,21] A short menstrual tampon may be inserted just comfortably inside the introitus; patients need to be instructed to use the tampon under dry conditions, which improves the adherence of the tampon. We instruct patients to use this "tampon trick" with a super tampon, and only on an occasional basis. The Conveen Continence Guard is available in some countries, and is a polyurethane foam cushion that is folded on its long axis and placed in the vagina. When moistened and partially unfolding, it acts as a backstop under the bladder neck. The device is available in three sizes and is worn for up to 18 hours then discarded. Several studies have documented good tolerance and significant reductions in urine loss with use.[22,23]

Non-supportive devices include a urethral insert (Rochester Medical, Inc.) and urethral suction caps,[24–26] (Uromed, not currently available.) Such urethral plugs and caps have been studied with some success, but do not seem to be popular in clinical practice. Examples of non-pessary devices for incontinence are shown in Figure 3.5.4.

Pessaries modified for use in incontinent women essentially are "bladder neck–supportive" during increased intraabdominal pressure. They include rings with knobs placed at the bladder neck, the Hodge pessary inserted backwards and upside down, the incontinence dish with support, PelvX ring, and the Suarez ring (Cook Urologic) (Fig. 3.5.5). Some women may wear these devices for activities only, whereas others need to wear them on a daily basis. Care of these pessaries is similar to that for supportive pessaries.

Although use of a short super tampon may be suggested on a temporary basis, we use the incontinence dish as our main incontinence pessary. Patients will immediately see the advantages (effective, no surgery) and disadvantages (small

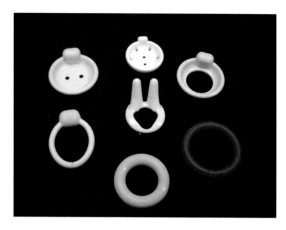

FIGURE 3.5.5. Incontinence pessaries. Clockwise from top: Incontinence dish with support (Mentor), Incontinence dish (Milex), PelvX ring (DesChutes Medical Products), Suarez ring (Cook Urologic), Incontinence ring with support (Milex), Incontinence dish with support (Milex); middle: Introl prosthesis (was Johnson and Johnson; currently not available).

amount of bother with insertion, need for continued use) associated with these devices.

Most studies evaluating the effectiveness of devices for stress incontinence are small in numbers and short in duration. In a prospective, randomized laboratory-based study,[21] 6 of 14 women were cured and 2 out of 14 improved while exercising wearing a super tampon in the vagina. Nine of 12 women had resolution of stress incontinence while wearing a contraceptive diaphragm during urodynamic testing[19] and 4 of 10 women wearing a contraceptive diaphragm for 1 week had improved continence.[27]

Of 190 women presenting to a tertiary care center with symptoms of stress or mixed UI who were offered pessary management,[13] 63% chose to undergo fitting and 89% achieved a successful fit in the office. Of the 106 women who took a pessary home, follow-up was available on 100. Fifty-five women used the pessary for at least 6 months as their primary method of managing urinary incontinence (median duration 13 months). Of the remaining 45 women who discontinued use before 6 months, most did so by 1 month.

Studies of intraurethral inserts showed that most women who use intraurethral devices are dry or improved (66%–95%) when the device is in place.[28–31] Not surprisingly, urinary tract

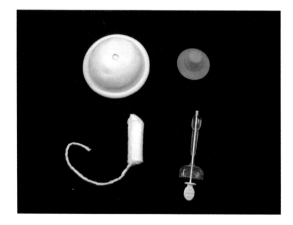

FIGURE 3.5.4. Examples of devices used to treat stress incontinence that are not pessaries. Top row: left, contraceptive diaphragm; right-urethral suction cap (was marketed by Uromed, currently not available). Bottom row: left, menstrual tampon; right-urethral insert (Rochester Medical, Inc).

infections are a common adverse event; however, the incidence of infections decreases after the first several months of use. In our practices, few women choose inserts as first line therapy, but some are very satisfied with them as an option when other therapies are unsuccessful.

Conclusion

Pessaries and other devices are an important part of the treatment armamentarium for POP and stress urinary incontinence. Further research is needed to determine which women are most likely to respond to devices. Long-term studies of both effectiveness and adverse events associated with various devices are essential to better understand the risk-benefit ratio.

References

1. Peschers UM, Gingelmaier A, Jundt K, et al. Evaluation of pelvic floor muscle strength using four different techniques. Int Urogynecol J Pelvic Floor Dysfunct. 2001;12(1):27–30.
2. Isherwood PJ, Rane A. Comparative assessment of pelvic floor strength using a perineometer and digital examination. BJOG. 2000;107(8):1007–1011.
3. Dumoulin C, Gravel D, Bourbonnais D, et al. Reliability of dynamometric measurements of the pelvic floor musculature. Neurourol Urodyn. 2004;23(2):134–142.
4. Wilson PD, Borland M. Vaginal cones for the treatment of genuine stress incontinence. Aust N Z J Obstet Gynaecol. 1990;30(2):157–160.
5. Olah KS, Bridges S, Denning J, et al. The conservative management of patients with symptoms of stress incontinence: a randomized, prospective study comparing weighted vaginal cones and interferential therapy. Am J Obstet Gynecol. 1990;162(1):87–92.
6. Versi E, Mantle J. Vaginal cones: a conservative method of treating genuine stress incontinence. Br J Obstet Gynaecol. 1989;96(6):752–753.
7. Herbison P, Plevnik S, Mantle J. Weighted vaginal cones for urinary incontinence. Cochrane Database Syst Rev. 2002;(1):CD002114.
8. Arvonen T, Fianu-Jonasson A, Tyni-Lenne R. Effectiveness of two conservative modes of physical therapy in women with urinary stress incontinence. Neurourol Urodyn. 2001;20(5):591–599.
9. Peattie AB, Plevnik S, Stanton SL. Vaginal cones: a conservative method of treating genuine stress incontinence. Br J Obstet Gynaecol. 1988;95(10):1049–1053.
10. Bo K, Talseth T, Holme I. Single blind, randomised controlled trial of pelvic floor exercises, electrical stimulation, vaginal cones, and no treatment in management of genuine stress incontinence in women. BMJ. 1999;318(7182):487–493.
11. Olsen AL, Smith VJ, Bergstrom JO, et al. Epidemiology of surgically managed pelvic organ prolapse and urinary incontinence. Obstet Gynecol. 1997;89(4):501–506.
12. Prashar S, Simons A, Bryant C, et al. Attitudes to vaginal/urethral touching and device placement in women with urinary incontinence. Int Urogynecol J Pelvic Floor Dysfunct. 2000;11(1):4–8.
13. Donnelly MJ, Powell-Morgan S, Olsen AL, et al. Vaginal pessaries for the management of stress and mixed urinary incontinence. Int Urogynecol J Pelvic Floor Dysfunct. 2004;15(5):302–307.
14. Clemons JL, Aguilar VC, Tillinghast TA, et al. Risk factors associated with an unsuccessful pessary fitting trial in women with pelvic organ prolapse. Am J Obstet Gynecol. 2004;190(2):345–350.
15. Cundiff GW, Weidner AC, Visco AG, et al. A survey of pessary use by members of the American urogynecologic society. Obstet Gynecol. 2000;95(6 Pt 1):931–935.
16. Wu V, Farrell SA, Baskett TF, et al. A simplified protocol for pessary management. Obstet Gynecol. 1997;90(6):990–994.
17. Pott-Grinstein E, Newcomer JR. Gynecologists' patterns of prescribing pessaries. J Reprod Med. 2001;46(3):205–208.
18. Sulak PJ, Kuehl TJ, Shull BL. Vaginal pessaries and their use in pelvic relaxation. J Reprod Med. 1993;38(12):919–923.
19. Clemons JL, Aguilar VC, Sokol ER, et al. Patient characteristics that are associated with continued pessary use versus surgery after 1 year. Am J Obstet Gynecol. 2004;191(1):159–164.
20. Suarez GM, Baum NH, Jacobs J. Use of standard contraceptive diaphragm in management of stress urinary incontinence. Urology. 1991;37(2):119–122.
21. Nygaard I. Prevention of exercise incontinence with mechanical devices. J Reprod Med. 1995;40(2):89–94.
22. Hahn I, Milsom I. Treatment of female stress urinary incontinence with a new anatomically shaped vaginal device (Conveen Continence Guard). Br J Urol. 1996;77(5):711–715.

23. Mouritsen L. Effect of vaginal devices on bladder neck mobility in stress incontinent women. Acta Obstet Gynecol Scand. 2001;80(5):428–431.
24. Brubaker L, Harris T, Gleason D, et al. The external urethral barrier for stress incontinence: a multicenter trial of safety and efficacy. Miniguard Investigators Group. Obstet Gynecol. 1999;93(6):932–937.
25. Bellin P, Smith J, Poll W, et al. Results of a multicenter trial of the CapSure (Re/Stor) Continence shield on women with stress urinary incontinence. Urology. 1998;51(5):697–706.
26. Tincello DG, Adams EJ, Sutherst JR, et al. A urinary control device for management of female stress incontinence. Obstet Gynecol. 2000;95(3):417–420.
27. Realini JP, Walters MD. Vaginal diaphragm rings in the treatment of stress urinary incontinence. J Am Board Fam Pract. 1990;3(2):99–103.
28. Nielsen KK, Walters S, Maegaard E, et al. The urethral plug II: an alternative treatment in women with genuine urinary stress incontinence. Br J Urol. 1993;72(4):428–432.
29. Peschers U, Zen Ruffinen F, Schaer GN, et al. [The VIVA urethral plug: a sensible expansion of the spectrum for conservative therapy of urinary stress incontinence?]. Geburtshilfe Frauenheilkd. 1996;56(3):118–123.
30. Staskin D, Bavendam t, Miller J, et al. Effectiveness of a urinary control insert in the management of stress urinary incontinence: early results of a multicenter study. Urology. 1996;47(5):629–636.
31. Sirls LT, Foote JE, Kaufman JM, et al. Long-term results of the FemSoft urethral insert for the management of female stress urinary incontinence. Int Urogynecol J Pelvic Floor Dysfunct. 2002;13(2):88–95.

3.6
Alternative Methods to Pelvic Floor Muscle Awareness and Training

Kaven Baessler and Barbara E. Bell

Key Messages

Pelvic floor muscles are increasingly incorporated into Yoga and Pilates classes and for lower back pain physiotherapy. Other techniques, including Feldenkrais physiotherapy, which develops pelvic floor awareness through movement and functional integration, and Cantienica, whereby a pelvic floor contraction is palpated externally after movements between the ischial tuberosities, greater trochanters, and the coccyx. Whole-body vibrations through biomechanical stimulation of the muscles increases metabolic power, might target type II muscle fibers, and is increasingly used in the training of athletes and the management of women with osteoporosis. Theoretically, this may lead to a more powerful muscle contraction.

Introduction

This chapter will give an overview of different approaches to pelvic floor awareness and training that have not been mentioned before. Awareness and voluntary control of the pelvic floor are the basis of any pelvic floor training. How a woman accomplishes this awareness may be individually different and may require the utilization of different techniques. Personal preferences on the woman's and the therapist's side will be a factor and will vary between therapists. The conscious, positive integration of the pelvic floor in other treatment or training regimes has gained popularity and recognition.

Pelvic Floor Incorporation in Yoga and Pilates Classes and Lower Back Pain Therapy

In some Yoga classes, the pelvic floor plays an active role, and Pilates schools have adopted the pelvic floor into their program, too. Programs based on the Pilates concept have been very successful with dancers and athletes as they focus on stability, balance, body alignment and awareness, and breathing and involve the deep abdominal muscles and the pelvic floor.[1,2] Pilates incorporates cognitive activation of the deep abdominal muscles before the performance of a task. As the transversus abdominis and the pelvic floor muscles (PFM) are part of the local stability system of the lumbopelvic region, certain movements and exercises will result in increased pelvic floor activity. A study using electromyographic (EMG) surface electrodes attached to the oblique and rectus abdominis muscles, and an intravaginal EMG probe to record pelvic floor muscle activity, demonstrated that, especially during the Pilates "clam" exercise, the pelvic floor muscles are activated (Fig. 3.6.1).[3] Pilates movements incorporating eccentric activity of the abdominal muscles generated greater PFM activity than concentric abdominal activity.[4]

FIGURE 3.6.1. Pilates "Clam exercise." The pelvic floor is activated concurrently.

The integration of the pelvic floor into Yoga and Pilates may range from simple coactivation to voluntary and active incorporation of the pelvic floor. It is not necessarily pelvic floor training, as it is likely to lack specificity and overload (see Chapter 3.1). With the changing view on how the pelvic floor and the abdominal and multifidi muscles work synergistically together, trunk stabilization exercise programs for lower back pain or sacroiliac joint dysfunction now increasingly include the pelvic floor muscles. There is also evidence that mind–body therapies like relaxation techniques, cognitive behavioral therapy, and biofeedback can be used as effective adjuncts to conventional medical treatment in the management of chronic lower back pain and osteoarthritis, for example.[5]

The Feldenkrais Approach to the Pelvic Floor

"Awareness through movement" is the goal of the so-called Feldenkrais method.[6] Moshé Feldenkrais (1904–1984) worked as a physicist and engineer. When he suffered from a knee-injury and adequate rehabilitation was not available, he developed a concept that enables a human being not only to feel but also to analyze even complex muscular–skeletal body movements, which subsequently leads to an increase in awareness. Based on this awareness, the individual can select the appropriate muscle action by optimizing the coordination between nervous system, muscles, and the skeleton. The Feldenkrais method is a pedagogical concept with the intention to train the brain to bring body movements into awareness.[6] This is especially effective in muscle pain and tension after prolonged misuse, e.g. as in neck and shoulder tensions of surgeons or typists and in pelvic floor re-education. The Feldenkrais approach offers benefits not only with increased pelvic floor awareness but also with the consciousness of how functional tasks such as weight lifting or coughing can be modified in order to minimize the unfavorable effects on the pelvic floor. In contrast to conventional physiotherapy, the Feldenkrais method does not directly approach the tensed muscles, but offers awareness of alternative body positions, which will be able to replace former habits. Although comparative literature is scarce, the Feldenkrais method was superior to conventional physiotherapy in non-specific musculoskeletal disorders.[7] The Feldenkrais method cannot replace conventional physiotherapy in general, but can be applied successfully in some individuals and might supplement physiotherapy.

Judy Pippen and Barbara Bell are trained physiotherapists in the Feldenkrais method and introduced the usefulness of this framework with pelvic floor education (www.pelvicpower.com). Initially, during "differentiation," one learns to sense the movement of the PFM more clearly. Lessons are organized in a slow progressive way so that attention is drawn to the ability to move the pelvic floor in parts. The front section may be differentiated from the back section, the right part of the pelvic floor from the left. Attention is also directed at sensing the difference between the pelvic floor tightening and that of the surrounding muscles, and then encouraging the integration/synergism with them. "Integration" is the second part of the Awareness Through Movement classes. It involves combination of movement of the whole body in an integrated way to support the functions of the pelvic floor. Moshe Feldenkrais said, "Awareness fits action to intention."[6] The pelvic floor is recruited in various degrees when a movement is intended (e.g. walking, breathing, singing, coughing) to keep continence, to keep our organs supported, and to be able to allow for relaxation and tightening during sexual activities. The third step is to include the pelvic floor in many activities throughout the day when needed, consciously or unconsciously, and to "empower" people to take control of improving function.

Within the Feldenkrais method, there is a particular focus on breathing. While trying to hold on to urine rushing to a toilet many people breathe in. The patient has to learn to breathe out while holding on; this helps negate the panic response, which only makes urgency more urgent. The forceful "out" breath (e.g. laughing, vomiting, coughing, or sneezing) is a challenging movement, especially for stress incontinent women. Using a functional approach, patients can learn in a graded way to breathe outwards and to organize the recruitment of the pelvic floor.

Breathing is also a focus because it can easily be brought under voluntary control, but is usually under involuntary control. Voluntary actions can change the state of overall "tonus" or tension and can change the subsequent rate, depth, and rhythm of the breath. In this meditative state between voluntary and involuntary where someone is very conscious and extremely focused, it can be sensed that breathing can be performed by the abdomen and the pelvic floor with coordination.

"Cantienica"

Benita Cantieni in Switzerland developed the concept of "sensual pelvic floor training" to exercise the "most important storey in the human building."[8] Many of her claims are questionable, and the efficacy of the method has not formally been researched, but the approach to achieve a pelvic floor contraction and how to verify it is resourceful and practical (Fig. 3.6.2). Women are

FIGURE 3.6.2. Three external ways to palpate activation of the pelvic floor: medial shift of the ischial tuberosities, muscle action over the major trochanters with slight external rotation and muscle pull at the inferior edge of the symphysis and movement of the coccyx.

asked to bend forward, palpate their ischial tuberosities and try and pull them together. A subtle shift of the ischial tuberosities can be felt externally, which probably is the action of the iliococcygeus and coccygeus muscles. A second method is to feel the gentle movement of the greater trochanters with a pelvic floor contraction, a movement that is probably generated by the obturator internus and piriformis muscles, which are innervated by the obturator nerve and pelvic branches from S1–S2, respectively. The third method to confirm a pelvic floor contraction is the palpation of a movement underneath the fingers on the pubic bone and the coccyx. The integration of the pelvic floor into daily life is encouraged and taught with numerous exercises.

Biomechanical Stimulation

Biomechanical stimulation or whole-body vibration therapy is a new neuromuscular training method that increases metabolic power and might specifically target type II muscle fibers.[9] It is thought to elicit muscular activity via stretch reflexes and allows the combination of voluntary and involuntary muscle work.[9] It has been used in the training of athletes[10] and in the management of women with osteoporosis[11] and lower back pain.[12] It has been shown to improve vertical jumping,[13,14] bone density,[15] and leg extension strength,[13] and to increase testosterone and growth hormone levels in men.[14] Standing with the knees slightly bent on a platform that vibrates between 5 and 30 Hz at an amplitude of up to 13 mm, all muscles that keep the body erect have to work (Fig. 3.6.3). The typical frequency is 25–30 Hz, forcing the muscle to contract and relax 25–30 times per second. Changing the position on the platform changes the focus of muscle work. Lower frequencies are supposed to be ideal for balance control, bone strengthening, and weaker individuals, whereas higher frequencies strengthen the muscles and are suitable for stronger individuals. Theoretically, this training can also improve the tone and strength of the PFM. Although this theory is currently being tested, no studies have been published yet. Without parallel pelvic floor education, it seems unlikely to prove better than current pelvic floor regimes because

FIGURE 3.6.3. Whole-body vibration therapy: The patient stands on a platform that tilts on a central axis with different frequencies and amplitudes. The musculature has to keep the body in its position and is forced to react to the oscillatory movements.

it lacks specificity. Women will still have to learn how to activate the PFM at the appropriate time, but after whole-body vibration they might be able to do so with a more powerful contraction.

Acknowledgment. The part on the Feldenkrais method was finalized by the assistance of Mrs. Erna Alig, Feldenkrais instructor, Switzerland.

References

1. Lake B. Acute back pain. Treatment by the application of Feldenkrais principles. Aust Fam Physician. 1985;14(11):1175–1178.
2. Hutchinson MR, Tremain L, Christiansen J, et al. Improving leaping ability in elite rhythmic gymnasts. Med Sci Sports Exerc 1998;30(10):1543–1547.

3. Sapsford R. Pilates and the pelvic floor. Proceedings, 2nd Biennal Excellence Down-under. School of Pysiotherapy, University of Melbourne, Australia; 2005.
4. Sapsford R, Kelly S, C.R. Pilates and the pelvic floor. Abstract Physiotherapy Conference; 2004.
5. Astin JA, Shapiro SL, Eisenberg DM, et al. Mind-body medicine: state of the science, implications for practice. J Am Board Fam Pract. 2003;16(2):131–147.
6. Feldenkrais M. Awareness through movement. Health exercises for personal growth. Hammondsworth, UK: Penguin Books; 1977.
7. Malmgren Olsson EB, Branholm IB. A comparison between three physiotherapy approaches with regard to health-related factors in patients with non-specific musculoskeletal disorders. Disabil Rehabil. 2002;24(6):308–317.
8. Cantieni B. Tiger feeling. The sensual pelvic floor training for her and him. 2000.
9. Rittweger J, Beller G, Felsenberg D. Acute physiological effects of exhaustive whole-body vibration exercise in man. Clin Physiol. 2000;20:134–142.
10. Bosco C, Colli R, Introini E, et al. Adaptive responses of human skeletal muscle to vibration exposure. Clin Physiol. 1999;19(2):183–187.
11. Rubin C, Recker R, Cullen D, et al. Prevention of bone loss in a post-menopausal population by low-level biomechanical intervention. Bone Min Res. 1998;23:1126.
12. Rittweger J, Just K, Kautzsch K, et al. Treatment of chronic lower back pain with lumbar extension and whole-body vibration exercise – a randomized controlled trial. Spine. 2002;27:1829–1834.
13. Torvinen S, Kannu P, Sievanen H, et al. Effect of a vibration exposure on muscular performance and body balance. Randomized cross-over study. Clin Physiol Funct Imaging. 2002;22:145–152.
14. Bosco C, Iacovelli M, Tsarpela O, et al. Hormonal responses to whole-body vibration in men. Eur J Appl Physiol. 2000;81:449–454.
15. Rubin C, Turner AS, Bain S, et al. Anabolism. Low mechanical signals strengthen long bones. Nature. 2001;412:603–604.

Part IV
Treatment: Condition-Specific Assessment and Approaches

4.1 Behavioral Treatment

Kathryn L. Burgio

Key Messages

- Behavioral treatments are usually comprised of several components and tailored to the needs of the individual woman.
- Active use of PFM is a key continence skill for avoiding both stress and urge incontinence.
- Altering voiding habits, fluid intake, and other aspects of life style promote improved continence status.
- Using a bladder diary enhances women's awareness of their incontinence and guides behavioral intervention.

Introduction

Behavioral interventions are a group of treatments that improve incontinence by changing women's habits or teaching them continence skills. Behavioral interventions include self-monitoring with a bladder diary, PFM training and exercise, active use of PFMs to prevent urine loss, urge suppression strategies, urge avoidance, scheduled voiding, delayed voiding, fluid management, weight loss, and other lifestyle changes. In general, these treatments are safe and without the risks and side effects of some other therapies. However, they require the active participation of a motivated woman and usually take some time and persistence to reach maximum benefit. Behavioral treatments have been recognized for their efficacy by the 1988 Consensus Conference on Urinary Incontinence in Adults[1] and the Guideline for Urinary Incontinence in Adults developed by the Agency for Health Care Policy and Research.[2] Although the majority of women are not cured with this approach, most can achieve significant improvement in continence status.

The Bladder Diary (See Chapter 2.6)

The bladder or voiding diary is a valuable clinical tool, both for the clinician and the woman. In the diagnostic phase, it provides information on the type and severity of urine loss and helps to plan appropriate components of behavioral intervention. During treatment, the diary can be monitored to determine the efficacy of various treatment components and guide the intervention. In addition to the value for the clinician, the self-monitoring effect of completing the diary enhances the patient's awareness of voiding habits and patterns of incontinence. It facilitates a woman's recognition of how her incontinence is related to her activities. In particular, clearly understanding the precipitants of urine leakage optimizes the woman's readiness to implement the continence skills learned through behavioral treatment.

Before initiating treatment, it is advisable to have the woman complete a bladder diary for 5 to 7 days.[3] At a minimum, the woman should record the time and volume of micturition, the time of each incontinent episode, its size, and the circumstances or reasons for the accident (Fig. 4.1.1). Through the process of reviewing the bladder diary, women can identify certain times when they are more likely to have incontinence and the activities that seem to trigger incontinence.

Date: *1-20-2006*

TIME URINATED IN TOILET	TIME OF SMALL LEAK	TIME OF LARGE LEAK	REASON FOR LEAKAGE
2:10 am			
4:30 am			
	6:40 am		Coughed in bed
8:09 am			
	8:45 am		Urge - Taking a shower
		10:10 am	Urge - Taking a walk
10:50 am			
1:35 pm			
	3:32 pm		Sneezed 3 times
5:55 pm			
		7:20 pm	Watching TV
9:25 pm			
10:30 pm			

<u>TIME</u> UP FOR THE DAY: ***8:00 am***   <u>TIME</u> TO BED FOR THE NIGHT: ***10:00 pm***
OF PADS USED TODAY: __4__   NUMBER OF LEAKAGES: __5__

FIGURE 4.1.1. Sample bladder diary.

PFM Training and Exercise

Pelvic floor muscle training and exercise are the foundation of behavioral treatment for stress urinary incontinence. This intervention has evolved as both a behavioral and a physical therapy, combining principles from both fields into a widely accepted conservative treatment for stress incontinence.

The goal of behavioral intervention for stress incontinence is to teach the patient how to improve urethral closure by voluntarily contracting her PFMs during physical activities that cause urine leakage, such as coughing, sneezing, or lifting. The first step in training is to assist the woman to identify her PFMs and to contract and relax them selectively. It is essential to confirm that patients have identified and isolated the correct muscles. Failure to find the correct muscles is perhaps the most common reason for failure of this treatment modality. Basics and advice on this subject are presented in Chapters 3.1 and 3.2.

It is best to begin treatment by ensuring that the woman understands which muscles to use. This can be accomplished by palpating the pelvic floor during pelvic examination and guiding her with verbal feedback to find the proper muscles. Pelvic floor muscle control can also be taught using biofeedback or by applying electrical stimulation.

Once the woman learns how to properly contract and relax the PFMs selectively, a program of daily practice and exercise is prescribed. The purpose of the daily regimen is not only to increase muscle strength but also to enhance motor skills through practice. The optimal exercise regimen has yet to be determined; however, good results are generally achieved using 45 to 50 exercises per day, over 2 to 3 sessions per day. It is important to encourage the woman to practice in various positions, so that she becomes comfortable using her muscles to avoid accidents in any position. (See Chapter 4.2 for more detail on PFM training.)

Using Muscles to Prevent Stress Accidents: Stress Strategies

Although exercise alone can improve urethral support and continence status, optimal results depend on the woman learning to use her muscles actively to prevent urine loss during situations of physical exertion. With practice and encouragement, the woman can develop the habit of first consciously and then automatically contracting PFMs to occlude the urethra before and during coughing, sneezing, or any other activities that have caused urine loss. This skill has been referred to varyingly as the stress strategy,[4,5] counterbracing, perineal blockage, precontraction and the "Knack."[6] Some women will benefit simply from learning how to control their PFMs and to use them to prevent accidents. Others will need a more comprehensive program of PFM rehabilitation to increase strength, as well as skill.

Using Muscles to Prevent Urge Accidents: Urge Suppression Strategies

In addition to its value in treating stress incontinence, this technique is now frequently used as a component in the treatment of urge incontinence as well.[7] In addition to using the PFMs to occlude the urethra, the woman learns to use PFM contraction and other urge suppression strategies to inhibit bladder contraction. Furthermore, the woman is taught a new way to respond to the sensation of urgency; instead of rushing to the toilet, which increases intraabdominal pressure and exposes her to visual cues that can trigger incontinence, she is encouraged to pause, sit down if possible, relax the entire body, and contract the PFMs repeatedly to diminish urgency, inhibit detrusor contraction, and prevent urine loss.[5] When urgency subsides, she can to proceed to the toilet at a normal pace. Behavioral training for urge incontinence has been tested in several clinical series utilizing prepost designs and in controlled trials using intention-to-treat models. Mean reductions of incontinence range from 60% to 80%.[7-10] (See Chapter 4.6 for more detail on urge suppression.)

Adherence and Maintenance

Exercising and using the PFMs requires the active participation of a motivated woman. It is often challenging to remember to use muscles strategically in daily life, as well as to persist over time in a regular exercise regimen to maintain strength and skill. This reliance on a woman's behavior change represents the major limitation of this treatment approach. In addition, progress with behavioral treatment is often gradual and progressive, usually evident by the fourth week of training and continuing for up to 6 months. Herein lies the challenge for behavioral treatment – to sustain the woman's motivation for a long enough time that she will experience noticeable change in her bladder control.

It is important in initiating behavioral treatment to make it clear to the woman that her improvement will be gradual and will depend on consistent practice and use of her new skills. Clinicians can provide support by scheduling follow-up appointments to track and reinforce her progress, make adjustments to the exercise regimen, and encourage persistence.

Voiding Habits and Schedules

Increasing Voiding Frequency

Many women have been advised by health care providers to increase frequency of urination as a way to prevent urgency and incontinence by avoiding a full bladder. Although immediate benefit can be observed, the long-term result is most likely counterproductive, because it removes opportunities for the bladder to be full, and the woman can lose the ability to accommodate bladder fullness. In addition, it feeds the cycle of urgency and frequency thought to perpetuate overactive bladder and urge incontinence in the long run.

Increasing the frequency of urination is generally reserved for women who clearly void below what is normal or in patients with dementia or who are otherwise cognitively impaired and incapable of learning new skills for bladder control. Women who have reduced bladder sensation may also benefit.

Decreasing Voiding Frequency: Bladder Training and Delayed Voiding

Bladder training is a behavioral intervention originally developed for the treatment of urge incontinence. The belief behind the bladder training method is that habitual frequent urination can reduce bladder capacity and lead to overactive bladder (OAB), which in turn causes urge incontinence.[11] The goal of the training is to break this cycle using consistent voiding schedules. The woman voids at predetermined intervals, and over time, the voiding interval is gradually increased (see Chapter 4.6).

Clinical series studies have demonstrated cure rates ranging from 44% to 90% using outpatient bladder training or a mixture of inpatient and outpatient intervention. The first randomized clinical trial of bladder training demonstrated an average 57% reduction of incontinence in older women.[12]

Lifestyle Changes

Fluid Management

Many behavioral clinicians recommend alterations in the volume or type of fluids that a woman consumes as a way to optimize her outcomes. It is not only helpful for bladder control but also good health advice to ensure that the woman is consuming an adequate amount of fluid each day. In an effort to cut down on urine loss, some patients restrict their fluid intake in general, or at particular times of day when they consider themselves to be at risk of incontinence. In some cases, this results in an inadequate intake of fluid and places them at risk of dehydration.

Fluid restriction is often appropriate in women who consume an abnormally high volume of fluid (e.g. >2,100 ml of output per 24 hours). Some people increase their fluid intake deliberately in an effort to "flush" their kidneys, lose weight, or avoid dehydration. In others it is simply a habit. In these cases, reducing excess fluids can relieve problems with sudden bladder fullness and resulting urgency. Reducing or eliminating fluids in the evening hours is often helpful for reducing nocturia.

Caffeine, in addition to being a diuretic, has also been shown to be a bladder irritant for many people. It is very difficult for most coffee drinkers to completely eliminate their morning coffee, but many will be willing to reduce caffeine intake at least partially. Though there is little data on the role of sugar substitutes in incontinence, there are clinical cases in which these substances appear to be aggravating incontinence, and reduction has provided clinical improvement.

Weight Loss

Obesity is a common health problem that has also been established as a risk factor for urinary incontinence. Women with high body mass index are not only more likely to develop incontinence but they also tend to have more severe incontinence than women with lower body mass index.[13]

A small amount of literature exists showing significant improvement in continence status accompanying weight loss of 45–50 kg after bariatric surgery[14] and of as little as 5% weight reduction with conventional weight loss programs.[15] Because this is an achievable goal for many overweight or obese women, it is reasonable to recommend weight loss as part of a comprehensive program to treat incontinence in overweight women.

Smoking

Smoking, adversely affects the continence mechanism and aggravates stress urinary incontinence by reducing lung function and vital capacity and producing a chronic cough with a corresponding rise in abdominal pressure. Therefore, it can be helpful to interrupt this cycle by stopping smoking and directing treatment toward the chronic cough.

Bowel Management

Constipation and fecal impaction have been sited as contributory factors to urinary incontinence, particularly in institutionalized populations. In severe cases, fecal impaction can obstruct normal voiding, precipitate overflow incontinence, or be an irritating factor in OAB. Disimpaction can resolve symptoms for some women, but it can recur unless a bowel management program is implemented. Bowel management may consist of instructions in normal fluid intake and dietary fiber (or supplements) to maintain normal stool consistency and regular bowel movements. When these measures are not adequate, enemas can be used to stimulate a regular daily bowel movement. Enemas should be timed after a regular meal such as breakfast to capitalize on postprandial motility.

Nocturia

Urinary incontinence is often accompanied by nocturia, which is waking up at night to void. Although getting up once per night is widely regarded as normal, getting up 2 or more times can be very bothersome when it results in sleep disruption or daytime fatigue, or increases the risk of falls. One practical approach to nocturia is to restrict fluid for 3 to 4 hours before bedtime. In patients who retain fluid during the day and have nocturia caused by fluid mobilization at night, interventions focus on managing daytime accumulation of fluid. This can be facilitated by wearing support stockings, elevating the lower extremities in the late afternoon, or using a diuretic. For patients who are already taking a diuretic, nocturia can often be improved by altering the timing of the diuretic (so that most of the effect has occurred before bedtime) or by using a long-acting diuretic.

In addition to the medical management of nocturia, behavioral training for urge incontinence has also been shown to reduce nocturia.[16] The woman is instructed to use the urge suppression strategy when she wakes up at night. If the urge subsides, she is encouraged to go back to sleep. If after a minute or two the urge to void has not remitted, she should get up and void so as not to interfere unnecessarily with her sleep.

References

1. Consensus conference. Urinary incontinence in adults. JAMA. 1989;261:2685–2690.
2. Fantl JA, Newman DK, Colling J, et al. Urinary incontinence guideline panel. March 1996. Urinary incontinence in adults: acute and chronic management. Clinical practice guideline. Rockville, MD: Agency for Health Care Policy and Research, Public Health Service, US Department of Health and Human Services.
3. Locher JL, Roth DL, Goode PS, et al. Reliability assessment of the bladder diary for urinary incontinence in older women. J Gerontol: Med Sci. 2001; 56A:M32–M35.
4. Goode PS, Burgio KL, Locher JL, et al. Effect of behavioral training with or without pelvic floor electrical stimulation on stress incontinence in women: a randomized controlled trial. JAMA. 2003; 290:345–352.
5. Burgio KL, Pearce KL, Lucco A. Staying dry: a practical guide to bladder control. Baltimore: The Johns Hopkins University Press; 1989.
6. Miller JM, Aston-Miller JA, DeLancey OL. A pelvic muscle contraction can reduce cough-related urine loss in selected women with mild SUI. JAGS. 1998;46:870–874.

7. Burgio KL, Whitehead WE, Engel BT. Urinary incontinence in the elderly: bladder/sphincter biofeedback and toileting skills training. Ann Intern Med 1985;103:507–515.
8. Burton JR, Pearce KL, Burgio KL, et al. Behavioral training for urinary incontinence in elderly ambulatory patients. J Am Geriatr Soc. 1988;36:693–698.
9. Burgio KL, Locher JL, Goode PS, et al. Behavioral vs drug treatment for urge urinary incontinence in older women. A randomized controlled trial. JAMA. 1998;280:1995–2000.
10. Burgio KL, Goode PS, Locher JL, et al. Behavioral training with and without biofeedback in the treatment of urge incontinence in older women: a randomized controlled trial. JAMA 2002;288:2293–2299.
11. Frewen WK. Role of bladder training in the treatment of the unstable bladder in the female. Urologic Clinics of North America. 1979;6:273–7.
12. Fantl JA, Wyman JF, McClish DK, et al. Efficacy of bladder training in older women with urinary incontinence. JAMA 1991;265:609–613.
13. Brown J, Grady D, Ouslander J, et al: Prevalence of urinary incontinence and associated risk factors in postmenopausal women. Heart & Estrogen/Progestin Replacement Study (HERS) Research Group. Obstet Gynecol 1999;94:66.
14. Bump R, Sugerman H, Fantl J, et al. Obesity and lower urinary tract function in women: effect of surgically induced weight loss. Am J Obstet Gynecol 1992;166:392.
15. Subak LL, Johnson CEW, Boban D, et al. Does weight loss improve incontinence in moderately obese women? Intl J Urogyn 2002;13:40.
16. Johnson TM, Burgio KL, Goode PS, et al. Effects of behavioral and drug therapy on nocturia in older incontinent women. J Am Geriatr Soc. 2005;53(5):846–850.

4.2
Stress Urinary Incontinence

Jo Laycock

Key Messages

- The knowledge of pelvic anatomy and physiology and causes of incontinence are necessary to facilitate pelvic floor re-education.
- The assessment of pelvic floor muscles (PFMs) includes their co-ordination, contraction, timing, and function during coughing.
- If there is no voluntary contraction, then teaching must be directed toward increasing pelvic floor awareness.
- If there is a weak pelvic floor contraction, then a strengthening program is necessary, and if PFMs are easily fatigued, then an endurance program is needed.
- Progression in the pelvic floor reeducation program involves teaching the active use of PFM exercises and stress strategies, combined – if necessary – with drug therapy such as estrogen or duloxetine.

Introduction

Stress urinary incontinence (SUI) is the involuntary loss of urine associated with an increase in intraabdominal pressure, such as with coughing or other physical activity (see Chapter 1.5). Continence during raised intraabdominal pressure is attributable to an integrated system of muscles, fascia, ligaments, and neural control (see Chapters 1.1 and 1.2). The connection of the levator ani to the arcus tendineus fascia pelvis permits active contraction of the PFMs to elevate and support the anterior vaginal wall.[1] In addition, continence also requires a competent urethral sphincter mechanism. Evidence suggests that a voluntary PFM contraction is accompanied by synergistic contraction of the urethral sphincter mechanism,[2] indicating that both components of the continence mechanism should respond to muscle training.

Assessment

Stress urinary incontinence has several causes (see Chapters 1.3, 1.4, and 1.5). In many cases, it does not occur in isolation, and women present with coexisting urological symptoms, such as urge incontinence, frequency, elevated postvoid residual urine, or urinary tract infection. Consequently, a holistic assessment is needed to identify and address all symptoms. In addition to the standard history and physical examination (see Chapter 2.1), the assessment should include a pain assessment of the pelvis, lower back, and pelvic floor, as pain in these regions may inhibit PFM recruitment (see Chapter 3.1).

A woman with SUI may have a defect in the fascial supportive tissues, but still have an intact urethral sphincter and PFMs, or, conversely, she may have weak muscles with intact fascial supports, or a combination of these deficits. Furthermore, inadequate muscular activity and weakness may be caused by disuse, which can be addressed by PFM exercises. However, weakness caused by partial denervation or muscles torn from their attachment is less likely to respond to muscle training (Fig. 4.2.1).

Pelvic Floor Muscle Re-education

Pelvic floor muscle re-education is an effective, low-risk intervention that can reduce incontinence significantly in varied populations, and should be considered as a first approach. The principles underlying this therapy are discussed in Chapter 3.1, and the effectiveness of treatment is described in Chapter 4.3. The following

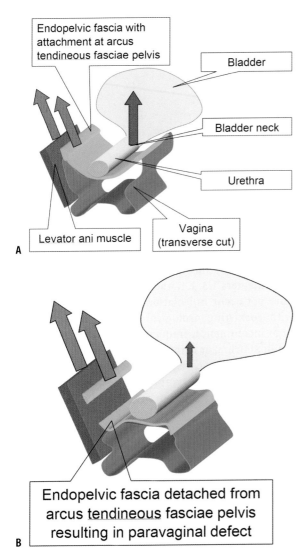

FIGURE 4.2.1. (A) With a sufficient pelvic floor contraction (orange arrows), the attached endopelvic fascia (via arcus tendineous fasciae pelvis; green) is elevated resulting in an elevation of the bladder neck (red arrow). (B) Detachment of the endopelvic fascia from the levator ani muscle precludes transmission of a pelvic floor contraction (orange arrows) onto the bladder neck (small red arrow).

step-by-step guide is recommended for the treatment of SUI.

Education

To optimize the usefulness of PFM training, the woman should understand the anatomy and physiology of the lower urinary tract.[3] This can be facilitated by the use of anatomical models and pictures. Realistic goal setting (depending on severity), and the need for the woman to participate actively in her treatment, is discussed. In this context, motivation and adherence to an exercise program lasting several months should be emphasized, and the consequences of noncompliance clearly stated. The importance of good posture and the avoidance of heavy lifting and activities that produce a great increase in intraabdominal pressure should be explained.

Many women find it difficult to exercise in general, and need to be convinced and motivated to adhere to their exercise regimen. Other women become excessive in their efforts and overexercise in the early days. Therefore, all women should be informed of the possibility of muscle soreness, which usually occurs the day after they begin. This soreness applies to both the PFMs and the lower abdominal muscles.

Many women will be wearing incontinence pads to avoid embarrassment and allow freedom of activities. Although necessary initially, the size and number of pads used should be monitored, and women should be encouraged to reduce pad use as symptoms improve.

All the above information can be provided in a group setting,[4,5] which can be both enjoyable and cost-effective, as well as enable women to realize that they are not alone with their problem. Finally, a simple handout explaining basic information should also be available to women.

PFM Assessment

Initially, the woman is told why the examination is to be carried out (to check whether she can contract the PFMs and how they perform), how the examination will be done (vaginal palpation), and alternative procedures if this is not acceptable (i.e. assessing the muscles by observation only or with a perineometer).

The periurethral muscles are palpated by placing the distal pad of the gloved, lubricated index finger on the anterior vaginal wall just inside the introitus and asking the woman to contract her muscles as if stopping urination, and then to relax the pelvic floor. The activity of the external urethral sphincter can thus be detected. The finger is moved to the right of the urethra

4.2. Stress Urinary Incontinence

and then to the left, and further contraction and relaxation requested. This should detect the contractility of the muscles lateral to the urethra (see Chapter 1.1). The resting tone of the muscles should also be noted because some women present with hypertonic PFMs caused by embarrassment, anxiety, or other factors (see Chapter 1.8). In this situation, a further contraction might not be possible or detectable, and an inexperienced clinician might misinterpret this as a weak or no contraction.

The index finger is then flexed and placed in the 8 o'clock and then the 4 o'clock positions, to enable palpation of the levator ani at rest and during a voluntary contraction. Once voluntary contractions have been established, a simple patient-specific exercise program is developed, depending on muscle strength and endurance.[6] (See Chapter 3.1 on Concepts of Neuromuscular Rehabilitation and Pelvic Floor Muscle Training).

While palpating the levator ani, the woman is also asked to cough, and the examiner monitors for the presence of cocontraction of the PFMs. The woman should then be instructed to tighten the PFM before and during a cough. Whether the woman is able to contract her muscles before the cough and hold this contraction should be assessed. Because many women have never used the PFMs during stress situations, this gives them immediate feedback, especially if urine leakage was avoided. For the therapist, the response can be utilized to guide the training program. In addition, this information will reveal those women who have a good PFM contraction with an initial elevation of the bladder neck, but descent of the urethra during coughing, while the levator ani is still contracted. This may be caused by fascial detachment, in which case the chance of success seems low, although there are no studies to corroborate this.

Initial Exercise Program

If the woman cannot produce a voluntary contraction or produces simply a flicker (grade 1 contraction), the techniques in Table 4.2.1 may be helpful. Women who can perform a voluntary PFM contraction of grade 2 or higher on the modified Oxford Scale are ready to embark on a muscle strength training program. If the maximum voluntary contraction is weak, then the woman is initially encouraged to squeeze harder and harder over the following weeks, to increase muscle awareness and improve contraction

TABLE 4.2.1. Techniques to assist voluntary contraction of the pelvic floor muscles

1. Apply firm pressure on the muscles (through the vaginal wall) to stretch the muscles and increase awareness. Many women are not aware of their PFMs and cannot locate them. Muscles often work better if stretched.
2. Use biofeedback via surface electromyography with a vaginal electrode (e.g. Periform™; Fig. 4.2.2) or vaginal or anal manometry to show the muscles working, however small the response. This helps the woman become aware of, and concentrate on, the appropriate muscles. It provides the feedback needed for operant conditioning (trial and error learning) and can inspire the woman to greater effort.
3. Teach contraction of transversus abdominis (TrA) (see Chapter 3.1). This may help to recruit the PFMs. The woman then tries to "join in" with her PFMs. Again, biofeedback is helpful. If biofeedback is not available, verbal feedback based on digital palpation is used to re-enforce the correct response. At this stage, the woman is requested to perform only short, gentle contractions. All too often, a woman will try too hard and use muscles not connected with the pelvic floor (e.g. tightening the shoulders or holding their breath). Exercising on an exercise ball or inflated cushion will introduce the importance of integrated muscle work for stability. Again, biofeedback can be used concurrently to demonstrate the effect on the PFMs.
4. If there is still no response, electrical stimulation can be used, providing there are no contraindications. This can be done with the Periform electrode (Fig. 4.2.2), as the movement of the indicator will help the woman to understand the action of a PFM contraction. The current is gradually increased until the indicator moves downwards (posteriorly). With subsequent electrically elicited contractions, the woman is instructed to "join in" and see the indicator move down further. The author uses the following electrical parameters:
   a. Frequency: 35 Hz
   b. Pulse-width (phase duration): 250 μs
   c. Duty cycle: 5 s on/10 s off
   d. Current intensity: Sufficient to produce maximum posterior movement of the indicator. The woman is encouraged to take the maximum current intensity that she can tolerate.

Alternate electrical stimulation and biofeedback is useful to re-enforce the correct action. The use of a home muscle stimulator, with the woman "joining in" with each contraction, may help to activate the muscles. Once the woman has located her PFMs, she should perform several voluntary contractions (i.e. without the stimulator) every hour, until she is ready for an exercise program.

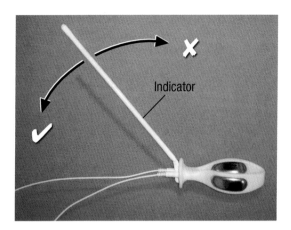

FIGURE 4.2.2. Periform vaginal electrode with indicator. A correct voluntary contraction of the PFMs produces downward (posterior) movement of the indicator; an incorrect voluntary muscle contraction or cough produces upward (anterior) movement of the indicator.

strength. If the woman demonstrates fatigue (short hold time and few repetitions), then the number of repetitions is increased and longer submaximal contractions are prescribed. The goal is to eventually hold a maximum voluntary contraction (MVC) for 10s and repeat 10 times, while breathing normally. All exercises should be done sitting and standing, or lying if the muscles are very weak.

Published trials have shown successful treatment of SUI by PFM exercise using a variety of exercise regimens. Therefore, it is advantageous to negotiate an individual regimen tailored to the circumstances and needs of each woman. For example, a woman in full-time employment may be limited to doing her exercises before and after work. It is best to be realistic in planning the exercise program, as too difficult a schedule may result in noncompliance.

To satisfy overload, it is necessary to evaluate the strength of a MVC, the endurance (hold time), and the number of repetitions the woman is capable of performing before the muscle output drops to below 50% of the MVC. This can be done via the aforementioned digital palpation using the P.E.R.F.E.C.T. scheme[6] or a perineometer (manometric or electromyographic). To keep the home exercise program simple, one should concentrate initially on either strength or endurance, whichever is the weakest component. For example, a woman with a weak PFM contraction, which she can hold for 5s and repeat 6 times, would initially concentrate on increasing strength by performing maximum short, 3s contractions. However, a woman with a strong MVC and poor endurance, e.g. 2-s hold and repeated 3 times, should concentrate initially on increasing endurance (hold time) and then the number of repetitions. This is best done by encouraging submaximal contractions and gradually increasing the hold time to 10s. On subsequent visits to the clinician, the muscle contractility is assessed and the exercise program advanced to satisfy overload.

Many women have difficulty in maintaining a normal breathing pattern during voluntary PFM exercises. When this occurs, biofeedback is a useful tool to help both the woman and the clinician to understand how the PFMs are behaving. First the woman is asked to breathe in; then, as she breathes out, she is instructed to gently tighten the PFMs (and transversus abdominis). This contraction is held while continuing to breathe in and out. As the technique is learned, which may take several days or weeks, the woman is encouraged to perform a stronger PFM contraction while maintaining normal breathing.

Exercise Progression and Active Use of Muscles

As PFM control improves, contractions generally become stronger and longer. Once a woman can sustain a PFM contraction for 5s, fast, 1-s MVCs may be added to the exercise program. In addition, when breathing control has been established, the woman can progress to holding a PFM contraction while balancing on an exercise ball and/or walking. Finally, the woman is taught how to prevent urine loss using a voluntary contraction of the PFMs before and during a cough or any increase in intraabdominal pressure that has precipitated an incontinent episode. This technique has been called counter-bracing, "The Knack,"[7] and the "stress strategy."[8]

Review and Follow-Up

Although the woman follows a home practice regimen, it is best to have regular follow-up visits to ensure that she is progressing, and to encourage her to adhere to her program. The number of treatment sessions depends on resources and the woman's progress. Published trials have used various protocols, and there is no agreement amongst clinicians regarding the minimum or optimum number of treatment sessions required. One approach is to see the woman once per week for 4 weeks and once per month for the next 3 months, with a final review visit after another 2 months. This entails a total of 8 treatment sessions over a period of 6 months.

A less expensive method is to see the patient in a group of women with similar problems. In the group model, the initial session includes education (see Education), followed by an individual session for PFM assessment. After this, if the woman has a voluntary PFM contraction, she attends a series of group sessions to strengthen the PFMs over a 4-month period. This is followed by an individual review 2 months later to assess both symptoms and muscle strength.

If there is no response after 6 months of treatment, despite convincing compliance, then a different approach may be needed. Perhaps the addition of biofeedback or electrical stimulation (if not given previously) may be warranted. If the woman admits to not doing sufficient exercises, alternative treatment or management should be sought.

One such technique is the use of an intravaginal tampon-like device, or a urethral insert. The vaginal devices are generally easier to use; they have been developed to produce upward pressure around the bladder neck, reducing mobility. Several urethral inserts have been tested[9] but discomfort, hematuria, and urinary tract infections have questioned the long-term safety of these devices. An example of a vaginal device – the Conveen Continence Guard – is a shaped foam tampon with a plastic applicator. The tampons are available in three sizes, and before use they are soaked in water, which doubles their size. Twenty-six women tested the Continence Guard,[10] with 41% reporting continence with the device in situ, and a further 45% reported significant improvement. After a further 12 months use by 19 women, 68% were subjectively dry and 26% were improved; furthermore, 86% were objectively improved on repeat pad testing and there were no serious complications reported.[11] These data have been confirmed in a larger study ($n = 126$) which demonstrated that 75% were continent or improved.[12] In addition to reducing/eliminating daily urine loss in women with moderate to severe SUI, these tampons are useful for women who only leak urine during occasional physical activities, such as tennis, and many report that they prefer to use a special continence tampon than to wear an incontinence pad. The studies quoted suffer from several problems, including the lack of blinding, lack of randomized design, and use of patients as their own controls; thus, further research is needed to ensure the safety and efficacy of these devices.

Pharmacological Therapy

Stress urinary incontinence is not a dichotomous condition; that is to say, a woman does not wake up one morning to find that she is suddenly leaking with every movement and every cough. In most cases, the incontinence begins with minor occasional loss of urine and increases in frequency and volume until it becomes a problem and she wishes treatment. Thus, there is a rationale for step-wise treatment of stress incontinence with medication and pelvic floor education, increasing to higher-risk treatments, such as surgery, only for those whose incontinence progresses in severity despite other treatments.

Alpha-Agonists

The rationale for the pharmacological treatment of stress urinary incontinence has focused on increasing the smooth muscle and striated sphincter muscle of the urethral closure mechanism. Alpha-agonists are a class of drugs that stimulate and increase the smooth muscle tone in all parts of the body. Phenylpropanolamine has the best data, and was found to cure SUI in 0–14% and reduce SUI in a further 19–60% of subjects (summary data from eight randomized controlled trials [RCTs], AHCPR 1005).[13]

The Cochrane Collaboration recently reviewed 15 randomized trials of alpha-agonists, including phenylpropanolamine (11 trials), midodrine (2 trials), and clenbuterol (2 trials), and found only weak evidence to suggest that adrenergic agonists were better than placebo. However, phenylpropanolamine was removed from the market in 2000 after it was recognized that there was an increased risk for hemorrhagic stroke.[14] Although the use of alpha-agonists for SUI is theoretically attractive, we await the development of more "uroselective" drugs.

Hormone Therapy

Estrogen therapy has been suggested for SUI because estrogen receptors do exist in the urethra. Thus, the hormone should improve mucosal coaptation, increase vascular tone, and maintain collagen content of the urethra. The Cochrane Collaboration recently reviewed estrogen therapy for incontinence. The analysis is made difficult by the small number of women included in trials specifically designed to look at incontinence, and the small number of RCTs with placebo. They concluded that estrogen had a positive effect on UI overall, with 50% cure or improvement seen in the 28 clinical trials compared to 25% seen with placebo, but that the effect was greater for urge than for stress incontinence. However, these conclusions are based on 374 women on estrogen versus 344 on placebo, including some studies with as few as 16 subjects in each arm. Two large American studies found that hormone replacement (HR) increased the risk of stress incontinence, but both were designed to study HR in relation to heart disease. Jackson and colleagues analyzed data from a large cohort of 1,584 white and black women, aged 70–79 years, and found a two-fold increased risk of stress incontinence in estrogen users (odds ratio (OR) 2.0, 95% confidence interval (CI) 1.3–3.1).[15] In a large RCT of combined estrogen/progesterone HR in over 2,700 women less than 80 years of age, incontinence improved in 26% of the women assigned to placebo compared with 21% of those on HR. Incontinence worsened in 27% of the placebo group versus 39% of the HR group.[16]

A major difficulty with the Cochrane analysis is that so few studies of topical estrogen therapy versus placebo were available. Now that it would appear that systemic HR is not useful to treat incontinence, more data are needed about the efficacy of vaginal estrogens. Most continence clinicians use local estrogens to improve atrophic symptoms, so that evidence is urgently needed regarding the effect of this therapy upon stress incontinence. The Cochrane report does not provide this subset analysis.

Tricyclic Antidepressants

Imipramine is a tricyclic antidepressant that has been used for many years in the treatment of stress and mixed incontinence. It has systemic anticholinergic side effects, which are not bladder specific, and also seems to inhibit the reuptake of norepinephrine and serotonin. There are no RCTs, and two open studies have demonstrated mild positive effects.[17,18] Imipramine should be used with caution in the elderly because of its well-known potential for arrhythmias.[19]

Selective Serotonin-Norepinephrine Reuptake Inhibitors

Duloxetine is a potent inhibitor of norepinephrine and serotonin reuptake that has been investigated for the treatment of stress incontinence.[20] In theory, both duloxetine and imipramine increase the activity of the urethral rhabdosphincter at the spinal cord level (Onuf's nucleus), thus, augmenting storage properties of the urethra without compromising emptying properties. The drug has been shown to have a moderate effect on improving stress incontinence when compared to placebo, regardless of the severity of the incontinence.[21,22] Duloxetine is the only drug approved for use in stress incontinence in Europe, but is not approved in the U.S.

Summary

There is a need for pharmacological treatment of stress urinary incontinence, as it offers an alternative to other conservative treatments that may be more acceptable in some individuals. However, only duloxetine has been shown to be effective in randomized trials, and is the only medication approved for use in stress incontinence.

References

1. DeLancey JOL, Starr RA. Histology of the connection between the vagina and levator ani muscles. J Reprod Med. 1990;35:765–771.
2. Bo K, Stein R. Needle EMG registration of striated urethral muscle wall and pelvic floor muscle activity patterns during cough, valsalva, abdominal, hip adductor and gluteal muscle contractions in nulliparous healthy females. Neurourol Urodyn. 1994;13:35–41.
3. Laycock J, Standley A, Crothers E, et al. Patient education and management. In: Clinical guide-

lines for the physiotherapy management of females aged 16 to 65 years with stress urinary incontinence. London: Chartered Society of Physiotherapy; 2001:19.
4. Bo K, Hagen RH. Pelvic floor muscle exercise in the treatment of female stress urinary incontinence; effects of two different degrees of pelvic floor muscle exercises. Neurourol Urodyn. 1990;9:489–502.
5. Demain S, Fereday Smith J, Hiller L, et al. Comparison of group and individual physiotherapy for female urinary incontinence in primary care. Physiotherapy 2001;87(5):235–242.
6. Laycock J, Jerwood D. 2001. Pelvic floor assessment: the P.E.R.F.E.C.T. scheme. Physiotherapy. 2001;87(12):631–642.
7. Miller JM, Ashton-Miller JA, DeLancey J. 1998. A pelvic muscle pre-contraction can reduce cough-related urine loss in selected women with mild SUI. J Am Geriatr Soc. 1998;46:870–874.
8. Burgio KL, Whitehead WE, Engel BT. Urinary incontinence in the elderly: bladder/sphincter biofeedback and toileting skills training. Ann Intern Med. 1985;103:507–515.
9. Tincello D. Prosthetic devices, inserts and plugs for the management of stress incontinence. In: Therapeutic management of incontinence and pelvic pain. Laycock J, Haslam J, editors. London: Springer-Verlag; 2002:91–94.
10. Thyssen H, Lose G. New disposable vaginal device (continence guard) in the treatment of female urinary stress incontinence. Design, efficacy and short term safety. Acta Obstet Gynecol Scand. 1996;75:170–173.
11. Thyssen H, Lose G. Long term efficacy and safety of a disposable vaginal device (continence guard) in the treatment of female urinary stress incontinence. Int Urogynecol J. 1997;8(133):130–132.
12. Hahn I, Milson I. Treatment of female urinary stress incontinence with a new anatomically shaped vaginal device (Conveen Continence Guard). Br J Urol. 1996;77:711–715.
13. Fantl JA, Newman DK, Colling J, et al. Urinary incontinence in adults: acute and chronic management. Clinical practice guidance, No. 2. 1996. Update; Rockville, MD: U.S. Department of Health and Human Services. Public Health Service, Agency for Health Care Policy and Research. AHCPR Publication No. 96-0682
14. Kernan WN, Viscoli CM, Brass LM, et al. Phenylpropanolamine and the risk of hemorrhagic stroke. N Engl J Med. 2000;343(25):1826–1832.
15. Jackson RA, Vittinghoff E, Kanaya AM, et al. Urinary incontinence in elderly women: findings from the Health, Ageing, and Body Composition Study. Obstet Gynecol. 2004;104(2):301–307.
16. Grady D, Brown JS, Vittinghoff E, et al. HERS Research Group. Postmenopausal hormones and incontinence: the Heart and Estrogen/Progestin Replacement Study. Obstet Gynecol. 2001;97(1):116–120.
17. Gilja I, Radej M, Kovacic M, et al. Conservative treatment of female stress incontinence with imipramine. J Urol. 1984;132(5):909–911.
18. Lin HH, Sheu BC, Lo MC, et al. Comparison of treatment outcomes for imipramine for female genuine stress incontinence. Br J Obstet Gynaecol. 1999;106(10):1089–1092.
19. Glassman AH. Cardiovascular effects of tricyclic antidepressants. Annu Rev Med. 1984;35:503–511.
20. Thor KB. Serotonin and norepinephrine involvement in efferent pathways to the urethral rhabdosphincter: implications for treating stress urinary incontinence. Urology. 2003;62(4 Suppl 1):3–9.
21. Dmochowski RR, Miklos JR, Norton PA, et al. Duloxetine versus placebo for the treatment of North American women with stress urinary incontinence. J Urol. 2003;170(4 Pt 1):1259–1263.
22. van Kerrebroeck P, Abrams P, Lange R, et al. Duloxetine versus placebo in the treatment of European and Canadian women with stress urinary incontinence. BJOG. 2004;111(3):249–257.

4.3
Evidence for the Effectiveness of Pelvic Floor Muscle Training in the Treatment and Antenatal Prevention of Female Urinary Incontinence

E. Jean C. Hay-Smith and Kate H. Moore

Key Messages

- Pelvic floor muscle training (PFMT) can cure or improve incontinence symptoms and reduce voiding frequency, leakage frequency, and amount of leakage.
- Women of all ages with stress, urge, or mixed incontinence can all benefit from PFMT.
- Clinicians should provide the most intensive PFMT supervision available.
- Supervision can be provided in individual or group settings.
- There is no additional benefit of adding biofeedback (BF) in the initial PFMT program.
- Antepartum PFMT can prevent postpartum incontinence, especially in women with increased bladder neck mobility.

Introduction

The purpose of the first part of this chapter is to review the evidence for the effectiveness of PFMT as a treatment for urinary incontinence in women, and discuss the implications for clinical practice. It is questioned whether PFMT is better than no treatment or control treatment, whether one approach to PFMT is better than another, and what the factors are that might affect treatment outcome. The second part of this chapter reviews the evidence for the effectiveness of antepartum PFMT for the prevention and treatment of urinary incontinence. It is considered whether PFMT during pregnancy can prevent antepartum or postpartum urinary incontinence in women without prior symptoms, and whether PFMT during pregnancy is an effective treatment for existing urinary incontinence. Because questions about the effectiveness of an intervention are best addressed in randomized controlled trials (RCTs), the chapter summarizes the evidence from existing systematic reviews of RCTs of PFMT.

Treatment of Urinary Incontinence

This section summarizes evidence from existing systematic reviews[1,2] that recruited women with existing stress, urge, or mixed urinary incontinence and used a PFMT program (but not PFMT in conjunction with a scheduled voiding regimen) in one or more branches of the study.

Is PFMT Better than No Treatment or Control Treatments?

PFMT has been compared to no treatment in 5 published, peer-reviewed trials.[3-7] Single trials found that women who received PFMT had significantly fewer leakage episodes and voids in 24 hours.[5,6] Data from a variety of pad tests suggest that the proportion of women who were cured or improved was greater, or the mean amount of leakage was less, in the PFMT group.[3-5,7] Most of the training parameters suggest that the PFMT programs were intended to increase muscle strength or endurance, although in one trial the aim was to improve the timing of a contraction in response to a rise in intraabdominal pressure (Table 4.3.1).

Six trials compared PFMT with control treatments, including placebo drug,[8] sham electrical stimulation,[9] offer of an anti-incontinence device,[10] advice on use of incontinence pads,[11] and written information about bladder function and PFMT.[12,13] Pooled data suggested that patients who received PFMT were about one and a half times more likely to report that they were cured or improved with treatment and had significantly fewer leakage episodes in 24 hours (approximately 2 less every 3 days). The studies recruited younger women with urodynamic stress incontinence (mean ages between 40 and 50 years) and older women with detrusor overactivity or mixed urinary incontinence (average age 50 to 70 years). All of the PFMT programs had characteristics of strength and/or endurance training, and two included the advice to contract the pelvic floor muscles to prevent leakage with raised intraabdominal pressure or with urgency (Table 4.3.1).

In summary, the data favored PFMT and suggested that women who received PFMT were more likely to report that their symptoms were cured or improved, and that voiding frequency, leakage frequency, and amount of leakage were significantly reduced. Women of all ages with stress, urge, or mixed incontinence showed benefit. Effect size was greater in some trials than others, but with so little data, it was not clear if the differences reflected methodological rigor, the sample characteristics (e.g. diagnosis and age), the training program, or other factors.

Is One Approach to PMFT Better than Another?

Trials have investigated the differences in exercise parameters, the amount of contact with health care professionals, the type of supervision, and the addition of BF, intravaginal resistance, or a cue to exercise.

The two studies that addressed differences in exercise parameters compared strength with endurance training,[14] and motor learning/strength training with motor learning alone,[15] in women with stress urinary incontinence. Neither found significant differences between the groups for leakage episodes; nor were there differences in self-reported cure or improvement in the latter study. These studies challenge the idea that the primary purpose of PFMT should be to strengthen

TABLE 4.3.1. Description of PFMT programs

Trial	Diagnosis	PFMT program
Aksac et al. 2003	USI	10 VPFMC (5 second hold, 10 second rest), 3 times a day, daily for 8 weeks. Progressed to 10 second hold and 20 second rest.
Bo et al. 1999	USI	8 to 12 high intensity VPFMC (6 to 8 second hold, 6 second rest), 3 times a day, daily at home, and a weekly PFMT exercise class for 6 months.
Burgio et al. 1998 and 2002	DO ± USI	15 VPFMC (10 second hold), 3 times a day, daily for 8 weeks. VPFMC with activities likely to cause leakage (e.g. urgency).
Burns et al. 1993	USI ± DO	20 VPFMC (10 fast with 3 second hold, 10 sustained with 10 second hold), 4 times a day, daily for 8 weeks. Progressed by 10 VPFMC per set, to a maximum of 200 contractions per day.
Goode et al. 2003	USI ± DO	15 VPFMC (2 to 4 second hold, equal period of rest), 3 times a day for 8 weeks Progressed to 10 second hold and 10 second rest. VPFMC with activities likely to cause leakage (e.g. cough).
Henalla et al. 1989	USI	5 VPFMC (5 second hold), 5 times an hour, daily for 12 weeks.
Lagro-Janssen et al. 1991	SUI	10 VPFMC (6 second hold), 5 to 10 times a day for 12 weeks.
Miller et al. 1998	SUI	VPFMC before cough and held until abdominal wall relaxed, for one week.
Yoon et al. 2003	UI	30 VPFMC (strength and endurance) daily for 8 weeks.

DO, detrusor overactivity; PFMT, pelvic floor muscle training; SUI, stress urinary incontinence; UI, urinary incontinence; USI, urodynamic stress incontinence; VPFMC, voluntary pelvic floor muscle contraction.

the muscles, but the width of the confidence intervals in these trials did not rule out potentially important differences.

In two small trials, women with stress urinary incontinence received PFMT, but women in one study arm had much more contact with the supervising physiotherapist.[16,17] Pooled data found that women in the intensive supervision group were 4 times more likely to report improvement. Two further studies compared 3 to 4 sessions of individual PFMT instruction with standard care (e.g. information on PFMT during postnatal class) in postnatal women with urinary incontinence symptoms 3 months after delivery.[18,19] Pooled data showed women receiving individual instruction were about a third more likely to report continence at 12 months postpartum than women receiving standard care. One large trial compared group versus individual supervision of PFMT in women with symptoms of urinary incontinence.[20] There were no significant differences in the number of women reporting cure or the number of leakage episodes in 24 hours at neither 12 weeks nor 9 months later. In summary, it is possible that more supervisory contact with a health care professional is better than less, but it might not matter whether this contact is given individually or in a group setting.

Four trials investigated the addition of home-based[21] or home- and clinic-based biofeedback[7,22,23] to a PFMT program for women with stress urinary incontinence. Pooled data found no significant difference between the BF and non-BF groups for self-reported cure or improvement.[21,23] Five trials compared PFMT and clinic BF with PFMT alone in women with stress urinary incontinence[6,24–26] or urge urinary incontinence.[27] There was no significant difference between the BF and non-BF groups for self-reported cure in pooled data (from 2 trials) or leakage episodes in 24 hours (from 3 trials). From these data, it appeared that PFMT with adjunctive home or clinic-based BF was no more effective than PFMT alone in the short term.

Finally, there were too few data from the 2 small trials investigating the effect of intravaginal resistance[28] or electronic cue to exercise[29] to draw any conclusions about the effects of either approach.

Summary

There is evidence that PFMT is an effective treatment for female urinary incontinence. Women with stress, urge, or mixed incontinence seem to benefit. Because follow up data (not presented here) are sparse and difficult to interpret, the longer-term benefits are less clear.

The limited evidence defining the best PFMT program suggests that: (a) clinicians should provide the most intensive PFMT supervision possible within service constraints, (b) supervision can be provided in individual or group settings, (c) there is no additional benefit of adding BF in the initial PFMT program. Although Table 4.3.1 describes a range of programs that have demonstrated effect, much more research is required to determine which type of exercise program or parameters are most beneficial, and for which groups of women. This review is restricted to RCTs, which provide the best evidence of effectiveness. In addition, there is also a large literature, which does not meet the strict criteria for this review, but which also helps establish the effectiveness of PFMT. Clearly, there is a growing body of evidence supporting PFMT and enough evidence to justify the recommendation that supervised PFMT is included in the treatment choices offered to women with urinary incontinence.

Antepartum Pelvic Floor Muscle Training to Prevent Postpartum Urinary Incontinence

This section summarizes the evidence from existing systematic reviews of RCTs of PFMT in pregnant women.[1,30,31] Studies that were included in these systematic reviews recruited pregnant women, with or without existing urinary incontinence, and used a PFMT program in one or more arms of the study.

There are three grades of prevention: primary (to remove the cause of a disease), secondary (to detect asymptomatic dysfunction and intervene to stop progression), and tertiary (treatment of existing symptoms to stop progression).[32] In this chapter, any study that recruited only pregnant women without existing urinary incontinence symptoms was considered to be a primary or secondary prevention study; these studies addressed the question of whether PFMT during pregnancy can prevent antepartum or postpartum incontinence. However, most studies to date have recruited pregnant women (some with and some without existing urinary incontinence), so they cannot answer questions about prevention. Although such studies have mixed prevention and treatment effects, they can address the question of whether PFMT during pregnancy reduces the prevalence of antepartum or postpartum urinary incontinence.

Can PFMT During Pregnancy Prevent Antepartum and/or Postpartum Urinary Incontinence in Women Without Prior Symptoms?

Three published, peer-reviewed trials compared supervised PFMT with standard antenatal care.[33–35] Although Mørkved and colleagues[33] and Sampselle and coworkers[35] included women with prior incontinence symptoms, data were also available for the subgroup of women without existing symptoms.

Reilly and colleagues[34] recruited primigravid women without a prepregnancy or current history of urinary incontinence, but with bladder neck hypermobility (more than 5 mm linear movement on perineal ultrasound) at 18 to 20 weeks of gestation. Mørkved et al.[33] and Sampselle et al.[35] also recruited primigravidae at 18 and 20 weeks gestation, respectively. The training parameters suggest that the purpose of the PFMT programs was to increase muscle strength, rather than endurance or the timing of a contraction (Table 4.3.2). In all three trials, the controls received usual antepartum care, which may have included advice on PFMT.

With regard to the effect of antepartum PFMT on preventing antepartum incontinence, one trial[34] found that women in the PFMT group were about half as likely to be incontinent at 36 weeks gestation. In contrast, another trial[35] found that urinary incontinence severity was not significantly different between the groups at 35 weeks.

Similarly, there was a mixed picture for the preventive effect of antepartum PFMT on postpartum incontinence. One study[34] found that women receiving PFMT were about half as likely to report urinary incontinence at 3 months postpartum, whereas another[33] did not find any significant difference. The third[35] found significantly less incontinence severity in the PFMT group at 6 weeks postpartum.

Medium- to long-term effects also varied. Sampselle et al. did not find any significant difference in incontinence severity between the groups at 6 or 12 months (the primary endpoint).[35] At four years, Reilly and colleagues[34] found that women from the training group were still less likely to report urinary incontinence, although only 100 of the original sample of 268 women responded.[36]

In summary, 3 trials of moderate to good quality, have addressed the effect of PFMT during pregnancy for prevention of antepartum and postpartum urinary incontinence. The findings suggest that women who are potentially at greater

TABLE 4.3.2. Description of antepartum PFMT programs*

Trial	Sample	PFMT program
Morkved et al. 2003	Primigravid women at 18 weeks gestation	8 to 12 near maximal VPFMC (6 to 8 seconds hold with 3 to 4 fast contractions at the end of each contraction, followed by 6 seconds rest), twice a day, and weekly exercise class (including PFMT) from weeks 20 to 36.
Reilly et al. 2002	Primigravid women, with bladder neck hypermobility, at 18 to 20 weeks gestation	8 to 12 VPFMC (6 second hold, 2 second rest), twice a day.
Sampselle et al. 1998	Primigravid women at 20 weeks gestation	Up to 30 near maximal VPFMC per day.

PFMT, pelvic floor muscle training; VPFMC, voluntary pelvic floor muscle contraction.
* Hughes et al does not appear here as their PFMT program was not described in the abstract.

risk of developing urinary incontinence (i.e. those with bladder neck hypermobility at 18 weeks of gestation) benefit from antepartum PFMT with half the risk of urinary incontinence at 3 months postpartum. This effect might last as long as 4 years. The findings from the other two trials are conflicting; it is not clear if antepartum PFMT reduces prevalence of urinary incontinence or incontinence severity (antepartum or postpartum) in pregnant women without incontinence symptoms at the onset of PFMT.

Can PFMT in Pregnant Women (Regardless of Continence Status) Reduce the Prevalence of Antepartum and/or Postpartum Urinary Incontinence?

Two published trials (see previous section),[33,35] and one that, so far, has only appeared as a published abstract of the International Continence Society,[37] were included. The trial by Hughes and co-workers recruited nulliparous women at 20 weeks gestation and randomized them to supervised antepartum PFMT (individual appointment with physiotherapist followed by one group session) or the usual community antenatal care.[37] There was no description of the PFMT program in the abstract.

Regarding the effect of PFMT during pregnancy on the prevalence of antepartum urinary incontinence, one trial[35] found that the PFMT group was approximately 30% less likely to experience urinary incontinence at 36 weeks gestation. Another[37] reported no statistically significant difference in the odds of stress, urge, or "spontaneous" incontinence (not associated with exertion or urgency) between the groups. The third[35] did not find any significant difference in incontinence severity at 35 weeks.

The effect on prevalence of postpartum incontinence also varied. One trial[33] reported that women in the PFMT group were approximately 40% less likely to be incontinent at 3 months postpartum, whereas another[37] did not find any difference in the odds of stress, urge, or unexplained urinary incontinence at 6 months postpartum. Analysis of unpublished data from the third sudy[35] found no significant differences in incontinence severity between the groups at 6 weeks, 6 months, or 12 months postpartum.

Of these three trials in samples of pregnant women (with and without incontinence symptoms when PFMT began), only one[33] found the prevalence of antenatal and postnatal incontinence to be significantly reduced in the women who received PFMT; the 95% confidence intervals suggested reduction in prevalence of 10–50% for antepartum incontinence and 10–60% for postpartum incontinence. Based on the description of PFMT, women in this positive trial had more regular contact with a health care professional while training because they attended a weekly exercise class in addition to PFMT at home. It might be that to reduce the prevalence of urinary incontinence in antepartum and postpartum women, intensively supervised programs and/or regular exercise classes, in addition to home training, are necessary.

Conclusion

Based on the limited evidence to date, it seems that antepartum PFMT can prevent postpartum urinary incontinence (at 3 months) in primiparous women with increased bladder neck mobility but no incontinence symptoms. This effect might last as long as 4 years. Furthermore, an antepartum PFMT program comprising a weekly exercise class, in addition to home training, might reduce the prevalence of antepartum (at 36 weeks) and postpartum urinary incontinence (at 3 months).

Thus, clinicians who are caring for childbearing women should consider first whether their current approach is targeting women who might benefit most (i.e. women with bladder neck hypermobility at 18 weeks of pregnancy, but no urinary incontinence), and second, whether it is sufficiently intensive.

References

1. Wilson PD, Hay-Smith J, Nygaard I, et al. Adult conservative management. In: Incontinence. Vol. 2. Management, Abrams P, Cardozo L, Khoury S, Wein A, editors. Health Publication, Ltd.; 2005: 855–964.

2. Hay-Smith EJC, Bo K, Berghmans LCM, et al., 2001. Pelvic floor muscle training for urinary incontinence in women. (Cochrane Review) In: The Cochrane Library, Issue 1. Update Software: Oxford.
3. Henalla SM, Hutchins CJ, Robinson P, et al. Nonoperative methods in the treatment of female genuine stress incontinence. J Obstet Gynaecol. 1989;9:222–225.
4. Miller JM, Ashton-Miller JA, DeLancey JOL. A pelvic muscle precontraction can reduce cough-related urine loss in selected women with mild SUI. J Am Geriatr Soc. 1998;46:870–874.
5. Yoon HS, Song HH, Ro YJ. A comparison of effectiveness of bladder training and pelvic muscle exercise on female urinary incontinence. Int J Nurs Stud. 2003;40:45–50.
6. Burns PA, Pranikoff K, Nochajski TH, et al. A comparison of effectiveness of biofeedback and pelvic muscle exercise treatment of stress incontinence in older community dwelling women. J Gerontol. 1993;48:M167–M174.
7. Aksac B, Aki S, Karan A, et al. Biofeedback and pelvic floor exercises for the rehabilitation of urinary stress incontinence. Gynecol Obst Invest. 2003;56:23–27.
8. Burgio KL, Locher JL, Goode PS, et al. Behavioral vs. drug treatment for urge urinary incontinence in older women: a randomized controlled trial. JAMA. 1998;280:1995–2000.
9. Hofbauer VJ, Preisinger F, Nurnberger N. Der stellenwert der physikotherapie bei der weiblichen genuinen streß-inkontinenz. Zeitschrift fur Urologie und Nephrologie. 1999;83:249–254.
10. Bo K, Talseth T, Holme I. Single blind, randomised controlled trial of pelvic floor exercises, electrical stimulation, vaginal cones, and no treatment in management of genuine stress incontinence in women. BMJ. 1999;318:487–493.
11. Lagro-Janssen TLM, Debruyne FMJ, Smits AJA, et al. Controlled trial of pelvic floor exercises in the treatment of urinary stress incontinence in general practice. Br J Gen Pract. 1991;41:445–449.
12. Burgio KL, Goode PS, Locher JL, et al. Behavioral training with and without biofeedback in the treatment of urge incontinence in older women: a randomized controlled trial. JAMA. 2002;288:2293–2299.
13. Goode PS, Burgio KL, Locher JL, et al. Effect of behavioral training with or without pelvic floor electrical stimulation on stress incontinence in women: a randomized controlled trial. JAMA. 2003;290:345–352.
14. Johnson VY. Effects of a submaximal exercise protocol to recondition the pelvic floor musculature. Nurs Res. 2001;50:33–41.
15. Hay-Smith EJC. Pelvic floor muscle training for female stress urinary incontinence. Dunedin School of Medicine. Dunedin, New Zealand: University of Otago; 2003.
16. Bo K, Hagen RH, Kvarstein B, et al. Pelvic floor muscle exercise for the treatment of female stress urinary incontinence: III. Effects of two different degrees of pelvic floor muscle exercise. Neurourol Urodyn. 1990;9:489–502.
17. Wilson PD, Al Samarrai T, Deakin M, et al. An objective assessment of physiotherapy for female genuine stress incontinence. Br J Obst Gynaecol. 1987;94:575–582.
18. Glazener CM, Herbison GP, Wilson PD, et al. Conservative management of persistent postnatal urinary and faecal incontinence: randomised controlled trial. BMJ. 2001;323:593–596.
19. Wilson PD, Herbison GP. A randomized controlled trial of pelvic floor exercises to treat postnatal urinary incontinence. Int Urogynecol J Pelvic Floor Dysfunct. 1998;9:257–264.
20. Janssen CC, Lagro-Janssen AL, Felling AJ. The effects of physiotherapy for female urinary incontinence: individual compared with group treatment. BJU Int. 2001;87:201–206.
21. Shepherd A, Montgomery E, Anderson RS. A pilot study of a pelvic exerciser in women with stress incontinence. J Obstet Gynaecol. 1983;3:201–202.
22. Aukee P, Immonen P, Penttinen J, et al. Increase in pelvic floor muscle activity after 12 weeks' training: a randomized prospective pilot study. Urology. 2002;60:1020–1023.
23. Mørkved S, Bo K, Fjørtoft T. Effect of adding biofeedback to pelvic floor muscle training to treat urodynamic stress incontinence. Obstet Gynecol. 2003;100:730–739.
24. Berghmans LCM, Frederiks CMA, de Bie RA, et al. Efficacy of biofeedback, when included with pelvic floor muscle exercise treatment, for genuine stress incontinence. Neurourol Urodyn. 1996;15:37–52.
25. Glavind K, Nohr SB, Walter S. Biofeedback and physiotherapy versus physiotherapy alone in the treatment of genuine stress urinary incontinence. Int Urogynecol J. 1996;7:339–343.
26. Pages I, Jahr S, Schaufele MK, et al. Comparative analysis of biofeedback and physical therapy for treatment of urinary stress incontinence in women. Am J Physical Med Rehabil. 2001;80:494–502.
27. Wang AC, Wang Y-Y, Chen M-C. Single-blind, randomized trial of pelvic floor muscle training,

biofeedback-assisted pelvic floor muscle training, and electrical stimulation in the management of overactive bladder. Urology. 2004;63:61–66.
28. Ferguson KL, P.L. M, Bishop KR, et al. Stress urinary incontinence: Effect of pelvic muscle exercise. Obstet Gynecol. 1990;75:671–675.
29. Sugaya K, Owan T, Hatano T, et al. Device to promote pelvic floor muscle training for stress incontinence. Int. J Urol., 2003;10:416–422.
30. Harvey MA. Pelvic floor exercises during and after pregnancy: a systematic review of their role in preventing pelvic floor dysfunction. JOGC. 2003;25:487–498.
31. Hay-Smith J, Herbison P, Morkved S, 2002. Physical therapies for prevention of urinary and faecal incontinence in adults. (Cochrane Review.) In: The Cochrane Library. Issue 3. Update Software: Oxford.
32. Hensrud DD. Clinical preventive medicine in primary care: background and practice: 1. Rationale and current preventive practices. Mayo Clinic Proceedings. 2000;75:165–172.
33. Mørkved S, Bø K, Schei B, et al. Pelvic floor muscle training during pregnancy to prevent urinary incontinence: a single-blind randomized controlled trial. Obstet Gynecol. 2003;101:313–319.
34. Reilly ET, Freeman RM, Waterfield MR, et al. Prevention of postpartum stress incontinence in primigravidae with increased bladder neck mobility: a randomised controlled trial of antenatal pelvic floor exercises. BJOG. 2002;109:68–76.
35. Sampselle CM, Miller JM, Mims BL, et al., Effect of pelvic muscle exercise on transient incontinence during pregnancy and after birth. Obstet Gynecol. 1998;91:406–412.
36. Udayakankar V, Steggles P, Freeman RM, et al. Prevention of stress incontinence by antenatal pelvic floor muscle exercises (PFE) in primigravidae with bladder neck mobility: A four year follow up [Abstract 219]. Int Urogynecol J. 2002;13:S57–58.
37. Hughes P, Jackson S, Smith P, et al. Can antenatal pelvic floor exercises prevent postnatal incontinence [Abstract 49]. Neurourol Urodyn. 2001;20:447–448.

4.4
Postpartum Management of the Pelvic Floor

Pauline E. Chiarelli

Key Messages

- Continence promotion in the postpartum period involves a multicomponent approach.
- Postpartum perineal management includes wound care, preparing for the first bowel movement, and minimizing perineal descent.
- Pelvic floor muscle (PFM) rehabilitation should be routine for all postpartum women, and should be tailored for those with incontinence.
- Continence promotion includes patient education, good bladder habits, PFM assessment and exercises, compliance aids, and follow-up contact.

Introduction

One of the most socially devastating sequelae of vaginal delivery is PFM damage and the subsequent loss of bladder or bowel control. The expulsive forces of the fetal head as it descends through the birth canal lead to overstretching of both nerve and muscle tissues, which do not always undergo full recovery. Such trauma can result in symptoms of urinary incontinence or anal sphincter incompetence, including incontinence of feces or flatus and fecal urgency.[1] Symptoms such as these are both embarrassing and debilitating, and although they may be transient following a first vaginal delivery, subsequent deliveries are likely to cause further deterioration.[2] Continence promotion in the postpartum period involves a multicomponent, multidisciplinary approach, which is described in this chapter.

Postpartum Perineal Management

In addition to trauma affecting the deeper structures of the pelvic floor, many women suffer painful trauma to the superficial structures of the perineum. Evidence supporting the morbidity associated with perineal repair includes wound breakdown, which is reported in 6% of women while in hospital and 12% at home up to 8 weeks postpartum.[3] Other morbidity, such as mild or moderate edema at 3 to 4 days postpartum, has been reported in 59% of women.[4]

In the presence of perineal discomfort, PFM contraction can be painful, and there is little likelihood that women might feel motivated to undertake PFM exercises routinely in the postpartum. Pain is known to inhibit muscle contraction and alter the timing of the onset of reflex muscle contractions.[5] In view of the prevalence of perineal trauma (sutured or otherwise) and its perceived impact on PFM exercise, it is reasonable to routinely inform women about the best management practices related to postpartum perineal trauma. Suggested care practices are shown in Table 4.4.1.

The discomfort of perineal suturing and PFM trauma might also have wider implications for women in the postpartum, and information about constipation and perineal support to minimize perineal descent during defecation might need to be included within any postpartum continence promotion program.

TABLE 4.4.1. Outline of suggested postpartum perineal care practices

Suggested Practice	Practice Details
Wound care Familiarization Use a hand mirror to view the perineum together with the woman	Point out: • the incision line • any visible sutures • lacerations • bruising • hemorrhoids Does the woman admit to the presence of perineal swelling?
Wound hygiene	Soap and water as preferred wound hygiene
Management of swelling /bruising	Rinse down following urination or bowel motion while sitting on the toilet. Suggest the use of a water-filled sports drink bottle. Pat dry with toilet paper or tissue. *Cryotherapy* Ice preferred and applied: • in a dampened cover • for maximum of 20 minutes • repeated fourth hourly *Pelvic floor muscle exercise* • Explain their role in reducing swelling (muscle pump action) • exercises started after 24 hours • performed within the limits of discomfort
Preparing the woman for her first postpartum bowel motion	Allay her fears related to "stitches bursting" during a bowel motion. Demonstrate the following while looking in the mirror: • gently support the perineum in an upward direction as she bears down to open her bowels. Advise her to cover her hand with paper tissue or a pad. • Her fingers can be parted to accommodate any suture line.
Management of hemorrhoids where present	Explain the need to protect any visible hemorrhoids from laceration during defecation using the above technique.
Minimizing perineal descent: Avoid constipation	Inquire about the: • time since her last bowel motion • usual frequency of her bowel motions • current fluid intake If the normal time between bowel motions has

4.4. Postpartum Management of the Pelvic Floor

TABLE 4.4.1. *Continued*

	been exceeded encourage her to seek such help as is routine/standard laxative care practice within the postnatal unit. Explain: • likely effect of increased fluid loss on her bowels • the need to increase her fluid intake • fluid loss is from the combined effects of lochia, breast milk production, and night sweats.
Use of correct toileting position	 The Body Position and Functional Pattern for Effective Defecation
Teach "The Knack" Explain the protective effect of precontracting the PFMs before coughing, sneezing or lifting	Using the mirror, check that the woman can contract her PFMs in an upward movement, away from the mirror. To activate TrA ask the woman to pull her belly button and bikini line toward her backbone. Teach cocontraction of TrA with PFMs using the following suggested instructions: "Breathe in, and as you let your breath out, pull up your PFM and pull in your belly button toward your backbone. Don't hold your breath." Demonstrate the effectiveness of the knack by asking the woman to cough. This is likely to be quite uncomfortable. Follow by asking the woman to precontract PFM and TrA and to repeat the cough. Advise the woman to use the knack before coughing or sneezing, as well as using it before ANY form of lifting, especially lifting the baby.

Postpartum PFM Rehabilitation

It is widely held that healthy PFMs make an important contribution to continence, and pelvic floor exercises (PFXs) should have top priority within any postpartum continence promotion program. Continence promotion should routinely include PFX for any woman who has had a baby. For women identified as having incontinence of urine, feces, or flatus, the PFX program should be individually tailored.

In view of the enormous demands and time constraints placed upon new mothers, it is often unrealistic to expect regular attendance at postpartum classes. It is important that any exercise program be simple, quick, and incorporated into the activities of daily living. To encourage widespread routine adoption of PFX by postpartum women, the exercise program needs to be carried out by the women unsupervised, without exercise adjuncts (such as cones, biofeedback, or electrical stimulation), and without ongoing input from healthcare professionals. Although many protocols include such adjuncts, they have not been shown to be necessary components of treatment.

An evidence-based protocol for the rehabilitation of the muscles of the pelvic floor following childbirth is presented in Table 4.4.2.[6-8]

Components of a Successful Postpartum Continence Promotion Program

Although PFXs are an important part of postpartum continence promotion, it is important to emphasize that continence promotion is more than PFX. Successful continence promotion is optimized by a comprehensive program. The author polled leading experts, including urologists, obstetricians, gynecologists, nurses, and physiotherapists, and designed such a program[9] (Fig. 4.4.1). A randomized controlled trial explored the continence promotion in postpartum women.[10] The components of this program were shown to be acceptable to women, and overall, the information components of the intervention were well received. Over 80% of women reported that they read the information booklet, and more than 30% used the stick-up dots as an aid to compliance. The intervention required 15–20 minutes to implement in the hospital and approximately 30 minutes for the postpartum intervention. It would appear that this postpartum continence promotion program is feasible for women who deliver in hospitals. In view of the costs to the healthcare system of female urinary incontinence, the costs of delivering this intervention might be seen as

TABLE 4.4.2. Summary of evidence-based exercise parameters for improving the functional capacity of striated pelvic floor muscles adapted for postpartum women

Exercise parameter	Suggestion supported by the literature	Adaptation for the postpartum intervention
Type of exercise	Isometric	Contraction of the PFMs with hold at end of range
Force	Maximum	Maximum or maximum against a cough
Number of repetitions per session	3 to 9	Starting number to be assessed from the per vaginam assessment of each woman, increasing as indicated by her own testing of her PFMs up to six repetitions
Length of hold	5 seconds	Starting length of hold to be assessed from the per vaginam assessment of each woman from one second or one hard cough up to five seconds or five hard, successive coughs
Rest between contractions	No	No
Speed	To address fast and slow twitch fibers	Squeeze and hold (slow) as well as squeeze, hold and cough (fast)
Specificity	Use the task usually performed by the muscle group	Addressed by using "the knack"
Sessions per day	One or more	Conducted individually with each woman, but aiming for two sessions per day
Sessions per week	Minimum of three	Daily to facilitate remembering

4.4. Postpartum Management of the Pelvic Floor

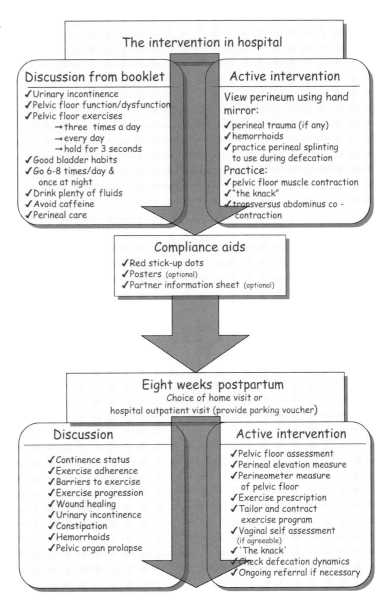

FIGURE 4.4.1. Components of the postpartum continence promotion program.

money well spent. In some cases, it might be necessary to prioritize interventions and limit them to members of a target group to achieve maximum impact with minimal resources.

Effectiveness of Postpartum Intervention

The evidence for effectiveness of PFM training during pregnancy has been explored in Chapter 4.3. Five studies relating to the efficacy of PFX for postpartum continence promotion examined the efficacy of PFX among postpartum women (not using exercise adjuncts) and measured continence outcomes.[11–15]

Three studies assessed urinary incontinence as well as PFM function as an outcome measure. Two studies found significant reductions in urinary incontinence postpartum, but no significant difference in measured PFM strength.[11,12] The study by Morkved and colleagues found significant improvement in both groups at 16 weeks postpartum.[13] Some improvement in measures of

PFM function were also noted in the control group (assumed to be the natural recovery of muscle function after childbirth), but measured changes in PFM function were significantly greater in the intervention group.

Chiarelli and colleagues adapted the protocol used by Bo,[8] and incorporated the precontraction of the PFMs in anticipation of a cough, as well as cocontraction of the transversus abdominis (TrA) and PFM during any activities likely to increase intraabdominal pressure (Table 4.4.1). This treatment consisted of a single visit by a physiotherapist to all eligible women (continent and incontinent). Women were given a continence promotion intervention, including a specifically designed information booklet. A single follow-up visit for further PFM assessment and exercise prescription followed. Although the program was effective in promoting PFXs and continence at three months postpartum,[10] there was no significant difference in continence status among the women in the intervention group at 12 months postpartum.[14] Further, continued adherence to PFXs at 12 months was predictive of continence at that time.

In another trial, incontinent postpartum women were seen initially at 3 months with follow-up visits at 6 months and 9 months postpartum. Significant improvements in continence status were noted, along with significantly higher reported performance of PFX.[15] Although one study tested continence promotion, while the other examined an intervention for incontinent women, the following observations might be made from these studies. First, extending the follow-up visits to 3, 6, and 9 months postnatally would seem more likely to provide consistent results in terms of continence. Second, PFX are significantly more likely to be performed at 12 months postpartum with even a minimal intervention, and continence status at 12 months is associated with the regular performance of PFX.

Adherence to PFX Programs

In the case of PFX programs, the most significant prognostic factor for success is the woman's adherence to the exercise program.[16] Studies of continence promotion programs show problems with adherence, with dropout rates of 15% for women who are not within the peripartum period,[16] and up to 52% for postpartum women.[12] Adherence to a treatment program cannot be assumed, and potential strategies for enhancing a woman's adherence to PFX regimens during and beyond the postpartum year must be considered an important component in any continence promotion program.

Studies of adherence to general exercise therapy show programs that include motivational strategies to be more effective. Key components of success include goal setting, negotiating treatment goals, formulating plans that are tailored to the individual needs of the woman's situation and daily routine, and finding specific events in a woman's daily routine to which program components might be anchored.[17] Lack of time is among the reasons given by postpartum women for difficulties in adhering to prescribed PFX protocols,[12] and women find it easier to adhere to programs that involve fewer lifestyle changes (Fig. 4.4.1).

Studies show that adherence levels deteriorate when any effective adherence-aiding strategy is withdrawn. Furthermore, women are more likely to comply with treatment programs that include specific positive-feedback program adherence. It is therefore important to check adherence at every consultation and maintain adherence interventions.

Integrating Continence Promotion into Routine Health Care

Continence promotion programs have been found to be acceptable to postpartum women.[17] The information components of the program are well utilized and there is good adherence with the recommended PFX regimen. Yet, in general, opportunities for continence promotion during pregnancy are not being utilized,[18] and studies of continence promotion programs show dropout rates of up to 52% for postpartum women.[12]

Because women's perceptions about their ability to contract these muscles are not always correct, well-designed PFX programs include vaginal preassessment of PFMs to ensure correct contraction.[19] Women appear to be receptive to such PFM function assessment in the postpartum period. The continence promotion study in postpartum women revealed that 74.5% of the women had attended their routine postnatal checkup and that 79.8% of these women had undergone a pelvic examination.[7] Moreover, few reported feeling embarrassed about PFM function testing during the examination.

Although opportunistic assessment of correct PFM contraction might seem to be a rational inclusion within a routine postnatal examination, only 27.5% of the women reported having their muscles tested in this manner.[7] Health professionals need to be informed and reassured that postpartum women are receptive to PFM function testing, and that postpartum visits present an ideal opportunity for such a procedure. Efforts should be made to ensure that all relevant healthcare professionals possess adequate knowledge about PFX.

Conclusion

There is little doubt that the conservative management of urinary and fecal incontinence is efficacious. Further, continence promotion among women in the early postpartum has been shown to prevent the development of urinary incontinence. Women benefit from education within a continence promotion program, and health care professionals should use the opportunity postpartum to assess the pelvic floor, to inform about pelvic floor dysfunction, and to encourage and instigate PFXs.

References

1. Sultan A, Kamm M, Hudson C, et al. Anal sphincter disruption during vaginal delivery. N Eng J Med. 1993;329:1905–1911.
2. Fynes M, Donnelly V, Behan M, et al. Effect of second vaginal delivery on anorectal physiology and faecal incontinence: a prospective study. Lancet. 1999;354:983–986.
3. Glazener C, Abdalla M, Stroud P, et al. Postnatal maternal morbidity: extent, causes, prevention and treatment. Br J Obstet Gynaecol. 1995;102:282–287.
4. Abraham S, Child A, Ferry J, et al. Recovery after childbirth: a preliminary prospective study. Med J Aust. 1990;152:9–12.
5. Petty N, Moore A. Function and dysfunction of muscle. In: Petty N, Moore A, editors. Principles of neuromusculoskeletal treatment and management. Edinburgh: Churchill Livingstone; 2004: 139–220.
6. Hay-Smith E, Bo K, Berghmans L, et al. Pelvic floor muscle training for urinary incontinence in women. The Cochrane Library 2003(1):Oxford: Update Software.
7. Chiarelli P. Female urinary incontinence in Australia: prevalence and prevention in postpartum women. [doctoral thesis]. Newcastle, Australia: University of Newcastle; 2001.
8. Bo K. Pelvic floor muscle exercise for the treatment of stress urinary incontinence: an exercise physiology perspective. Int Urogynecol J Pelvic Floor Dysfunct. 1995;6:282–291.
9. Chiarelli P. What the experts said: developing a postpartum continence promotion programme. Aust Continence J. 1999;5(2):41–42.
10. Chiarelli P, Cockburn J. Promoting urinary continence in women following delivery: randomised controlled trial. Br Med J. 2002;324(25 May):1241–1247.
11. Sampselle C, Miller J, Mims B, et al. The effect of pelvic muscle exercise on transient incontinence during pregnancy and after birth. Obstet Gynecol. 1998;91:406–412.
12. Wilson P, Herbison G. A randomised control trial of pelvic floor muscle exercises to treat postnatal urinary incontinence. Int Urogynecol J Pelvic Floor Dysfunct. 1998;9:257–264.
13. Morkved S, Bo K, Schei B, et al. Pelvic floor muscle training during pregnancy to prevent urinary incontinence: A single-blind randomised controlled trial. Obstet Gynecol. 2003;101(2):313–319.
14. Chiarelli P, Murphy B, Cockburn J. Promoting urinary continence in postpartum women:12 month follow-up data from a randomised controlled trial. Int Urogynecol J Pelvic Floor Dysfunct. 2004;15:99–105.
15. Glazener C, Herbison G, Wilson P, et al. Conservative management of persistent postnatal urinary and faecal incontinence: Randomized controlled trial. Br Med J. 2001;323:1–5.
16. Lagro-Janssen T, Debruyn F, Smits A, et al. Controlled trial of pelvic floor exercises in the treatment of urinary stress incontinence in general practice. Br J Gen Pract. 1991;9(3):284–289.
17. Hallberg J. Teaching patients self care. Nurs Clin North Am. 1970;5:223–231.
18. Chiarelli P, Campbell E. Incontinence during pregnancy. Prevalence and opportunities for continence promotion. Aust New Zealand J Obstet Gynaecol. 1997;37(1):66–73.
19. Morkved S, Bo K. The effect of postpartum pelvic floor muscle exercise in the prevention and treatment of urinary incontinence. Int Urogynecol J Pelvic Floor Dysfunct. 1997;8:217–222.

4.5
Role of a Perineal Clinic

Ranee Thakar

Key Messages

- The perineal clinic offers access to multidisciplinary care for a variety of postpartum problems, including dyspareunia, perineal pain, wound breakdown, and anal and urinary incontinence.
- Women who sustain obstetric anal sphincter injury should be seen within 3 months postpartum.
- Pregnant women with a prior injury can be evaluated and counseled regarding mode of delivery.

Introduction

More than 85% of women sustain perineal trauma,[1] and up to two thirds of women need suturing. This can have a devastating effect on family life and sexual relationships.[2] Perineal pain and discomfort affects up to 42% of women 10 days postpartum, and in 10% of women these problems persist at 18 months.[3] Fifty-eight percent of women experience superficial dyspareunia 3 months after delivery.[4] Urinary incontinence affects 32% of women 3 months postpartum,[5] although this improves in two thirds within 1 year.[6] Incontinence to flatus and feces affects 45% and 10% of women, respectively, within 3 months of delivery.[7]

Although there have been great advances in antenatal and intrapartum care, postnatal maternal morbidity has remained largely neglected. Encountering such unexpected problems after childbirth can leave women feeling inadequate and distraught. As these problems are usually of a sensitive nature, women should ideally be seen in a dedicated clinic instead of a busy general clinic. Furthermore, this environment facilitates childcare and breastfeeding. The establishment of a dedicated one-stop clinic enables provision of evidence-based quality care by experienced professionals. A dedicated perineal clinic also provides women an opportunity to be given an explanation of the circumstances under which the perineal injury occurred and counseled appropriately regarding modes of subsequent delivery. Women attending such a dedicated clinic have reported high satisfaction rates and found the service very valuable.

Structure of the Clinic

The perineal clinic should be staffed by a consultant urogynecologist (competent at urodynamics, anal manometry, and ultrasound) and a trained nurse/midwife. There should be easy access to a physiotherapist, a continence nurse specialist, colorectal nurse specialist, colorectal surgeons, and psychosexual counselor. Integration of multidisciplinary professionals promotes a holistic approach to pelvic floor and perineal problems. Furthermore, this clinic should accept self-referrals and should be easily accessible to general practitioners and midwives to allow fast tracking.

This clinic should be restricted to childbirth-related problems and include conditions such as dyspareunia, perineal pain, wound breakdown,

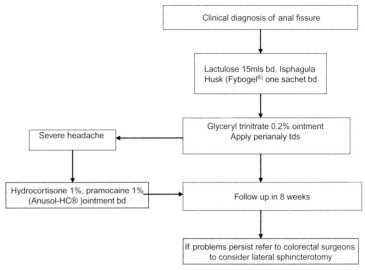

FIGURE 4.5.1. Pathway for management of anal fissure.

infection, prolapse, and anal and urinary incontinence. Women who have sustained obstetric anal sphincter injury (OASI) should be seen within 3 months postpartum. In addition, pregnant women with previous OASI can be evaluated and counseled regarding mode of delivery. Ideally, the clinic should be equipped with facilities for endoanal ultrasound scans and for anal manometry, to facilitate a one-stop approach.

Treatment for Conditions Encountered in the Perineal Clinic

Anal fissures, perineal pain, and OASI can be managed according to the pathways in Figures 4.5.1, 4.5.2, and 4.5.3.[8] All women with urinary and anal incontinence should be instructed to do pelvic floor muscle exercises and/or biofeedback

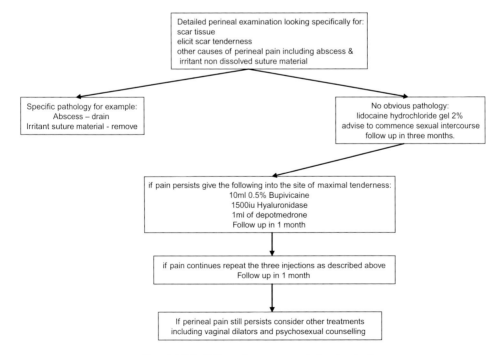

FIGURE 4.5.2. Pathway for management of perineal pain.

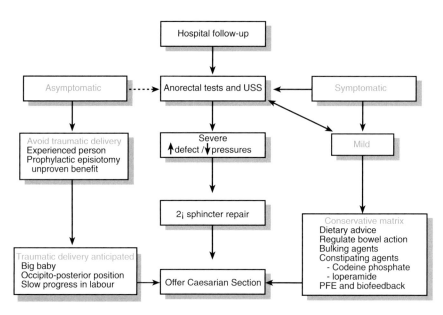

FIGURE 4.5.3. Pathway for management after 3°/4° anal tear.

should be initiated. Women with perineal wound breakdown should have careful inspection of their perineum, swabs taken, antibiotics given as appropriate, and advice given on perineal hygiene.

Our Experience

In Mayday University Hospital, Croydon, UK, we have a one-stop perineal clinic. Over a 21-month period, between September 2002 and May 2004, 249 new referrals were seen. The mean age was 30 years (range 16–42) with a median parity of 1 (range 1–5). The reasons for referral are shown in Table 4.5.1.

All women with anal fissures, perineal pain, and dyspareunia had resolution of symptoms when managed by protocol. On direct questioning, 7 (5.2%) with OASI had symptoms of anal incontinence. Twenty (15.6%) had ultrasound defects at follow-up. Nineteen (7.6%) women attending the clinic required directed pelvic floor muscle exercises and 24 (9.6%) had biofeedback. Five women have subsequently had surgery, two for urinary incontinence and prolapse, and 3 had secondary anal sphincter repairs.

TABLE 4.5.1. Reasons for referral

	n	%
OASI follow up	134	54.0
Dyspareunia	8	3.2
Perineal pain	12	4.8
Wound breakdown	2	4.8
Prolapse	9	3.6
Urinary incontinence	20	8.0
Vulval hematoma	3	1.2
Anal incontinence	16	6.4
Previous OASI	15	6.0
Vaginal discharge	7	2.8
Traumatic vaginal delivery	7	2.8
Miscellaneous	6	2.4

$n = 249$.

Conclusion

Postpartum problems are clearly an integral part of care of childbearing women, and such clinics should be available to mothers. This kind of service would lead to women being seen promptly and to the development of local expertise in managing perineal problems. More research is needed in the management of perineal problems, including the management of women with OASI in subsequent pregnancies, and a perineal clinic is an ideal setting for this.

References

1. Sleep J, Grant A, Garcia J, et al. West Berkshire Perineal Management Trial. BMJ. 1984;289:587–590.
2. Kettle C, Hills RK, Jones P, et al. Continuous versus interrupted perineal repair with standard or rapidly absorbed sutures after spontaneous vaginal birth: a randomised controlled trial. Lancet. 2002:359;2217–2223.
3. Carroli G, Belizan J, Stamp G. Episiotomy policies in vaginal births. (Cochrane Review). In: Cochrane Library. Issue 2. Oxford; Update Software; 1998. Updated quarterly.
4. Barrett G, E Pendry, J Peacock, C, et al. Women's sexuality after childbirth: A pilot study. Arch Sex Behav. 1999; 28(2):179–191.
5. Morkved S, Bo K, Schei B, et al. Pelvic floor muscle training during pregnancy to prevent urinary incontinence: A single-blind randomised controlled trial. Obstet Gynecol. 2003;101:313–319.
6. Arya L, Jackson N, Myers DL, et al. Risk of new-onset urinary incontinence after forceps and vacuum delivery in primiparous women. Am J Obstet Gynecol. 2001;185:1318–1324.
7. MacArthur C, Glazener CMA, Wilson PD, et al. Obstetric practice and faecal incontinence three months after delivery. BJOG. 2001;108:678–683.
8. Thakar R, Sultan AH. Management of obstetric anal sphincter injury. Obstet Gynaecol. 2003;5(2):72–78.

4.6
Overactive Bladder

Kathryn L. Burgio, Dudley Robinson, and Linda Cardozo

Key Messages

- Two main approaches to OAB are behavioral interventions and pharmacologic therapies.
- Bladder training focuses on reducing voiding frequency using incremental voiding schedules and techniques for coping with urgency.
- Behavioral training focuses on teaching patients a new response to urgency and urge suppression strategies, including how to use pelvic floor muscle (PFM) contractions to voluntarily suppress detrusor contraction.
- Pharmacological therapy includes oxybutynin, propiverine, tolterodine, trospium, solifenacin, darifenacin, imipramine, and desmopressin.
- Combining behavioral and drug therapy can optimize patient outcomes.

Introduction

Overactive bladder (OAB) is defined as urgency, with or without urge incontinence, usually with frequency and nocturia.[1] It is a common condition with a significant impact on the quality of life.[2] There are two primary approaches to the treatment of OAB; behavioral interventions and pharmacologic therapies. Both have been shown to produce significant reduction of OAB symptoms in large numbers of patients.

Behavioral Interventions

Historically, behavioral interventions have focused on the treatment of urge urinary incontinence, rather than the broader spectrum of OAB symptoms. However, over time they have also proven useful for reducing the symptoms of urgency, frequency, and nocturia. Several behavioral techniques have been described, including scheduled toileting (timed voiding, delayed voiding, bladder drill, bladder training, and prompted voiding), PFM training and exercise, biofeedback, use of PFMs for urethral occlusion, urge inhibition training, urge suppression strategies, urge avoidance, self-monitoring with bladder diaries, and fluid and diet modification. The goal of each of these methods is to reduce urinary incontinence by changing patient behavior or by teaching continence skills.

In clinical practice, the most successful behavioral programs appear to be multicomponent programs that combine various behavioral elements into a program individualized to the patient. There are two fundamentally distinct approaches to behavioral management of urge incontinence and OAB that incorporate one or more of these techniques: bladder training, which targets voiding habits, and behavioral training, which focuses on teaching patients new skills using methods such as PFM training and urge suppression strategies to control detrusor contraction.

Bladder Training

First described in the 1970s, bladder training has long been viewed as a standard behavioral treatment for urge incontinence.[3] The goal of bladder training is to modify bladder function by altering voiding habits. The training focuses on reducing

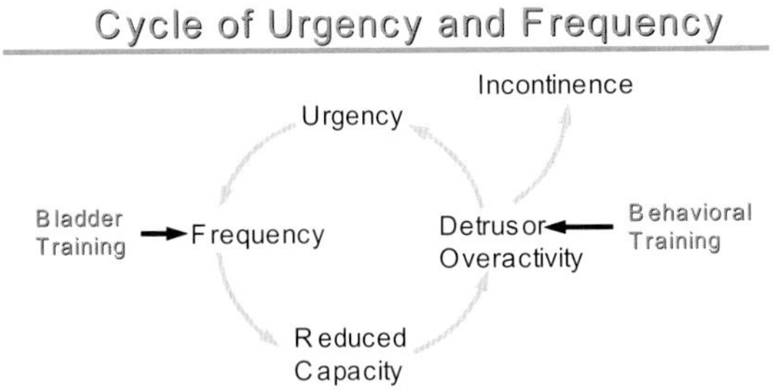

4.6. Overactive Bladder

Cycle of Urgency and Frequency

Bladder Training → Frequency
Urgency → Incontinence
Detrusor Overactivity ← Behavioral Training
Reduced Capacity

FIGURE 4.6.1. Cycle of urgency and frequency, depicting the two points where bladder training and behavioral training are thought to break the cycle. (*Source*: Burgio KL. Current perspectives on management of urgency using bladder and behavioral training. Supp J Am Acad Nurs Pract. 2004;16:4–7.)

ity are thought to contribute to detrusor overactivity. Uncontrolled detrusor contraction produces incontinence and contributes to urgency, completing a self-perpetuating cycle of OAB symptoms.

Bladder training breaks the cycle of urgency and frequency using consistent incremental voiding schedules. It begins with the patient completing a voiding diary, which shows the clinician when and how often the patient is voiding. The diary is reviewed with the patient to determine the longest time interval between voids that is comfortable, and this baseline interval becomes the starting point. Patients are given instructions to void according to this schedule rather than in response to urgency. Their daily schedule includes voiding upon waking, each time the interval passes during the day, and just before bed. To comply with this regimen, patients must resist the sensation of urgency and postpone urination. To help the patient cope with their urgency, they can be instructed to use distraction or relaxation techniques or self-affirming statements to get them to the next scheduled voiding time. Over time, the voiding interval is increased at comfortable intervals to a maximum of every 3 to 4 hours. See Figure 4.6.2 for guidelines for implementing bladder training.

voiding frequency to increase bladder capacity and eliminate detrusor overactivity.

Figure 4.6.1 presents a model of OAB in which urgency is a sensation that drives frequency. Unlike the ordinary sensation of fullness, urgency is a strong sensation that is difficult to defer. In this model, urgency promotes the increased frequency, which over time, can lead to reduced bladder capacity. In turn, the habit of frequent voiding and a smaller functional bladder capac-

	Guidelines for Bladder Training
Step 1:	Review bladder diary with the patient and identify varying voiding intervals.
Step 2:	With the patient, select the longest voiding interval that is comfortable for her.

Step 3:	Provide written patient instructions:
	Empty your bladder…
	  o First thing in morning
	  o Every time your voiding interval passes during the day
	  o Just before you go to bed

Step 4:	Teach patient coping strategies that she can use when she has an urge to void before her interval has passed: ➢ Self-statements (e.g., "I can wait until it is time to go.") ➢ Distraction to another task ➢ Deep breathing and relaxation ➢ Urge suppression using pelvic floor muscle contraction
Step 5:	Gradually increase the voiding interval… ➢ When patient is comfortable on her schedule for at least 3 days ➢ By 30-minute intervals, more or less as determined by the patient's confidence and clinician judgment

FIGURE 4.6.2. Guidelines for bladder training.

The effectiveness of bladder training for reducing incontinence has been demonstrated in several studies.[4–6] The most definitive is a randomized controlled trial demonstrating in older women an average 57% reduction in incontinent episodes and 54% reduction in the quantity of urine loss.[6]

Behavioral Training with PFM Rehabilitation

The second basic approach to the behavioral treatment of urge incontinence and OAB is behavioral training. The goal of behavioral training is to improve bladder control by teaching the patient how to voluntarily suppress detrusor contractions. This involves a new response to urgency and learned use of the PFM to control urgency and detrusor contraction (Fig. 4.6.3).

PFM training is a central element of behavioral training. Its use is based on the premise that voluntary contraction of the PFM not only can occlude the urethra, but also can inhibit or abort detrusor contractions. It is a skill that can be accomplished by most patients and provides significant reduction of incontinence.

The first step in behavioral training for OAB is to assist patients to identify the PFMs and to contract and relax them selectively without increasing pressure on the bladder or pelvic floor. Among the techniques shown to be successful for this step are biofeedback,[7] and verbal feedback using vaginal or anal palpation.[8] Once patients have learned to properly control the PFM in the clinic, they are given instructions for daily progressive exercises to strengthen their motor skills, as well as muscle strength (see Chapter 4.2).

One exercise found to be helpful for patients with urge incontinence is to interrupt or slow the urinary stream during voiding once per day. Not only does this provide practice in occluding the urethra and interrupting detrusor contraction, it does so in the presence of urge sensation, when patients with OAB need the skill most. Some clinicians have concerns that repeated interruption of the stream may lead to incomplete bladder emptying in certain groups of patients. Therefore, caution should be used when recommending this technique for patients who may be susceptible to voiding dysfunction.

Another cornerstone of behavioral training is teaching patients a new way to respond to urgency: the urge suppression strategy.[9] Ordinarily, OAB patients feel compelled to rush to the nearest bathroom when they feel the urge to void. In behavioral training, they learn how this natural response is actually counterproductive, because it adds physical pressure on the bladder, enhances the sensation of fullness, exacerbates urgency, triggers detrusor contraction, and increases the risk of an incontinent episode. Although the new response is counterintuitive at first, patients can learn instead to stop what they are doing, sit down if possible, and contract the PFM repeatedly to suppress the detrusor contraction. They concentrate on voluntarily inhibiting the urge sensation and wait for the urge to subside before they walk at a normal pace to the toilet. See Figure 4.6.4 for patient instructions for using the urge suppression strategy.

The effectiveness of behavioral training has been established in several studies.[7,8,10,11] In the first randomized controlled trial, behavioral training reduced incontinence episodes significantly more than drug treatment, and patient perceptions of improvement and satisfaction with their progress were higher.[11] A subsequent study demonstrated that the results of behavioral training using biofeedback versus verbal feedback based on vaginal palpation did not differ significantly,[8] indicating that careful training with either method can achieve good results.

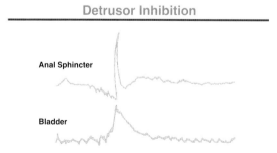

FIGURE 4.6.3. Polygraph tracings showing the mechanism of the urge suppression strategy: immediate suppression of detrusor contraction using active, voluntary contraction of PFMs (anal sphincter).

FIGURE 4.6.4. Patient instructions for the urge suppression strategy. (*Source*: Burgio KL, Pearce KL, Lucco A. Staying Dry: A Practical Guide to Bladder Control. Baltimore: Johns Hopkins University Press, 1989.)

	Patient Instructions for the Urge Suppression Strategy
	When you experience a strong urge to urinate…
Step 1:	Stop and stay still. Sit down if you can.
Step 2:	Squeeze your pelvic floor muscles quickly 3 to 5 times and repeat as needed.
Step 3:	Relax the rest of your body. Take several deep breaths.
Step 4:	Concentrate on suppressing the urge feeling.
Step 5:	Wait until the urge subsides.
Step 6:	Walk to the bathroom at a normal pace.

Urgency, Frequency, and Nocturia

Because most research on bladder training and behavioral training has been conducted with incontinence as the primary outcome measure, less is known about their effects on urgency or daytime frequency and nighttime urination. In studies of bladder training, frequency is not usually viewed as an outcome measure because it is the mechanism by which incontinence is reduced. In studies of behavioral training, reductions in daytime and nighttime urination have been observed incident to treatment for urinary incontinence.[11,12] Thus, there is evidence that both bladder training and behavioral training can be used to successfully to treat the broad spectrum of OAB symptoms.

Pharmacological Therapy

Pharmacological therapy continues to play an important role in the management of women with an overactive bladder.

Drugs with Mixed Actions

Oxybutynin

Oxybutynin is a tertiary amine with a mixed action consisting of both an antimuscarinic and a direct muscle relaxant effect. In addition it has local anesthetic properties, which is important when given intravesically. Oxybutynin has been shown to have a high affinity for muscarinic receptors in the bladder and a higher affinity for M_1 and M_3 receptors over M_2.

The effectiveness of oxybutynin in the management of patients with OAB is well documented in placebo controlled trials, although as many as 80% of patients complain of significant adverse effects, principally dry mouth or dry skin.[13,14] Oxybutynin has been shown to be more effective than previous antimuscarinic agents, such as propantheline, and as efficacious as propiverine. However, the antimuscarinic adverse effects of oxybutynin are often dose limiting.[15,16] Using an intravesical or intrarectal route of administration, higher local levels of oxybutynin can be achieved while limiting the systemic adverse effects.

More recently, controlled release oxybutynin (Ditropan XL) preparations using an osmotic system (OROS) have been developed and shown to have efficacy comparable to immediate release oxybutynin, although with fewer adverse effects.[17] In one study, incidence of moderate to severe dry mouth was 23%, and only 1.6% of participants discontinued because of adverse effects.[18]

An oxybutynin transdermal delivery system has also been developed and compared with extended release tolterodine. Both agents significantly reduced incontinence episodes, increased volume voided and improved quality of life in patients with mixed incontinence. The most common adverse event with the oxybutynin patch was application site pruritis in 14%. The incidence of dry mouth was lower (4.1%) compared to tolterodine (7.3%).[19]

Propiverine

Propiverine has been shown to combine anticholinergic and calcium channel–blocking actions and is the most popular drug for OAB in Germany, Austria, and Japan. Efficacy has been demonstrated in open studies and in a placebo-controlled trial of its use in neurogenic detrusor overactivity, it was shown to significantly increase bladder capacity and compliance. Dry mouth was experienced by 37% in the treatment group compared to 8% in the placebo group, with dropout rates of 7% and 4.5%, respectively.[20]

Antimuscarinic Drugs

Tolterodine

Tolterodine is a competitive muscarinic receptor antagonist with relative functional selectivity for bladder muscarinic receptors, and although it shows no specificity for receptor subtypes, it does appear to target the bladder over the salivary glands. Several controlled trials on patients with idiopathic or neurogenic detrusor overactivity have demonstrated significant reductions in incontinent episodes and frequency[21] and confirmed the safety of tolterodine, with an incidence of adverse events no different than placebo.[22] Data from several trials indicate that the clinical efficacy of tolterodine is comparable to that of oxybutynin, and is associated with fewer adverse events, dose reductions, and patient withdrawals.[23]

More recently, tolterodine has also been developed as an extended release once daily preparation (Detrusitol XL). When compared to immediate release tolterodine and placebo, this preparation was found to be significantly more effective for reducing urge incontinent episodes and had better tolerability, with a 23% lower incidence of dry mouth.[24]

Extended release oxybutynin and extended release tolterodine were compared in the OPERA trial (Overactive bladder: Performance of Extended Release Agents), which involved 71 centers in the United States. Improvements in episodes of urge incontinence were similar for the two drugs, but oxybutynin was significantly more effective for reducing frequency of micturition. Significantly more women taking oxybutynin were also completely dry (23% vs. 16.8%; P = 0.03), although they experienced significantly more dry mouth.[25]

Trospium

Trospium chloride is a quaternary ammonium compound that is nonselective for muscarinic receptor subtypes and shows low biological availability. It has produced significant clinical improvement, as well as increases in maximum cystometric capacity and threshold volume for detrusor contraction. The frequency of adverse events was similar to placebo.[26] Compared to oxybutynin, both agents showed significant improvement in bladder capacity and compliance and were not significantly different from each other. Those taking trospium had a lower incidence of dry mouth (4% vs. 23%) and were less likely to withdraw (6% vs. 16%).[27]

Solifenacin

Solifenacin is a potent M_3 receptor antagonist that has selectivity for the M_3 receptors over M_2 receptors and has much higher potency against M_3 receptors in smooth muscle than in salivary glands. Despite expressing a higher potency than darifenacin in a model of inhibition of M_3 receptor–mediated calcium ion mobilization in guinea pig colonic smooth muscle cells,[28] it has been shown to be 40-fold less potent than oxybutynin and 79-fold less potent than tolterodine in its inhibition of salivary secretion.[29] In the S.T.A.R. study (Solfenacin in a flexible dose regime with tolterodine XL as an active comparator in a double-blind, double-dummy, randomized OAB symptom trial) on 1,177 patients, solifenacin was more effective than tolterodine in producing continence (59% vs. 49%) using a three-day diary. There was a 31% greater reduction in pad usage with solifenacin than tolterodine (P = 0.0023) and symptoms of urgency were also decreased. The suggested dosage is 5 mg daily, with an increased dose to 5 mg bd if required.[30]

Darifenacin

Darifenacin is a highly selective M_3 receptor antagonist that has been found to have a 5-fold higher affinity for the human M_3 receptor, relative to the M_1 receptor. Darifenacin is equipotent with atropine in the ileum and bladder and 6-times less potent at inhibiting muscarinic receptors in the salivary gland. Salivary responses are inhibited at doses 6–10-fold higher than those required to inhibit bladder responses. Furthermore, a pilot study has demonstrated its ability to reduce the number, maximum amplitude, and duration of uninhibited bladder contractions.[31]

Antidepressants: Imipramine

Imipramine has been shown to have systemic anticholinergic effects and blocks the reuptake of serotonin. Some authorities have found a significant effect in the treatment of patients with OAB[32] although others report little effect. In light of this evidence and the adverse effects associated with tricyclic anti-depressants, their role in OAB remains of uncertain benefit although they are often useful in patients complaining of nocturia or bladder pain.

Antidiuretic Agents: Desmopressin

Desmopressin (1-desamino-8-D-arginine vasopressin; DDAVP) is a synthetic vasopressin analogue. It has strong antidiuretic effects without altering blood pressure. The drug has been used primarily in the treatment of nocturia and nocturnal enuresis in children and adults. More recently, nasal desmopressin has been reported as a "designer drug" for the treatment of daytime urinary incontinence.[33] Desmopressin is safe for long-term use, however it should be used with care in the elderly because of the risk of hyponatremia.

Combining Behavioral and Drug Therapy

One of the drawbacks of both behavioral treatment and pharmacological therapy is that the majority of patients do not achieve full continence. On average, only 20% to 30% become dry. Considering that these therapies may work by different mechanisms, combining them may be one strategy to enhance outcomes. Early studies provide evidence for the effectiveness of combined therapy. In one study, bladder training significantly increased the effects of tolterodine for reducing voiding frequency and increasing volume voided.[34] In another study of women treated initially with behavioral training or immediate-release oxybutynin alone, crossover to combined treatment improved outcomes significantly over those achieved with the single therapy.[35] Thus, combining behavioral and drug therapy may be one way to optimize outcomes for patients with OAB.

References

1. Abrams P, Cardozo L, Fall M, et al. The standardisation of terminology of lower urinary tract function. Report from the standardisation subcommittee of the International Continence Society. Urology. 2003;61:37.
2. Stewart WF, Van Rooyen JB, Cundiff GW, et al. Prevalence and burden of overactive bladder in the United States. World J Urol. 2003;20:327
3. Frewen WK. Role of bladder training in the treatment of the unstable bladder in the female. Urol Clin North Am. 1979;6:273–277.
4. Jarvis GJ, Millar DR. Controlled trial of bladder drill for detrusor instability. Br Med J. 1980;281: 322–1323.
5. Pengelly AW, Booth CM. A prospective trial of bladder training as treatment for detrusor instability. Br J Urol. 1980;52:463–466.
6. Fantl JA, Wyman JF, McClish DK, et al. Efficacy of bladder training in older women with urinary incontinence. JAMA. 1991;265:609–613.
7. Burgio KL, Whitehead WE, Engel BT. Urinary incontinence in the elderly: bladder/sphincter biofeedback and toileting skills training. Ann Intern Med 1985;103:507–515.
8. Burgio KL, Goode PS, Locher JL, et al. Behavioral training with and without biofeedback in the treatment of urge incontinence in older women: a randomized, controlled trial. JAMA. 2002;288:2293–2299.
9. Burgio KL, Pearce KL, Lucco A. Staying dry: A practical guide to bladder control. Baltimore: The Johns Hopkins University Press; 1989.
10. Burton JR, Pearce KL, Burgio KL, et al. Behavioral training for urinary incontinence in elderly ambulatory patients. J Am Geriatr Soc. 1988;36:693–698.
11. Burgio KL, Locher JL. Goode PS, et al. Behavioral vs. drug treatment for urge urinary incontinence in older women: a randomized controlled trial. JAMA. 1998;280:1995–2000.
12. Johnson TM, Burgio KL, Goode PS, et al. Effects of behavioral and drug therapy on nocturia in older incontinent women. J Am Geriatr Soc. 2005; 53:846–850.
13. Moore KH, Hay DM, Imrie AE, et al. Oxybutynin hydrochloride (3 mg) in the treatment of women with idiopathic detrusor instability. Br J Urol. 1990;66:479–485.

14. Tapp AJS, Cardozo LD, Versi E, et al. The treatment of detrusor instability in post menopausal women with oxybutynin chloride: a double blind placebo-controlled study. Br J Obstet Gynaecol. 1990;97:479–485.
15. Thuroff JW, Bunke B, Ebner A, et al. Randomized, double-blind, multicentre trial on treatment of frequency, urgency and incontinence related to detrusor hyperactivity: oxybutynin versus propantheline versus placebo. J Urol. 1991;145;813–817.
16. Madersbacher H, Halaska M, Voigt R, et al. A placebo-controlled, multicentre study comparing the tolerability and efficacy of propiverine and oxybutynin in patients with urgency and urge incontinence. BJU Int. 1999;84:646–651.
17. Anderson RU, Mobley D, Blank B, et al. Once daily controlled versus immediate release oxybutynin chloride for urge urinary incontinence. OROS Oxybutynin Study Group J Urol. 1999;161:809–1812.
18. Gleason DM, Susset J, White C, et al. Evaluation of a new once-daily formulation of oxybutynin for the treatment of urinary urge incontinence. Ditropan XL Study Group. Urology. 1999;54:420–423.
19. Dmochowski RR, Sand PK, Zinner NR, et al. Transdermal Oxybutynin Study Group. Comparative efficacy and safety of transdermal oxybutynin and oral tolterodine versus placebo in previously treated patients with urge and mixed urinary incontinence. Urology. 2003;62;237–242.
20. Stoher M, Madersbacher H, Richter R, et al. Efficacy and safety of propiverine in SCI-patients suffering from detrusor hyperreflexia: a double-blind, placebo-controlled clinical trial. Spinal Cord. 1999;37:196–200.
21. Millard R, Tuttle J, Moore K, et al. Clinical efficacy and safety of tolterodine compared to placebo in detrusor overactivity. J Urol. 1999;161:1551–1555.
22. Rentzhog L. Stanton SL, Cardozo LD, et al. Efficacy and safety of tolterodine in patients with detrusor instability: a dose ranging study. Br J Urol. 1998;81:42–48.
23. Abrams P, Freeman R, Anderstrom C, et al. Tolterodine, a new antimuscarinic agent: as effective but better tolerated than oxybutynin in patients with an overactive bladder. Br J Urol. 1998;81:801–810.
24. Van Kerrebroeck P, Kreder K, Jonas U, et al. Tolterodine Study Group. Tolterodine once-daily: superior efficacy and tolerability in the treatment of overactive bladder. Urology. 2001;57:414–421.
25. Diokno AC, Appell RA, Sand PK, et al. OPERA Study Group. Prospective, randomised, double blind study of the efficacy and tolerability of the extended-release formulations of oxybutynin and tolterodine for overactive bladder: results of the OPERA trial. Mayo Clin Proc. 2003;78:687–695.
26. Cardozo LD, Chapple CR, Toozs-Hobson P, et al. Efficacy of trospium chloride in patients with detrusor instability: a placebo-controlled, randomized, double-blind, multicentre clinical trial. BJU Int. 2000;85:659–664.
27. Madersbacher H, Stoher M, Richter R, et al. Trospium chloride versus oxybutynin: a randomized, double-blind, multicentre trial in the treatment of detrusor hyperreflexia. Br J Urol 1995;75:452–456.
28. Ikeda K, Kobayashi S. Effects of YM905, Oxybutynin and Darifenacin on carbachol induced intracellular calcium mobilization by dispersed longitudinal smooth muscle cells of guinea pig colon. Yamanouchi Pharmaceutical Co. (1998). Registration No. D199803320–02.00. International Study ID: 905-PH-006.
29. Ikeda K, Kobayashi S. Effects of YM905, tolterodine and Oxybutynin on M3 receptor mediated cytosolic free Ca2+ mobilization in acutely dissociated cells of guinea pig urinary bladder smooth muscle and murine submandibular gland. Yamanouchi Pharmaceutical Co. (1998). Registration No. D199803217–02.000. International Study ID: 905-PH-005.
30. Chapple C, Martinez-Garcia R Selvaggi L, et al A comparison of the efficacy and tolerability of Solifenacin Succinate and extended release Tolterodine at treating overactive bladder syndrome: results of the S.T.A.R trial. Eur Urol. 2005; 48:464–470.
31. Rosario DJ, Leaker BR, Smith DJ, et al. A pilot study of the effects of multiple doses of the M3 muscarinic receptor antagonist darifenacin on ambulatory parameters of detrusor activity in patients with detrusor instability. Neurourol Urodyn. 1995;14:464–465.
32. Castleden CM, Duffin HM, Gulati RS. Double-blind study of imipramine and placebo for incontinence due to bladder instability. Age Ageing. 1986;15:299–303.
33. Robinson D, Cardozo L, Akeson M, et al. Women take control; Desmopressin – A drug for daytime urinary incontinence. Neurourol Urodyn. 2002;21: 385–386.
34. Mattiasson AJ, et al. Simplified bladder training augments the effectiveness of tolterodine in patients with an overactive bladder. BJU Int. 2003;91:54–60.
35. Burgio KL, Locher JL, Goode PS. Combined behavioral and drug therapy for urge incontinence in older women. J Am Geriatr Soc. 2000;48:370–374.

4.7
Sexual Dysfunction and the Overactive Pelvic Floor

Wendy F. Bower

Key Messages

- Dyspareunia and overactive pelvic floor (OAPF): stretching techniques of the pelvic floor muscle (PFM), stimulation of painful regions.
- PFM training may improve orgasmic response.
- Sexual dysfunction, including coital incontinence, is common in incontinent women.
- Management of OAPF includes postural correction, joint mobilization, and reduction of strain on pelvic fascia.
- Subcutaneous restrictions, trigger points, or scar tissue are treated with deep massage and friction techniques, augmented by the use of ultrasound or heat.

Introduction

Female sexual dysfunction is defined as a disorder of sexual desire, arousal, orgasm, and/or sexual pain.[1] Loss of arousal is marked by vaginal lubrication problems, whereas orgasmic dysfunction includes infrequent or nonorgasmic response.[2] Pain associated with sexual activity may be either dyspareunia, vaginismus, or non-coital sex pain.[3] Up to 63% of community-dwelling women and 76% of women with pelvic pain may have some form of sexual dysfunction.[1,4-6] Personal distress, negative experience in sexual relationships, and adverse effects on quality of life are common in women with sexual dysfunction.[5]

Sexual arousal depends on an intact sympathetic nervous system and can be overridden by the expectation of pain. The anticipation of discomfort induces anxiety and limits the sexual arousal response, thus, preventing vaginal lubrication.[6] Pain medication, estrogen deficiency, alcohol, and antidepressant, anticholinergic, and antihistamine medications also have a negative effect on vaginal lubrication.[2] Poor lubrication of vaginal tissues leaves them susceptible to trauma with penetration and subsequent pain.

Symptoms and Findings

Dyspareunia

Dyspareunia, or pain on penile, digital, or other methods of vaginal penetration, ranges from localized introital tenderness to diffuse deep soreness and can be sustained for up to 3 days after sexual activity.[7] Common causes in women with pelvic pain include endometriosis, adhesions, organ pathology, acute or chronic inflammation, contact with the cervix, vaginismus, atrophy, and urethral or bladder wall trauma.[2,8] Dyspareunia has been reported in up to 41% of patients with incontinence and 57% of patients with OAPF. It is generally attributed to stretching of the shortened pelvic floor, stimulation of painful regions, or organ dysfunction/adhesions.[1,4,9,10]

Initially, treatment should involve the identification and management of associated pathological conditions of the bladder, urethra, and vagina.

Sexual counseling is appropriate if psychological or relationship problems are evident or there is a disclosure of previous sexual abuse. When deep dyspareunia is present, women should be educated to limit deep thrusting during intercourse until they are highly aroused, the vaginal apex has expanded, and the uterus has moved upward.[2] Sexual positions that minimize discomfort include the woman being astride, partners side by side, or vaginal entry from behind. Techniques to desensitize the introitus using the woman's own finger, her partner's fingers, or a dilator/vibrator may prove helpful.

Vaginismus

Vaginismus is the involuntary contraction of muscles in the outer third of the vagina, which limit or prevent vaginal penetration. The muscle spasm is considered to be a conditioned response to events that have previously caused pain or discomfort and may have its origin in prior vaginal infection, sexual trauma, or unresolved psychological conflict.[2,8,11] As vaginismus is situation specific, pelvic muscle spasm may not always be evident on examination. However, in most such cases, careful palpation of the pelvic floor reproduces the woman's pain.[2]

There is known to be a high prevalence of vaginismus in patients with both urethral syndrome and vulvar vestibulitis,[12] suggesting a link with impaired vasodilatory arousal response and dyspareunia.[2] Although women with vaginismus would be expected to have altered muscle mechanics, van der Velde and Everaerd reported no difference in the quality of pelvic floor relaxation or contraction when compared with healthy volunteers.[13]

Therapy is multimodal and involves exploration of physical causes, identification of significant psychological issues, and reconditioning of the learned response to sexual contact. Patient education involves the woman looking at her vulva and coming to understand the vaginismic response. This is followed by muscle awareness of contraction and relaxation in other regions of the body, with eventual progression to the perineal and PFM groups. The woman and/or her partner gradually and progressively introduce a finger, tampon, dilator, or EMG electrode into the vagina while maintaining a lax introitus, even breathing, and relaxation. If biofeedback is used, the woman has immediate visual feedback of vaginal pressure and relaxation. At all times the patient is in control of the situation, and progression to penetration will occur only when she chooses, and when she has trust in her sexual partner.[2,12]

Orgasmic Dysfunction

Female orgasm follows sufficient sensory stimulation and induces repeated 1-s motor contractions of the pelvic floor, succeeded by repeated uterine and vaginal smooth muscle contractions.[1] Poor pelvic floor positioning, tone, and strength are associated with decreased intensity of orgasm.[14] In women with pelvic pain, a lack of orgasm is frequently associated with a learned inhibitory elevation of the pelvic floor, based on prior painful encounters.[2] Low arousal, inadequate stimulation, dyspareunia, anxiety, vaginismus, and sexual inhibition can all be primary factors driving a nonorgasmic response. Although there is no correlation between sexual function and PFM strength in the normative muscle,[15] one third of women who reported difficulty in reaching climax improved after pelvic floor rehabilitation treatment.[9] To date, no studies have demonstrated improvement in pelvic pain with pelvic floor rehabilitation.

Sexual Dysfunction and Incontinence

Sexual dysfunction, and particularly the aspects of hypoactive desire and sexual pain, are significantly more common in women with incontinence than in those who are continent.[15,16] Although it is unclear whether urinary symptoms may be the cause or result of sexual dysfunction, patients with mixed urinary incontinence appear more vulnerable than those with isolated overactive bladder or stress incontinence.[16–18] Women with an overactive bladder report the need to void and fear of coital leakage as major causes of hypoactive sexual desire.[15]

The prevalence of incontinence during sexual activity ranges from 2% to 10% in a community sample and from 10% to 56% in women with lower urinary tract disorders.[7,9,17,19] Coital leakage is more prevalent in women with severe rather than moderate symptoms and peaks around 50 years of age.[20,21] Unless a clinician specifically asks about incontinence during intercourse, women are unlikely to volunteer the symptom.[22] Aside from coital leakage, wetness during the night, the need to wear pads to bed, odor, and nocturnal enuresis have all been associated with a decrease in the frequency of sexual activity.[9]

Women with urinary incontinence or lower urinary tract symptoms have a significantly higher rate of sexual pain than women without these symptoms.[15] Women with an overactive bladder are more likely to experience pain or orgasmic difficulties during intercourse than are women with isolated stress incontinence; however, the latter group reported greater interference with sexual activity.[15,16] Women with recurrent cystitis and voiding dysfunction may be at the most risk for sexual pain disorders, possibly because of impaired blood flow to the pelvic region, inflammation, or underlying OAPF.[15]

Surgical treatment of incontinence and urogenital prolapse, although curative, may have adverse effects on sexual function. Alteration of local blood flow, vaginal architecture, sensation, and connective tissue structure can diminish libido and arousal, cause dyspareunia, and inhibit orgasm.[8,23]

Conservative Management of the OAPF

Musculoskeletal

The basic premise behind any treatment of the OAPF is that, as a consequence of its attachment to the bony pelvis and fascial structures, dysfunction often extends beyond the fibers of the levator ani. Accordingly, a holistic therapy approach begins with postural correction, joint mobilization, and gait modification. Spinal and hip joints need full range of movement to reduce strain on the pelvic fascia and its attachment to the OAPF. In particular, limitation of pelvic movement at the coccygeus or sacrum can impair the rehabilitation of the PFMs.[24] Because joints are anchored by their capsules and connective tissue, stretching, strengthening, and correction of muscle imbalances around the trunk and hip joints is often necessary.[25] If a woman has a significant diastasis of the rectus, optimal function of the pelvic floor cannot be achieved, and she should be taught to support the abdominal wall and begin rehabilitation of the apposed muscle.[26,27]

Psychological

Because there is a known association between prior sexual and physical abuse and functional disorders (e.g. dysfunctional voiding, defecation difficulties) in women with pelvic pain, clinicians should be sensitive to possible disclosures. Aside from showing calm concern and care, a referral for psychotherapy or other mental health intervention should be considered. Intervention may be modified by the need to proceed more slowly to techniques involving vaginal contact, and perhaps to substitute external therapies for internal techniques.

Box #1
Assessment of the woman with an OAPF muscle

1. Visualize the perineum at rest: note size of genital hiatus and position of perineal body.
2. Observe PFM contraction, relaxation, and bearing down; note movement, recruitment patterning, timing, proprioception, fatigue, and accessory muscle activity.
3. Check sensation/neurological integrity; note anal wink reflex.
4. Palpate the external tissues: note tissue quality, variation in color, sensation, temperature, tenderness, possible trigger points, and sites of pain referral.
5. Palpate the vagina/rectum with a single, well-lubricated digit; note presence of pain (site, localized or diffuse), PFM tone (compressibility); relaxation after a contraction (absent, partial, full), spasm, and contractile activity (+/− surface EMG or manometry).

> **Box #2**
> **Point-form treatment of dyspareunia patient with short pelvic floor muscles**
>
> - Goal setting of 4–6 weeks before tissue normalization
> - Home introital desensitization (+/− vaginal dilator) and exploration of arousal techniques
> - Connective tissue work on pelvic and abdominal region +/− heat and ultrasound
> - Treatment of any overactive bladder symptoms or bowel-emptying dysfunction
> - Stretching and fascial release of vaginal wall and anomalies identified on palpation
> - Contract–relax techniques immediately after fascial therapy
> - Posttreatment electrostimulation via surface or vaginal electrode for pain relief or muscle normalization
> - Explanation of sexual positioning to minimize discomfort

General

Women are encouraged to initiate general fitness activities, such as walking or swimming, to stimulate circulation, decrease stress and depression, and assist in pain relief.[25] From the outset, activities and postures (e.g. excessive posterior pelvic tilt in sitting) that increase tension on the shortened OAPF or generate pain should be avoided.[25,27] Vigorous PFM exercise is not an appropriate first-line therapy for the woman with OAPF.

Manual Therapy

Connective tissue of the pelvic and abdominal region may be noted to have subcutaneous restrictions, trigger points, or scar tissue.[27] Such limitations are treated with deep massage and friction techniques, augmented by the use of ultrasound or heat, until there is unrestricted mobility in all directions.[24,26] The PFMs are assessed by digital vaginal palpation for fibrous bands, areas of soreness, trigger points, and decreased tissue mobility.[25] Tissue abnormalities are treated with techniques to stretch and release fascia and elongate the underlying muscle.[28] Women may voluntarily contract the pelvic floor after stretching to effect the more complete relaxation and lengthening that follows muscle contraction. Adjunctive neuromuscular electrical stimulation (NMES) can be used to increase local circulation, thereby removing metabolites. This facilitates optimal neural drive, allows greater tissue extensibility, and provides a posttreatment analgesic effect.[26] Resolution of trigger-point pain and restoration of tissue homogeneity may take from 6 to 10 weeks.[24,27]

Muscle Rehabilitation

Extensive muscle awareness training is implemented and may involve the use of perineal or vaginal biofeedback. Initially, the OAPF will have a high resting tone and will be weak in its shortened range. Therapy aims to retrain resting muscle tension by bringing muscle tension to a conscious level. Biofeedback is particularly useful for enhancing this awareness.[25] Women may use

> **Box #3**
> **Point-form treatment of pelvic pain patient with coexisting shortened pelvic floor**
>
> - patient education and counseling
> - lifestyle and physical stress reduction techniques
> - postural correction to rectify imbalances generated by pain-protective postures
> - spinal and pelvic mobilization to address any secondary limitations in range of movement at hips, spinal or pelvic joints
> - general exercise regimen to combat physiological de-conditioning
> - specific PFM awareness and relaxation training (+/− surface biofeedback)
> - deep friction or myofascial techniques/trigger point pressure
> - active and passive stretching of PFM with emphasis on achieving descent postcontraction and restoring pain-free range of movement
> - heat/NMES
> - home program of PFM awareness and active relaxation
> - strength training to regain normal contractile capability

home EMG biofeedback or neuromuscular electrical stimulation to train proprioception and functional recruitment, respectively, in the pelvic floor rehabilitation phase.

A home program of PFM exercise should not be given until the muscle has resumed its normal length-tension ability. PFM exercise can be accompanied by bladder training, bowel management, and defecation techniques to address associated symptoms. Other techniques that promote general body relaxation, such as focused breathing, ambient environment, visualization, and progressive active relaxation, are helpful for the woman with OAPF.[29]

References

1. Munarriz R, Kim NN, Goldstein I, et al. Biology of female sexual function. Urol Clin N Am. 2002;29:685–693.
2. Bachmann GA, Phillips NA. Sexual dysfunction. In: Steege JF, Metzger DA, Levy BS, editors. Chronic pelvic pain. Philadelphia: WB Saunders; 1998:77–90.
3. Basson R, Berman J, Burnett A, et al. Report of the international consensus development conference on female sexual dysfunction: definitions and classifications. J Urol. 2000;163(3):888–893.
4. FitzGerald MP, Kotarinos R. Rehabilitation of the short pelvic floor: background and patient evaluation. Int Urogynecol J. 2003;14:261–268.
5. Laumann EO, Paik A, Rosen RC. Sexual dysfunction in the United States: prevalence and predictors. JAMA. 1999;281:1174.
6. Marthol H, Hilz MJ. Female sexual dysfunction: a systematic overview of classification, pathophysiology, diagnosis and treatment. Fortschr Neurol Psychiatr. 2004;72(3):121–135.
7. Salonia A, Munarriz RM, Naspro R, et al. Women's sexual dysfunction: a pathological review. BJU Int. 2004;93:1156–1164.
8. Pauls RN, Berman JR. Impact of pelvic floor disorders and prolapse on female sexual function and response. Urol Clin N Am. 2002;29:677–683.
9. Beji NK, Yalcin O. The effect of pelvic floor training in sexual function of treated patients. Int Urogynecol J. 2003;14:234–238.
10. Meadows E. Treatment for patients with pelvic pain. Urolog Nurs. 1999;19(1):33–35.
11. Shafik A, El-Sibai O. Study of the pelvic floor muscles in vaginismus: a concept of pathogenesis. Eur J Obstet Gynecol Reprod Biol. 2002;10:105(1):67–70.
12. Gibbons JM. Vulvar vestibulitis. In: Steege JF, Metzger DA, Levy BS, editors. Chronic pelvic pain. Philadelphia: WB Saunders; 1998:181–187.
13. Van der Velde J, Everaerd W. Voluntary control over pelvic floor muscles in women with and without vaginismus. Int Urogynecol J. 1999;10:230–236.
14. Shafik A. The role of the levator ani muscle in evacuation, sexual performance and pelvic floor disorders. Int Urogynecol J Pelvic Floor Dysfunct. 2000;11:361–376.
15. Salonia A, Zanni G, Nappi RE, et al. Sexual dysfunction is common in women with lower urinary tract symptoms and urinary incontinence: Results of a cross-sectional study. Europ Urol. 2004;45:642–648.
16. Walters MD, Taylor S, Schoenfeld LS. Psychosexual study of women with detrusor instability. Obstet Gynecol. 1990;75:22–26.
17. Shaw C. A systematic review of the literature on the prevalence of sexual impairment in women with urinary incontinence and the prevalence of urinary leakage during sexual activity. Eur Urol. 2002;42:432–440.
18. Gordon D, Groutz A, Sinai T, et al. Sexual function in women attending a urogynecology clinic. Int Urogynecol J Pelvic Floor Dysfunct. 1999;10(5):325–328.
19. Amarenco G, Le Cocquen A, Bosc S. Stress urinary incontinence and genitor-sexual conditions. Prog Urol. 1996;6:913–919.
20. Moller LA, Lose G, Jorgensen T. the prevalence and bothersomeness of lower urinary tract symptoms in women 40–60 years of age. Acta Obstet Gynecol Scand. 2000;79:298–305.
21. Nygaard I, Milburn A. Urinary incontinence during sexual activity: prevalence in a gynecologic practice. J Women's Health. 1995;4(1):83–86.
22. Hilton P. Urinary incontinence during sexual intercourse: a common, but rarely volunteered, symptom. Br J Obstet Gynaecol. 1988;95(4):377–381.
23. Maaita M, Bhaumik J, Davies AE. Sexual function after using tension-free vaginal tape for the surgical treatment of genuine stress incontinence. BJU Int. 2002;90:540–543
24. Lukban JC, Whitmore KE. Pelvic floor muscle re-education treatment of the overactive bladder and painful bladder syndrome. Clin Obstet Gynecol. 2002;45(1):273–285.
25. Shelly B, Knight S, King P, et al. Treatment of pelvic pain. In: Laycock J, Haslam J, editors. Therapeutic management of incontinence and

pelvic pain. London: Springer-Verlag; 2002:177–189.
26. Costello K. Myofascial syndromes. In: Steege JF, Metzger DA, Levy BS, editors. Chronic pelvic pain. Philadelphia: WB Saunders; 1998:251–266.
27. FitzGerald MP, Kotarinos R. Rehabilitation of the short pelvic floor: treatment of the patient with the short pelvic floor. Int Urogynecol J. 2003;14:269–275.
28. Weiss JM. Pelvic floor myofascial trigger points: manual therapy for interstitial cystitis and the urgency–frequency syndrome. J Urol. 2001;166(6):2226–2231.
29. Baker PK. Musculoskeletal problems. In: Steege JF, Metzger DA, Levy BS, editors. Chronic pelvic pain. Philadelphia: WB Saunders; 1998:215–240.

4.8
Anal Incontinence and Evacuation Difficulties

Christine Norton

Key Messages

- Treatment often involves several conservative components tailored to the patient's needs.
- Patients often benefit from education about normal bowel function and what has gone wrong in their case.
- If possible, establish a consistent bowel pattern with complete evacuation at the same time each day, preferably after meals.
- Proper diet, fluids, and physical activity facilitate normal stool consistency and bowel habits.
- Evacuation training addresses regular habit, posture, breathing, pushing without straining, relaxing the back passage, and avoiding laxatives.
- Pelvic floor muscle retraining, bowel retraining, and biofeedback can improve bowel control.

Introduction

Many patients with fecal incontinence (FI), anal incontinence (AI), which includes loss of flatus or mucus, or evacuation difficulties can be helped by conservative measures.[1,2] This often involves several different elements, rather than a single definitive intervention, and in clinical practice it makes sense to combine approaches in a retraining program to maximize patient benefit. Uncomplicated mild-to-moderate constipation will often respond to simple measures, such as a regular habit, increasing fiber and fluid intake, and getting more exercise, which should always be tried first. Anal incontinence will likewise often improve with attention to diet, bowel habits, and some retraining and exercises.

Patient Education

Bowels are a mystery to most members of the public. From the time of mastery of control until things go wrong, most people think little about their bowels – like any other skill which has become "incorporated," the body functions on a semiautomatic level, and little conscious thought goes into either holding on when an urge or gas is felt, or into evacuating once sitting in the correct position. This can make it very difficult to regain control if FI or evacuation difficulties develop. Patients often benefit from education about normal bowel function and learning why it has gone wrong in their individual case. This can be achieved verbally, by using diagrams or models, or through written information (www.bowelcontrol.org.uk).[3-5] Often, a combination of methods, tailored to each person's needs, will enable the best understanding.

Bowel Routine

If it is possible to establish a predictable bowel pattern, with complete evacuation at the same time each day, then most bowel problems are much more manageable. For most people, the bowel is relatively inactive while asleep and maximum motility occurs in the first few hours after getting up in the morning. Motility is enhanced by stimulating the gastrocolic response by eating and drinking at breakfast. Approximately 20–30 minutes after eating is the most likely time for mass movements in the colon to result in propulsive waves of peristalsis, delivering a bolus of stool to the rectum. Unfortunately, with modern life styles, this is often the time of day when we are traveling to work or taking the children to school. If the urge to defecate is ignored, stool may be propelled away from the anus by retrograde peristalsis. If this is done repeatedly, the stool becomes harder as more water is absorbed through the colonic mucosa and the urge will fade and even disappear. Deliberately ignoring the call to stool slows colonic transit,[6] and many constipated people report that they never feel an urge, possibly a result of a long-term habit of ignoring this sensation.

This is why patients are advised very strongly to make time for breakfast and their bowels in the morning, even if this involves getting up 30 minutes earlier or adjusting their morning routine. There is no evidence that it matters what is eaten and drinking any type of fluid seems to be almost as effective as drinking caffeine in stimulating motility.[7]

Patients are instructed to eat something, consume two drinks, and then attempt to use the toilet 20–30 minutes later, allowing 5–10 minutes of uninterrupted privacy for this, every day at the same time if it is at all feasible. If this proves impossible, an evening routine may be substituted.

Physical Activity, Diet, and Fluids

Exercise is often recommended as a means of promoting a regular bowel habit with complete evacuation, as it increases propulsive activity in the colon.[8] The epidemiological literature is consistent in showing less constipation in people who report that they exercise.[9] Excessive exercise can even cause diarrhea. However, there have been no studies evaluating the efficacy of increasing exercise in people with long-standing or severe constipation.[10]

Clinically, patients with slow-transit constipation often report less discomfort and bloating when they reduce, rather than increase, the fiber content of their diet. If transit is slow, adding fiber, particularly unrefined wheat bran, will simply add to what is sitting in the colon and is unlikely to promote peristalsis. More soluble forms of fiber, such as in fruit and vegetables, may be better tolerated. It is certainly worth it for each person to experiment, as some people seem to have an intolerance, and to find that their constipation worsens with certain food groups or combinations. It is unwise to restrict whole food groups in the long term without good dietary advice, as deficiencies can develop.

Standard advice for constipated people is to drink more, and this may work well if the patient is clinically dehydrated and the body is trying to preserve fluid by increasing colonic fluid uptake. However, if already adequately hydrated, drinking more will simply cause increased urine, not stool, production. Drinking fluids may have a limited effect on motility via the gastrocolic response, but it will not generally soften the stool.[10]

People with incontinence are more prone to leakage or accidents if the stool is soft or loose, and many people modify their diets considerably in an attempt to avoid FI. In these cases, the goal is to keep the stool formed, while avoiding hard stool or constipation. Diet has a major role in this. Some patients need more fiber to bind stool together, and adding supplements such as psyllium or gum arabic to the diet has been found to improve FI.[11] Others achieve a firm stool by restricting their fiber intake.

Evacuation Training

Patients with both slow transit constipation and evacuation difficulties respond equally well to retraining techniques, often described in the literature as "biofeedback," whether or not feedback equipment is utilized.[12] Table 4.8.1 gives an example of the basic instructions given to patients on evacuation technique. The elements are outlined for all patients, but with different emphasis, depending on the presenting problems and findings on assessment. Patients are advised that, although each element sounds simple, putting it all together takes time, practice, and determination. Routine and daily practice is emphasized.

4.8. Anal Incontinence and Evacuation Difficulties

TABLE 4.8.1. An example of the basic instructions given to patients on evacuation technique

Getting into a Regular Habit

- Most people's bowels respond best to a regular habit. Some of us are too busy to make time for our bowels. Others live a very irregular lifestyle, which makes a habit difficult.
- The bowel usually goes to sleep at night and wakes up in the morning. Eating, drinking, and moving around all stimulate the bowel. The most likely time for a bowel action is about 30 minutes after the first meal of the day.
- This makes it important not to skip breakfast. Try to eat something for breakfast and have two warm drinks. Try to make 5–10 minutes of free uninterrupted time about 30 minutes later. This is not always easy if your house is busy in the morning, so you may need to plan ahead or get up a little earlier while you retrain you bowel.

Sitting Properly

- The way you sit on the toilet can make a big difference to ease of opening your bowels. The "natural" position (before toilets were invented) is squatting. Countries where squat (hole in the floor) toilets are still common seem to have fewer problems with constipation.
- Whereas actually squatting is not very practical, many people find that adopting a "semisquat" position helps a lot. One of the footstools that toddlers use to reach a sink is ideal, 8–12 inches high (20–30 cm). Position this just in front of your toilet and rest your feet flat on the stool, keeping your feet and knees about 1 foot (30 cm) apart. Lean forward, resting your elbows on your thighs. Try to relax.

Breathing

- It is important not to hold your breath when trying to open your bowels. Many people are tempted to take a deep breath in and then hold their breath while trying to push. Try to avoid this. Sit on the toilet as described above, relax your shoulders, and breathe normally. You may find it easiest to breathe in through your nose and out through your mouth.
- If you hold your breath and push, this is STRAINING, which tends to close your bottom more tightly. Also, when you hold your breath, you are limited in how long you can hold this, and when you have to take the pressure off and breathe you are back at square one.
- If you find that you cannot help straining and holding your breath, try breathing out gently or humming or reciting a nursery rhyme.

Pushing Without Straining

- The best way to open your bowels is by using your abdominal (stomach) muscles to push. Leaning forward, supporting your elbows on your thighs and breathing gently, relax your shoulders. Make your abdominal muscles bulge outwards to "make your waist wide." Now, use these abdominal muscles as a pump to push backwards and downwards into your bottom. Keep up the gentle but firm pressure.

Relaxing the Back Passage

- The final part of the jigsaw is to relax the back passage. Many people with constipation actually tighten the back passage when they are trying to open the bowels, instead of relaxing, without realizing what they are doing. This is like squeezing a tube of toothpaste while keeping the lid on!
- To locate the muscles around the back passage, first squeeze as if you are trying to control wind.
- Now imagine that the muscle around the anus is a lift. Squeeze to take your lift up to the first floor. Now relax, down to the ground floor, down to the basement, down to the cellar.

Putting it all Together

This is a bit like learning to ride a bike. The above instructions tell you WHAT to do, but do not tell you HOW to do it. It sounds simple, but coordinating everything takes practice, and you have to work it out for yourself. Some people find it easier than others.
- Sit properly
- Breathe normally
- Push from your waist downwards
- Relax the back passage

Keep this up for about 5 minutes, unless you have a bowel action sooner. If nothing happens, don't give up. Try again tomorrow. It often takes several weeks of practice until this really starts to work.

Posture

For some patients, posture on the toilet seems to be very important, and evacuation difficulty is largely resolved by adopting a "semisquat" posture, with feet raised on a 20–30-cm-high footstool placed in front of the toilet. Patients are instructed to lean forward with the feet apart and support the elbows on the knees, allowing the abdominal muscles to relax between the knees.

Balloon Expulsion Training

If the patient has difficulty grasping the correct technique, balloon expulsion can be used to teach the coordination of pushing. A rectal balloon, filled with approximately 50 ml of air or water, has been found useful when teaching correct technique.[2] Placed just inside the rectum near the anal margin, this can give the patient the sensation of rectal fullness or the urge to defecate and, if pushing correctly, they experience the proprioception of stool moving in the right direction in response to pushing. If technique is good, the balloon may actually be expelled. By holding the end of the balloon just outside the anus, the therapist can assess whether propulsive effort is effective and improving.

Success Rates

Studies report that 50–80% of patients are improved with these techniques.[2] Patients may have unrealistic expectations, and it is important to dispel myths about the necessity to have a bowel movement every day. A more appropriate goal is that evacuation will be comfortable and without straining. Patients with a symptomatic rectocele or intraanal intussusception often achieve moderate symptom relief, and some symptoms even resolve completely.

Laxatives

Many authors have recommended allowing no oral laxatives (prescribed or self-purchased) during the retraining period, even if previous doses were high and dependence is long standing.[2] The exception is in patients with neurological disorders and children, who may need 6–12 months of regular laxatives to establish a regular bowel habit.[13] Many patients are frightened by the prospect of coming off laxatives. However, if the purpose is to get the bowel functioning normally again then this habit must be broken. Clinical experience suggests that it is better to stop laxatives completely than to attempt a gradual weaning.[2] Glycerin suppositories may be used as a rescue if no stool has been passed for five days. Often, abdominal discomfort and bloating are improved simply by stopping the use of laxatives, as these can be side effects of the medication rather than a result of the constipation.

Some patients fail the retraining program and may need to use laxatives long-term. The logical laxative is a stimulant for slow transit constipation and a softener for evacuation difficulties. Table 4.8.2 gives a summary of laxative classes.

TABLE 4.8.2. Laxatives and evacuants and their uses

Class	Examples	Main use	Notes
Bulking agents	Ispaghula Sterculia Methylcellulose	Stool softening when diet not tolerated	Introduce slowly and with adequate fluid intake. May cause bloating and discomfort. Avoid if patient is severely constipated or impacted. Can take several days to work.
Osmotic laxatives	Magnesium salts Lactulose Macrogols	Can adjust dose to produce desired stool consistency	Often work in 8–12 hours. Can cause bloating (especially lactulose). Macrogols can also be used for disimpaction.
Stimulant laxatives	Bisacodyl Senna Sodium picosulfate Docusate sodium Danthron (terminal illness only)	Stimulate peristalsis in slow transit constipation	Often work in 12–24 hours. Can cause abdominal cramping. No evidence of harm with long-term use, but may cause discolouration of colonic mucosa (melanosis coli).
Rectal agents	Glycerol suppositories	To initiate defecation or to evacuate completely	Often work within 5–20 minutes. Can be more predictable and controllable than oral preparations.
	Bisacodyl suppositories	Stimulant, to initiate defecation if glycerol does not work	Not acceptable to all patients.
	Tap water enemas	Can use low volume to initiate or higher volume to irrigate	Patients with disabilities may find them difficult to use independently.
	Microenemas (e.g. sodium citrate)	Stimulate rectal contraction	
	Phosphate enemas	When micro enemas are ineffective	

Unfortunately, there is very little research on which laxative is best, and there is really no substitute at present for trial and error to find the best preparation (most effect with least inconvenience/side-effects) for each individual. Almost all oral laxatives seem to become less effective with time, so dependence on a single preparation is best avoided. Rotating a few effective preparations and keeping doses as infrequent and low as possible is important for long-term users, as is maximizing the use of nondrug interventions, such as diet and habit. Some patients find rectal evacuants more predictable in the timing of effect and, thus, more convenient.

Exercises for Anal Incontinence

Pelvic floor muscle retraining is widely used for treatment of AI. Yet, there is no consensus about the optimum frequency or intensity for an exercise program. Indeed, there are very few reports of exercises being used without the additional use of biofeedback.[14,15] In clinical practice, exercise instructions are very similar to those given for urinary incontinence. Patient instructions are available at www.bowelcontrol.org.uk. One study found no additional benefit of exercises, over that which was gained from advice, information, and deferment techniques.[16] However, other studies have shown benefit, with women performing exercises for urinary incontinence after childbirth reporting much less AI at 12 months postpartum.[17]

Bowel Retraining

Patients with urgency will often benefit from an urge resistance program akin to bladder retraining for urge urinary incontinence. It may only take a single episode of major FI in public to set up a vicious circle of anxiety, which stimulates colonic peristalsis, leading to urgency, which engenders panic, and a flight to the toilet with an accident on the way. Hypersensitivity to rectal contents becomes a self-protective mechanism with constant monitoring of sensation. Any arrival of stool or gas results in a sensation of urgency and running to the toilet "just in case." This cycle can be broken, but it requires considerable effort and may take several months. It can be helped by rectal balloon distension to help re-learn that the urge will wear off if resisted, or by loperamide to dampen motility (see next section).

Medication for Fecal Incontinence

The aim of medication for AI is to firm the stool, making it less likely to leak, and to slow gut motility. Loperamide is the drug of first choice for this as it has a rapid onset of action and is safe in a range of doses.[1] Up to 16 mg per day is commonly used, and is adjusted to achieve the desired formed, but not hard, stool. Syrup formulation is available if tablets are too constipating. This allows for the accurate adjustment of dose to individual needs. It is most effective if taken about 30 minutes before eating, as it also dampens the gastrocolic response. Development of side effects, tolerance, or addiction is unusual with loperamide. Codeine phosphate is an alternative, but can be associated with drowsiness and eventual addiction. Early work with alpha-adrenergic stimulants to raise anal resting pressure has not so far lived up to early promise of clinical efficacy.[18]

Biofeedback for Anal Incontinence and Constipation

Biofeedback is often cited as the management of first choice for AI.[19] However, unlike the literature on urinary incontinence, in the colorectal and gastrointestinal literature, several different interventions and complex packages of care have been grouped under the term "biofeedback." Therefore, when reading the literature on success rates it is crucial to determine what intervention was actually used in the name of "biofeedback."

As originally described, biofeedback was used to enhance rectal sensitivity to distension.[20] By using a rectal balloon, filling it to the threshold of sensation, and then attempting to "teach" the patient to detect smaller and smaller volumes, it was intended to enable the patient to detect rectal contents sooner. Patients were then taught to respond appropriately to resist the urge to defecate by squeezing the anal sphincter to resist the drop in anal pressure with rectal distension (the rectoanal inhibitory reflex). This technique remains the most commonly used in the USA, particularly in gastroenterology settings. The same equipment, or a simpler manometric or EMG anal probe or surface electrodes can be used to teach anal sphincter exercises and to give the patient feedback about the progress of muscle training.

Biofeedback for evacuation difficulties usually involves displaying sphincter activity and training relaxation to coordinate with pushing.

Does Biofeedback Make a Difference?

Clinical series have almost always reported positive results using biofeedback for AI or constipation, with two-thirds of patients reporting improvement of FI.[14,21] There is also evidence that certain neurological patients (e.g., spina bifida, multiple sclerosis) can benefit from retraining techniques.[22,23] Some have reported a differential response between constipation and FI, whereas others have found good responses in both. Useful clinical changes do not necessarily depend on changes in the physiological parameter being "trained."[24] Improved bowel function has been found to be associated with a measurable improvement in autonomic gut function.[25]

There is a small but growing body of evidence from controlled studies that the biofeedback may not be the crucial element, with equally good results obtained from well-supervised conventional management in children and adults.[16,26–28] It has been shown that a computer display of sphincter EMG does not enhance the response of constipated patients when compared with balloon expulsion in the absence of visual feedback.[29] This does not mean that biofeedback techniques are not helpful with individual patients, but it does mean that it is very feasible to set up a service for these patients without major capital investment.

Electrical Stimulation

Electrical stimulation could in theory be helpful for improving striated muscle function, improving sensory function and ability to exercise the correct muscle, and/or inhibiting unwanted rectal motility. Anal electrical stimulation has been reported as helpful in a few case series, but was not helpful in others. In the absence of good randomized studies, it is impossible to determine if this is likely to be useful.[30]

Conclusion

Conservative management for AI and constipation often yields clinically useful improvement. However, it is not clear which elements are most effective for which patients, and more research is needed to refine techniques.

Cases

Patient 1: Ms. JD

26-year-old office worker with a history of severe constipation, which she reports started when she left home at age 16, after prolonged disputes with her parents. Onset probably coincided with chaotic lifestyle and erratic eating habits, but symptoms persist now despite a settled lifestyle and a happy relationship with a supportive partner. Opens her bowels once in 10–14 days by taking large doses of sodium picosulphate, with excessive straining and bloating. Hard pellet stool with bright red bleeding.

Assessment

An intelligent, motivated patient with no other health concerns, no medications, a balanced diet, good fluid intake and a reasonable level of exercise. Transit study showed slow colonic transit on all 3 marker sets. Digital examination was normal. Attempted balloon expulsion showed no paradoxical anal contraction, but no relaxation and poor propulsive effort with straining (Valsalva).

Intervention

Intervention included detailed patient education, demonstration of good evacuation posture and technique while breathing, and a daily routine (Table 4.8.1). All laxatives were stopped and glycerine suppositories allowed only if no stool was produced in 5 days. She was advised that symptoms might worsen initially, but the importance of retraining the bowel to act without laxatives was stressed.

Outcome

After 5 monthly sessions of advice, encouragement, and repeat balloon expulsion at sessions 2 and 3, she was opening her bowels once in 3–4 days with less effort, no laxatives, and no bleeding. She was content with this outcome.

Patient 2: Mrs. SF

A 56-year-old woman with constipation, which she felt had started after her abdominal hysterectomy 9 years previously. This changed 18 months ago, and she developed frequency, with some bleeding and FI. Mrs. SF is obese (BMI 32), has type 2 diabetes, and eats a nutritionally poor diet. She also reports stress urinary incontinence. She currently opens her bowels 2–4 times per day to a variable stool consistency with urgency, rare urge FI, and soiling most days. Evacuation is sometimes easy, but she often has to strain and has a feeling of incomplete evacuation. In the past she has had 3 vaginal deliveries; the largest baby was 4.2 kg and was delivered using forceps.

Assessment

In view of her change in bowel habit, a full colonoscopy was ordered, which was normal. Anal ultrasound showed intact, but atrophic, anal sphincters. Manometry showed reduced resting and squeeze anal pressures and slightly blunted anorectal sensation. Proctogram showed a large (4 cm) rectocele, which did, however, empty completely. There was major pelvic floor descent with evacuation. Digital examination found inability to localize anal squeeze and rapid fatigue. Inspection revealed no obvious atrophic vaginitis.

Intervention

A combination of anal sphincter exercises to address perineal descent and sphincter weakness, urge resistance training to combat urgency, with evacuation training and advice to try gentle digital vaginal pressure to assist complete emptying of the rectocele. Detailed advice was also given on diet and avoidance of artificial sweeteners and caffeine, and she was encouraged to join a weight reduction and exercise class. Her family doctor was contacted and her diabetic oral medication was changed to one less likely to cause loose stool. She initially found the program challenging and was seen twice weekly to help maintain motivation. This was increased to monthly visits once she was progressing.

Outcome

After a total of 6 sessions she was improved, but symptoms were still troublesome. She was referred to a counselor to address apparent depressive symptoms and motivation with weight loss and exercise. Another 6 months later she was still making progress and had lost weight, seeming a lot more confident. At this point, home anal electrical stimulation (35 Hz, 20 minutes daily) was added and, after 4 months, she was fully continent.

References

1. Norton C, Whitehead WE, Bliss DZ, et al. Conservative and pharmacological management of faecal incontinence in adults. In: Abrams P, Khoury S, Wein A, Cardozo L, editors. Incontinence (Proceedings of the Third International Consultation on Incontinence). Plymouth: Health Books; 2004.
2. Horton N. Behavioural and biofeedback therapy for evacuation disorders. In: Norton C, Chelvanayagam S, editors. Bowel continence nursing. Beaconsfield: Beaconsfield Publishers; 2004.
3. Chiarelli P, Markwell S. Let's get things moving: overcoming constipation. East Dereham: Neen Healthcare; 1992.
4. Norton C, Kamm MA. Bowel control – information and practical advice. Beaconsfield: Beaconsfield Publishers; 1999.
5. Heaton KW. Understanding your bowels. London: Family Doctor Publications, British Medical Association; 1995.
6. Klauser AG, Voderholzer WA, Heinrich CA, et al. Behavioural modification of colonic function – can constipation be learned? Dig Dis Sci. 1990;35:1271–1275.
7. Brown SR, Cann PA, Read NW. Effect of coffee on distal colon function. Gut. 1990;31:450–453.

8. Cheskin LJ, Crowell MD, Kamal N, et al. The effects of acute exercise on colonic motility. J Gastrointest Motility. 1991;4:173–177.
9. Everhart JE, Go VL, Johannes RS, et al. A longitudinal survey of self-reported bowel habits in the United States. Dig Dis Sci. 1989;34:1153–1162.
10. Muller-Lissner SA, Kamm MA, Scarpignato C, et al. Myths and misconceptions about chronic constipation. Am J Gastroenterol. 2005;100(1): 232–242.
11. Bliss DZ, Jung H, Savik K, Lowry AC, et al. Supplementation with dietary fiber improves fecal incontinence. Nurs Res. 2001;50(4):203–213.
12. Chiotakakou-Faliakou E, Kamm MA, Roy AJ, et al. Biofeedback provides long-term benefit for patients with intractable, slow and normal transit constipation. Gut. 1998;42:517–521.
13. Clayden GS, Hollins G. Constipation and faecal incontinence in childhood. In: Norton C, Chelvanayagam S, editors. Bowel continence Nursing. Beaconsfield: Beaconsfield Publishers; 2004.
14. Norton C, Kamm MA. Anal sphincter biofeedback and pelvic floor exercises for faecal incontinence in adults – a systematic review. Aliment Pharmacol Ther. 2001;15:1147–1154.
15. Norton C, Hosker G, Brazzelli M. Biofeedback and/or sphincter exercises for the treatment of faecal incontinence in adults (Cochrane review). The Cochrane Library; 2002
16. Norton C, Chelvanayagam S, Wilson-Barnett J, et al. Randomized controlled trial of biofeedback for fecal incontinence. Gastroenterol. 2003;125:1320–1329.
17. Glazener CM, Herbison P, Wilson PD, et al. Conservative management of persistent postnatal urinary and faecal incontinence: randomised controlled trial. Br Med J. 2001; 323: 593–596.
18. Cheetham M, Kamm MA, Phillips RK. Topical phenylephrine increases anal canal resting pressure in patients with faecal incontinence. Gut. 2001;48:356–359.
19. Whitehead WE, Wald A, Norton N. Treatment options for fecal incontinence: consensus conference report. Dis Colon Rectum. 2001;44:131–144.
20. Engel BT, Nikoomanesh P, Schuster MM. Operant conditioning of rectosphincteric responses in the treatment of faecal incontinence. N Eng J Med. 1974;290:646–649.
21. Heymen S, Jones KR, Ringel Y, et al. Biofeedback treatment of fecal incontinence: a critical review. Dis Colon Rectum. 2001; 44:728–736.
22. Wald A. Biofeedback for neurogenic faecal incontinence: rectal sensation is a determinant of outcome. J Pediatr Gastroenterol Nutr. 1983;2:302–306.
23. Wiesel PH, Norton C, Roy AJ, et al. Gut focused behavioural treatment (biofeedback) for constipation and faecal incontinence in multiple sclerosis. J Neurol Neurosurg Psychiatry. 2000;69(2):240–243.
24. Ko CY, Tong J, Lehman RE, et al. Biofeedback is effective therapy for fecal incontinence and constipation. Arch Surg. 1997; 132(8):829–833.
25. Emmanuel AV, Kamm MA. Successful response to biofeedback for constipation is associated with specifically improved extrinsic autonomic innervation to the large bowel. Gastroenterol. 1997;112: A729.
26. Solomon MJ, Pager CK, Rex J, et al. Randomised, controlled trial of biofeedback with anal manometry, transanal ultrasound, or pelvic floor retraining with digital guidance alone in the treatment of mild to moderate fecal incontinence. Dis Colon Rectum. 2003;46(6):703–710.
27. Loening-Baucke V, Desch L, Wolraich M. Biofeedback training for patients with myelomeningocele and faecal incontinence. Dev Med Child Neurol. 1988;30:781–790.
28. Whitehead WE, Parker L, Bosmajian L, et al. Treatment of fecal incontinence in children with spina bifida: comparison of biofeedback and behavior modification. Arch Phys Med Rehabil. 1986;67(4):218–224.
29. Koutsomanis D, Lennard-Jones JE, Roy A, et al. Controlled randomised trial of visual biofeedback versus muscle training without a visual display for intractable constipation. Gut. 1995;37:95–99.
30. Hosker G, Norton C, Brazzelli M. Electrical stimulation for faecal incontinence in adults (Cochrane review). The Cochrane Library; 2002.

4.9 Incontinence During Sports and Fitness Activities

Alain P. Bourcier

Key Messages

- Pelvic floor disorders are a significant problem for women who exercise, especially those involved in high impact sports
- Recommendations for exercise incontinence include vaginal devices and pelvic floor muscle (PFM) training, with emphasis on PFM precontraction (perineal blockage) to prevent urethral descent and leakage during exertion
- Techniques of biofeedback include training in different positions, simulation of daily activities, and PFM practice while performing real physical activities
- The goal is to provoke a reflex PFM contraction in response to physical stress

Introduction

Pelvic floor disorders are widely regarded as problems affecting older, postmenopausal, multiparous women. Many do not realize that it is also a significant problem for female exercisers of all ages. The prevalence of urinary stress incontinence in young female athletes and women who exercise ranges from 8% to 47%.[1-6] Physically active women are more likely than sedentary women to experience incontinence,[3] and the problem is most common in high-impact sports. Because stress incontinence occurs during physical exertion, this condition represents a major problem for females participating in fitness or sports activities.

Etiology

Repeated increases in intraabdominal pressure on the pelvic floor structures may overload and weaken pelvic floor muscles (PFM) over time and lead to loss of support to the bladder neck and urethra. Fatigue of the PFMs might be the reason why women with symptoms of stress incontinence often describe that leakage starts only after several incidents, such as repetitive jumps, indicating that the continence control system weakens after repetitive stresses.

In addition to this mechanical aspect, a combination of weak connective tissue and/or partial pudendal denervation may worsen the problem.[2,7] It has also been suggested that chronic repetition of intraabdominal pressure may damage the cardinal and uterosacral ligaments and connective tissue, as well as neuromuscular structures, resulting in loss of neural control and weakness of the PFM.[8]

High-impact movements result in impact forces up to 3 or 4 times a person's body weight. Investigators have found that the flexibility of the foot arches in incontinent athletes was significantly lower than in the continent athletes,[9] indicating that the way in which impact forces are absorbed may be another etiological factor.

Pelvic Floor Rehabilitation

For female athletes with "exercise incontinence," recommendations include vaginal devices – either tampons or pessaries – to elevate the bladder neck and help avoid leakage during activity. But even those who use such devices should also do PFM exercises, which can dramatically relieve the problem.

Most rehabilitation programs teach women to contract PFMs before and during increased intraabdominal pressure, particularly with strenuous effort.[10] This technique is called the "stress strategy," "the perineal blockage before stress technique," or "the Knack." A quick, strong, well-timed PFM contraction can compress the urethra (increasing urethral pressure), prevent urethral descent, and prevent leakage during an abrupt or sustained increase in intraabdominal pressure. Eventually, women will be conscious of a constant contraction of the levator ani and can use the technique of perineal blockage before physical stress. PFM strength and control are also important in the control of urge incontinence associated with sports.[11] (See Chapter 4.6).

Techniques of Biofeedback

Urinary stress incontinence and pelvic organ prolapse are primarily related to the erect posture. In the vertical position, the urethra leaves the bladder at the point of maximum combined intraabdominal pressure and gravity force. Intraabdominal pressure in the standing position is two or three times greater than that in the supine position. One technique of applied biofeedback uses electromyography (EMG) of the levator ani combined with synergistic and antagonistic muscle activity in a standing position (Fig. 4.9.1). The method consists of teaching PFM

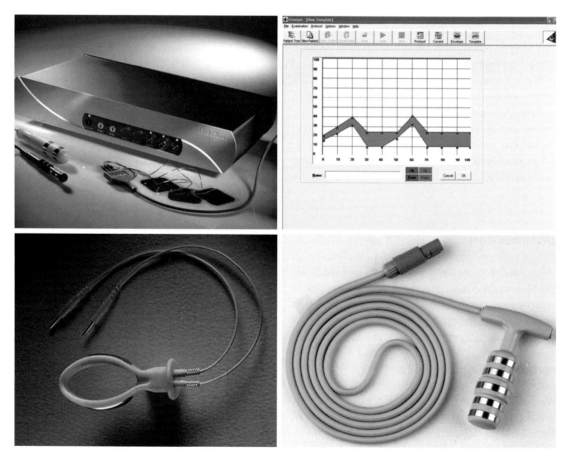

FIGURE 4.9.1. Multichannel system used for biofeedbock (Laborie Inc.) Different vaginal probes are available for easy insertion and possibility of standing position for pelvic floor muscle training program.

contraction during the abrupt increasing intraabdominal pressure that occurs during standing and exercising.[12]

Equipment is used that allows women to assume different positions as they learn to use the PFM.[10,12] The trainer helps women change their responses by establishing targets and assisting them to develop new habits and modify their physical activities. The simulation of daily activities is a very important stage, in which selected "home stresses" or physical task are used to access the woman's ability to perform a real-life activity. The woman is asked to perform various tasks, very similar to domestic activities, such as carrying a baby basket or lifting items from the floor. In general, one should start with easy tasks and progressively make the activities more difficult and more functional.

Successful recovery of the ability to perform daily activities without incontinence is most likely to occur when the PFM strength is combined with a refined control of the functional activity. For this purpose, equipment, such as a video monitor connected to a computerized unit, is used to monitor EMG activity of muscle groups (Fig. 4.9.2).[5] It is optimal for women to be standing when they perform these exercises, but if the woman has some difficulty in performing the exercises, it is suggested that she practice in different positions, such as lying and sitting.

Sportswomen should use a treadmill, which provides an opportunity to combine speed and endurance. The underlying concept is that the circumstances and precipitants of incontinence must be taken into account during treatment sessions to provoke a reflex PFM contraction in response to physical stress. Accuracy of speed (from 2.5mph to 6mph) and slope (from 5% to 12%) is assumed for a precisely controlled workout. The introduction of the treadmill or other sports equipment into the program for sportswomen is one way to allow the woman to perform real physical activities in a medical environment. Recently, another technique has been introduced with telemetry (wireless biofeedback), in which the infrared connection allows more freedom for the woman while practicing different physical activities.

FIGURE 4.9.2. Applied biofeedback during exercises. A short vaginal sensor and electrodes wire set for abdominal muscles are used for the exercise program. The woman practices the perineal blockage technique while performing movement.

Conclusion

Stress incontinence is surprisingly high in groups of physically active women. More research is needed to clearly understand how impact from different exercises affects pelvic organs and PFMs. Based on current knowledge, specific PFM training programs are recommended as the first choice of treatment. The perineal blockage technique is a very adjunctive modality and can be proposed for active women.

References

1. Nygaard IE, DeLancey JOL, Arnsdorf L, et al. Exercise and incontinence. Obstet Gynecol. 1990; 75:848–851.

2. Nygaard IE, Thompson FL, Svengalis SL, et al. Urinary incontinence in elite nulliparous athletes, Obstet Gynecol. 1994;84:183–187.
3. Bo K, Maehlum S, Oseid S, et al. Prevalence of stress urinary incontinence amongst physically active and sedentary female students. Scand J Med Sci Sports. 1989;11:113–116.
4. Bourcier AP, Juras JC. Conservative treatment of stress incontinence in sportswomen. Neurourol Urodyn. 1990;9:232–234.
5. Fall M, Frankenenberg S, Frisen M. 456 000 svenskar kan ha urininkontinens. Endasr var fjaerde soker hjelp for besvaeren. Laekartidningen. 1985; 82:2054–2056.
6. Jacquetin B, Lambert T, Grumberg P, et al. Incontinence urinaire de la femme sportive. In: Le Pelvic Féminin, Statistique et Dynamique. Paris: Masson; 1993:142.
7. Norton P, Baker J, Sharp H, et al. Genito-urinary prolapse. Relationship with joint hypermobility. Neurourol Urodyn. 1991;9:225–228.
8. Nichols D Milley P. Functional pelvic floor anatomy: the soft tissue supports and spaces of the female pelvic organs. In: The human vagina. Amsterdam: Elsevier/North Holland Biomedical Press; 1978;21–37.
9. Nygaard IE, Glowaski C, Saltzman L. Relationship between foot flexibility and urinary incontinence in nulliparous varsity athletes. Obstet Gynecol. 1996;87:1049–1051.
10. Bourcier AP, Juras JC, Jacquetin B. Urinary incontinence in physically active and sportswomen. In: Appell RA, Bourcier AP, La Torre F, editors. Pelvic floor dysfunction: Investigations and conservative treatment. Rome: CESI; 1999:9–17.
11. Burgio KL, Locher JL, Goode PS, et al. Behavioral versus drug treatment for urge incontinence in older women: A randomized clinical trial. JAMA. 1998;23:995–2000.
12. Bourcier AP, Burgio KL. Biofeedback therapy. In: AP Bourcier, EJ McGuire, P Abrams, editors. Pelvic floor disorders. Philadelphia: Sauders-Elsevier; 2004:297–311.

4.10
Pelvic Organ Prolapse – Pessary Treatment

Jane A. Schulz and Elena Kwon

Key Message

Pessary treatment of prolapse is one of the oldest remedies in medicine and is an important conservative treatment that is particularly valuable for the physically frail. Pessaries can be used for diagnosis and treatment of prolapse, for voiding dysfunction and urinary incontinence and for the management of incontinence or retention during pregnancy. The guidelines for pessaries and the role of the woman in taking care of her pessary is emphasized. The main types of pessary and the specific indications are reviewed – choice will depend on the type of prolapse and the vaginal anatomy. The success rate and the complications and their management are outlined. The role of pelvic floor exercises and supportive garments are reviewed. The importance of future randomized control trials and establishment of clinical guidelines is emphasized.

Introduction

The lifetime risk for pelvic organ prolapse (POP) or incontinence surgery for a female by the age of 80 years old is 11.1%. Up to 30% of women will require repeat prolapse surgery, and up to 10% of women will require repeat continence surgery.[1] Treatment of prolapse depends on several factors, including the patient's wishes for management, the severity of prolapse and its symptoms, the woman's general health, and whether childbearing is completed. In the past, conservative treatment of prolapse has been reserved for those with mild prolapse, those who are too frail or unwilling to have surgical management, or for those who wish to have more children. However, because evidence indicates that we still do not have good durability of prolapse repairs, and with women living longer, conservative management options must be considered for all as a method of treatment.

Historical Perspective

Mechanical devices as a conservative management tool for POP have been used for many centuries. They were described as far back as the time of Hippocrates. Multiple variations have been described, such as pomegranates, bone, sea sponges, and various external braces (see Fig. 4.10.1). Other conservative methods included repositioning of the prolapse, leg binding, douching, herbal remedies, and the use of leeches.

Pessaries gained popularity in the 1800s for the management of uterine retroversion. All were precursors to our current pessaries and were used very frequently because of the high surgical morbidity and mortality. However, with advances in anesthesia and surgical techniques, they fell out of favor. More recently, with newer pessaries, and a wide range of styles, the longer lifespan of women, and the realization of the impermanence of surgery, mechanical devices for POP are experiencing a rebirth in popularity of use.[2]

Research in this area is still lacking. The recent Cochrane review of mechanical devices for POP in women found no eligible, completed, published

FIGURE 4.10.1. Cup and stem pessary with belt. (*Source*: Reproduced with permission from Milex Products Inc., Chicago, Illinois.)

or unpublished randomized controlled studies; therefore, no data collection or analysis was possible.[3]

Types of Pelvic Organ Prolapse and Evaluation

Conservative Management Options for POP

Pessaries

An extensive range of mechanical devices has been described for the management of pelvic floor disorders, and they are listed in Chapter 3.4. Because these devices are often underutilized, they are covered separately in this chapter, where we will consider their use for specific indications of POP. These mechanical devices consist mainly of pessaries. Pessaries are primarily made of medical grade silicone covering surgical steel. The advantages of silicone are that it has a longer lifespan for use, it can be autoclaved, it does not absorb odors or secretions, and it is an inert material. Pessaries come in a wide range of shapes and sizes (Fig. 4.10.2). They may be used to prevent prolapse from becoming worse, to decrease the frequency or severity of symptoms of prolapse, and to avert or delay surgical management. Pessaries can also be used as a diagnostic tool. Examples of their use for diagnosis include whether pessary insertion corrects the patient's symptoms of prolapse, and whether associated symptoms such as voiding dysfunction and urinary incontinence are corrected by pessary insertion. Pessaries are believed to work by creating an artificial shelf of levator support to reduce the prolapse. Incontinence pessaries also work by elevating the bladder neck back to the normal anatomic position, and by some degree of obstructive effect on the urethra.[4] Pessaries need to be fitted by a health care professional. A nurse-run clinic for pessary fitting is a good option as a time- and cost-saving measure.[5]

Indications for Pessary Fitting

Pessaries can be used for all types and all stages of POP. Pessaries can also be used for stress, urge, mixed, and overflow urinary incontinence. Although, historically, incontinence type pessa-

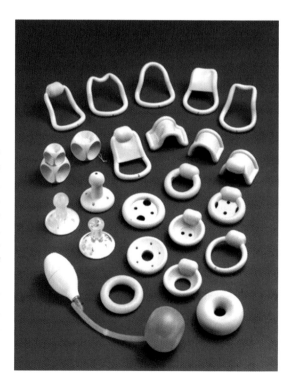

FIGURE 4.10.2. Milex pessaries. (*Source*: Reproduced with permission from Milex Products Inc., Chicago, Illinois.)

ries have been used for stress urinary incontinence, there have been reports of success with urge incontinence in 64–67% of patients.[6,7] Pessaries have been used for the diagnosis and management of latent stress urinary incontinence.[8] Hextall et al. found stress incontinence was unmasked in 27% of women in their unit with prolapse that were being investigated with urodynamics.[9] The use of a pessary before surgery is also useful to predict whether women will achieve relief of their prolapse symptoms, and whether urinary symptoms, such as urgency and voiding dysfunction, will resolve.[10] Pessaries are a valuable tool for the management of the pregnant woman who has urinary incontinence, POP, or urinary retention secondary to uterine retroversion or incarceration. In pregnancy, the size of the pessary may have to be changed with advancing gestation. Hodge pessaries work best for uterine incarceration with associated voiding dysfunction. Once the pregnant uterus moves up out of the pelvis in the second trimester, symptoms often improve.

There are few contraindications to pessary fitting. Active infections of the pelvis or vagina, such as vaginitis or pelvic inflammatory disease, preclude the use of a pessary until the infection has resolved. Allergies to the pessary are very uncommon, especially since now most are made of silicone. However, any allergic response to a vaginal pessary would be a contraindication to fitting. The only other caution is with patients who are not likely to be compliant with pessary care and follow-up; these patients should not be fitted with a device.[19]

Guidelines for Pessary Fitting

An adequate amount of time should be allotted for pessary fitting. A clinical setting in a nurse-run pessary clinic is ideal. In postmenopausal women, pretreatment with local estrogen therapy for at least 6 weeks is helpful to optimize successful fitting.[6] A postvoid residual should be checked before pessary fitting, as pessaries can cause obstruction of urinary flow. To fit a pessary, size the vaginal vault by examining the vagina with two fingers. Start with a covered ring pessary, or the appropriate design for the diagnosis. The pessary should be lubricated on the end and then inserted and tilted up behind the pubic symphysis (Fig. 4.10.3). A fingerbreadth should fit between the pessary and the vaginal mucosa. Once the pessary has been fitted, the patient should walk around and exercise in the clinic to ensure it will not immediately fall out. It is necessary to ensure that patients are able to void and are given appropriate education before leaving the clinic with their new pessary. If possible, patients should be taught to remove, clean, and replace their pessaries themselves. If pessaries are difficult to remove, fishing wire or dental floss may be attached to the pessary to aid in removal. There are few guidelines for pessary removal and cleaning, and the recommendations that do exist are variable. Current Canadian practice advises any woman who is able to remove her own pessary to remove, wash, and replace it once per week. The patient is advised to wash the pessary in soap and warm water. If she cannot remove her pessary, she should have it removed, cleaned, and replaced every 3 months by a health care professional.[7] Similar guidelines are followed by family physicians and gynecologists.[13,19] Women may have intercourse with their pessary in place; however, many elect to remove it.

Figure 4.10.3. Pessary fitting. (*Source*: Reproduced with permission from Milex Products Inc., Chicago, Illinois.)

Proper pessary fitting may require trials of several styles and sizes. Difficulty with pessary fitting may arise if there is a large posterior wall defect, poor perineal body support, or a shortened vagina.[11] Many pessaries rely on good perineal support to remain in place. Patients that have had prior radiation or multiple pelvic surgeries may also encounter difficulties with pessary fitting because of a scarred or shortened vagina.[11,12] Peri- or postmenopausal women with significant vaginal atrophy may have significant discomfort when pessary fitting is attempted. In this situation, 4 to 6 weeks of local estrogen therapy is often helpful to increase the success rate of pessary fitting.[5]

Types of Pessaries

Ring pessaries are the most widely available and most commonly used;[13] they are available in open and covered forms. The covered ring pessary has drainage holes to allow the vaginal secretions to escape; it is useful in patients that still have a uterus, to prevent the cervix from slipping through the ring. Open and covered ring pessaries are best used in POPQ stage I and II prolapse.[14] The Shaatz pessary is a stiffer circular pessary that is used when more support is required for management of the prolapse. Shaatz pessaries can be used if the rings fall out, or if there is still protrusion of the prolapse beyond a ring pessary. Doughnut pessaries are shaped like their namesake; they are used for more significant uterine prolapse, especially if accompanied by anterior and posterior wall descent. A variation of the doughnut pessary is the Inflatoball; this is made of latex and must be deflated daily for removal and cleaning. The Inflatoball pessary is used in patients with a narrow introitus but a capacious upper vagina. The Regula is a newer pessary for mild prolapse. Its unique bridge-shaped design helps to prevent expulsion.[14]

The Gellhorn, or stem, pessary is indicated for more advanced stage III or IV prolapse. It is often useful in reducing a complete procidentia or vaginal vault eversion. Like the other pessaries, the Gellhorn creates an artificial levator shelf, but also creates a suction to provide a little more support. The stem helps to prevent the pessary from shifting position. The Gellhorn is more difficult to remove, and cannot be used if a patient is sexually active, unless she is able to remove the pessary herself. To remove the Gellhorn, the suction must be broken at the dish of the device; occasionally, a Kelly clamp is required to pull on the stem and assist with removal.

The cube or tandem cube pessaries are used when other pessaries are unsuccessful, or when there is very poor pelvic tone. They work using suction to the vaginal walls, as all their sides are concave. Older cube pessaries did not have drainage holes and required daily removal and cleaning. However, newer versions do have some drainage holes that allow them to be left for up to a week. There is a string attached to the cube pessary; however, this is to assist with locating the pessary and is not for traction for removal.

There are now a variety of lever pessaries that are all modifications of Hodge's original design from the 1860s.[15] These include the Hodge, the Smith-Hodge, the Risser, and the Gehrung. The Hodge pessary has been used traditionally for uterine retroversion and incompetent cervix. Variations of this pessary are for variations in pubic arch anatomy. Traditionally, the Gehrung has been used for women with both a cystocele and rectocele, although it is sometimes difficult to keep this pessary in position.[14]

Incontinence pessaries are variations on the other forms of pessaries with an elevated knob to support the bladder neck. There are incontinence ring and dish pessaries, and now also incontinence versions of some of the lever pessaries. If a patient with prolapse develops stress incontinence after being fit with one of the other styles of pessaries, switching to an incontinence pessary may address both problems.

Success Rates with Pessaries

The reported success rates for pessaries vary by diagnosis. Vierhout reported a 63% subjective improvement or cure rate with pessary use for stress urinary incontinence.[16] In a prospective review by Clemons et al. of 100 women being fitted for a pessary for POP, 73% had a 2-week successful pessary-fitting trial.[17] Of the group that had successful pessary fitting, almost all had complete resolution of their prolapse symptoms, 50% had improvement of their urinary symptoms, and 92% were satisfied with their pessary. Dissatisfaction with pessary fitting was associated with occult stress incontinence.

In a retrospective review of 1,216 women in a tertiary care gynecology unit, 86% of women were able to be fit with pessaries, and of these 71% were able to wear them successfully. Successful fit was achieved in 83% of patients with uterine prolapse, 82% of patients with cystocele, 69% of patients with vault prolapse/enterocele, and 66% of patients with cystocele/rectocele.[6]

There is some suggestion that the use of a pessary may prevent the progression of POP.[18] However, there is still significant study required in this area.

Pessary Complications

Overall pessary complications are uncommon and affect less than 10% of patients.[6,7]

Vaginal Discharge

Vaginal discharge is one of the more common complaints with pessary fitting. With insertion of a foreign body into the vagina, it is normal to see an increase in the vaginal discharge, especially if local estrogen treatment is also being used, such as in the postmenopausal population. However, if there is patient concern, or if there are other symptoms such as foul smell, bleeding, or pruritis, the discharge should be investigated. A vaginal examination and culture can be completed. If there is a yeast infection or bacterial vaginosis, the pessary should be left out for a week while the appropriate antibiotic or antifungal treatment is used. Concern about recurrent vaginal infections is a common patient concern. However, this is unusual; in the postmenopausal population this is best prevented with local estrogen use. The use of Trimosan cream, which is provided with the Milex pessaries, has been recommended to help decrease the amount of odor and discharge, although it has not been studied in clinical trials.[13,14,19]

Vaginal Erosions

Erosions of the vaginal mucosa usually start as an area of redness or abrasion where the pessary is resting. They may present with vaginal bleeding or a change in discharge. If left untreated, they may progress to ulcers. In patients with a uterus still in place, other causes of abnormal vaginal bleeding must be ruled out. These areas may also become secondarily infected, leading to further tissue breakdown. Erosions occur in 2% to 8.9% of patients.[6,17] They usually respond well to local estrogen therapy; addition of an antibiotic cream may also be required if secondary infection has occurred. Diligent pessary care and inspection of the vaginal tissues every 6 to 12 months helps to prevent erosions. Pessary size may also have to be adjusted to prevent further erosions from developing.

Fistulas

One of the keys to long-term pessary care is ensuring that the patient takes adequate precautions to prevent the more serious complications. Fistulas, although very rare, are among the most serious complications of neglected pessaries.[20] They can be rectovaginal, vesicovaginal, or urethrovaginal. An impacted pessary can develop erosions that break down, or get infected, leading eventually to fistula. It is very important that pessaries are regularly removed, washed, and replaced, and that the vagina inspected for any signs of infection or erosion. In a cognitively impaired patient, it is imperative that a caregiver be committed to ensuring ongoing pessary care and cleaning.

Pelvic Floor Physiotherapy

Pelvic floor prolapse is an anatomical defect associated with functional changes. These may include urinary incontinence (urge, stress, and overflow), defecatory dysfunction, and pelvic pressure. There is evidence that pelvic floor exercises are helpful for some of the resultant conditions and functional changes associated with POP. These include pelvic floor exercises and bladder retraining for urinary incontinence.[21–22]

However, for the direct treatment of pelvic floor prolapse as an anatomical defect, there is little documentation regarding the effect of pelvic floor physiotherapy.[23] For mild prolapse there is a perceived benefit;[24] however, more severe prolapse is unlikely to be corrected by exercises alone. Defects such as stress urinary incontinence and POP have been associated with electromyographic changes that may represent either motor unit loss or failure of central activation,[25] and this would certainly impact the potential success of pelvic floor therapy for these conditions. Randomized clinical trials, and the establishment of clinical and referral guidelines, are required in this area.

Fembrace[26]

For many centuries, conservative management of POP relied primarily on the use of pessaries and pelvic floor exercises. Since the marketing of a new V-brace™ support garment in New York in the fall of 2000, however, there exists an alternative for POP symptom relief in patients who cannot use a pessary for various reasons (Fig. 4.10.4). The V-brace™ garment is a panty with a padded double crotch and cross elastic straps that acts to reduce the symptoms of pelvic organ prolapse by providing support and pressure to the vaginal area. The creators of the V-brace™ garment suggest that even women who use a pessary can alternate and also use the V-brace™ every other day when not wearing the pessary, for ultimate relief of prolapse symptoms. The garment is also recommended for reducing varicose veins on the vulva, as well as for reducing hip, leg, or pelvic pain.

Other Alternatives

Other options include the use of tight bicycle shorts as a perineal support in women that are unable to fit a pessary and unable to have surgery. Some women have used their contraceptive diaphragms or tampons to attempt to reduce their prolapse, or to provide relief for their urinary incontinence. However, a common complaint of women with moderate to severe prolapse is the inability to retain a tampon. Desperate patients that have come to our clinic have also described the use of sticky tape across the vaginal opening.

Summary

Pelvic organ prolapse is a prevalent condition that impacts quality of life. As women live longer, further research is needed to study conservative options to treat prolapse. Pessaries are currently the main conservative management tool.[27] Other options that exist include pelvic floor physiotherapy and the Fembrace support.

References

1. Deval B, Haab F. What's new in prolapse surgery? Curr Opin Urol. 2003; 13(4):315–323.
2. Davila GW. Vaginal prolapse: management with nonsurgical techniques. Postgrad Med. 1996; 99(4):171–81.
3. Adams E, Thomson A, Maher C, et al. Mechanical devices for pelvic organ prolapse in women (Cochrane Review). In: The Cochrane Library. Issue 2. Chichester, UK: John Wiley and Sons Ltd.; 2004.
4. Bhatia NN, Bergman A, Gunning JE. Urodynamic effects of a vaginal pessary in women with stress urinary incontinence. Am J Obstet Gynecol. 1983; 147(8): 876–884.
5. Tam F, Schulz J, Flood CG, et al. Factors affecting the success of pessary fitting in a nurse-run clinic. Int Urogynecol J. 2000;11:S17.
6. Hanson L, Schulz JA, Flood CG, et al. Vaginal pessaries in managing women with pelvic organ prolapse and urinary incontinence: patient characteristics and factors contributing to success. Int Urogynecol J Pelvic Floor Dysfunct. 2006;17(2): 155–159. Epub 2005 Jul 26.

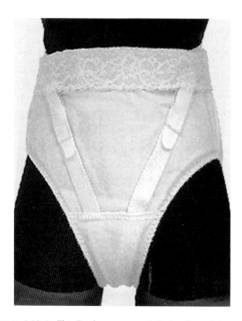

FIGURE 4.10.4. The Fembrace support device for pelvic organ prolapse.

7. Wu V, Farrell SA, Baskett TF, et al. A simplified protocol for pessary management. Obstet Gynecol. 1997; 90(6): 990–994.
8. Liang CC, Chang YL, Chang SD, et al. Pessary test to predict postoperative urinary incontinence in women undergoing hysterectomy for prolapse. Obstet Gynecol. 2004; 104(4): 795–800.
9. Hextall A, Boos K, Cardozo L, et al. Videocystourethrography with a ring pessary in situ. A clinically useful preoperative investigation for continent women with urogenital prolapse? Int Urogynecol J Pelvic Floor Dysfunct. 1998;9(4): 205–209.
10. Lazarou G, Scotti RJ, Mikhail MS, et al. Pessary reduction and postoperative cure of retention in women with anterior vaginal wall prolapse. Int Urogynecol J Pelvic Floor Dysfunct. 2004;15(3):175–178. Epub 2004 Feb 14.
11. Clemons J, Aguilar V, Tillinghast T, et al. Risk factors associated with an unsuccessful pessary fitting trial in women with pelvic organ prolapse. Am J Obstet Gynecol 2004; 190: 345–350.
12. Farrell SA, Singh B, Aldakhil L. Continence pessaries in the management of urinary incontinence in women. J Obstet Gynaecol Can. 2004 Feb; 26(2): 113–117.
13. Pott-Grinstein E, Newcomer J. Gynecologists' patterns of prescribing pessaries. J Repro Med 2001; 46(3): 205–208.
14. Milex (Chicago, Illinois). Website. http://www.milexproducts.com/products/pessaries. Accessed January 1, 2005.
15. Miller D. Contemporary use of the pessary. In: Sciarra JJ, editor. Gynecology and Obstetrics: Clinical Gynecology. Vol 1. Philadelphia: Lippincott; 1999:1–13.
16. Vierhout ME, Lose G. Preventive vaginal and intra-urethral devices in the treatment of female urinary stress incontinence. Curr Opin Obstet Gynecol. 1997;9:325–328.
17. Clemons JL, Aguilar VC, Tillinghast TA, et al. Patient satisfaction and changes in prolapse and urinary symptoms in women who were fitted successfully with a pessary for pelvic organ prolapse. Am J Obstet Gynecol. 2004;190:1025–1029.
18. Handa VL, Jones M. Do pessaries prevent the progression of pelvic organ prolapse? Int Urogyne J. 2002;13:349–532.
19. Viera A, Larkins-Pettigrew M. Practical use of the pessary. Am Fam Physician. 2000; 61(9): 2719–2726.
20. Chow S, LaSalle M, Rosenberg G. Urinary incontinence secondary to a vaginal pessary. Urology. 1997;49(3):458–459.
21. Proceedings of the international consultation on incontinence. Abrams P, Cardozo L, Khoury S, Wein A, editors. 2004.
22. Borello-France D, Burgio K. Nonsurgical treatment of urinary incontinence. Clin Obstet Gynecol. 2004;47(1):70–82
23. Wallace SA, Roe B, Williams K, et al. Bladder training for urinary incontinence in adults. (Cochrane Review). In: The Cochrane Library. Issue 1. Chichester, UK: John Wiley and Sons Ltd.; 2004.
24. Hagen S, Stark D, Cattermole D. A United Kingdom-wide survey of physiotherapy practice in the treatment of pelvic organ prolapse. Physiotherapy. 2004;90:19–26.
25. Cundiff G, Addison A. Management of pelvic organ prolapse. Obstet Gynecol Clin N America. 1998;25(4):907–921.
26. Weidner A, Barber M, Visco A, et al. Pelvic muscle electromyography of levator ani and external anal sphincter in nulliparous women and women with pelvic floor dysfunction. Am J Obstet Gynecol. 2000;183:1390–1401.
27. Fembrace Support Garment. Advertisement. Website. http://www.fembrace.com/. Accessed 18 Aug 2004.

Part V
What to Do if Physiotherapy Fails?

5.1 Stress Urinary Incontinence: Choice of Surgery

Stuart L. Stanton

Key Messages

After an adequate trial of physiotherapy, surgery should be offered, which may either be minimally or majorly invasive. It is important to be aware of the cure rates and complications, and an adequate follow-up is always needed to assess these.

Introduction

When stress urinary incontinence, despite conservative treatment, remains an intrusive symptom, it is necessary to consider surgery. Preoperative urodynamic studies, including an MSU for culture and sensitivity (Chapter 2.3.1), are necessary to confirm the diagnosis and exclude an overactive bladder and voiding. Cystometry and uroflowmetry are the basic and most helpful studies, which may be combined with video cystourethrography (radiological control – VCU) to demonstrate urinary loss and any anatomical abnormality (Chapter 2.1.6).

Surgical treatment may be classified as follows:

1. Simple
   - urethral bulking
   - midurethral tape
2. Complex
   - colposuspension – open or laparoscopic
   - sling
   - artificial urinary sphincter
   - urethral closure
   - urethral diversion

Minimally Invasive Procedures

Bulking Agents

Urethral bulking agents (e.g. collagen, Uretex, Zuidex, Macroplastique, and Durasphere) act either by preventing premature bladder neck opening or improving the urethral "seal." Their success rate varies enormously – most clinicians accept 70% as a subjective improvement and 50% as an objective improvement, but only up to 1 year.[1]

Because of these rather indifferent results, bulking agents are used as follows:

1. for mild SUI
2. as an adjunct to enhance the cure rate of continence surgery
3. where child bearing is incomplete and a procedure is needed at the moment
4. where other continence surgery is contraindicated.

The procedure is simple and quick and can be performed as a day case under local anesthetic with or without sedation. Complications include short-term voiding disorders, local (periurethral) abscess formation, systemic allergic reaction with collagen and particle migration (Teflon, silicone, and carbon-coated beads).

The bulking agent is conventionally injected at the bladder neck, but recent studies have suggested bulking should be carried out more distally.[2]

Midurethral Tape

This innovative procedure was described by Petros and Ulmsten in 1990[3] as the intravaginal sling and was refined until it became commercially available as the tension-free vaginal tape in 1997[4]: all subsequent tapes, whether retropubic or transobturator, are derived from this work. The tapes are polypropylene and may either be monofilament (preferably) or multifilament. There are now at least 20 commercially available varieties.

Midurethral tape is believed to stabilize and support the midurethral segment on effort, when the proximal and distal urethral portions of the urethra descend, leaving the midurethra supported and leading to kinking at this point (rather similar to how a garden hose can be bent to interrupt the flow).

The indications are either primary or secondary SUI at any age. Contraindications include previous tape or sling erosion of the midurethra or a fistula. Previous irradiation is a relative contraindication.

Complications include short-term voiding difficulty (11%), retention requiring resection of the tape (1–2.8%), retropubic hemorrhage, and bladder perforation at the time of surgery (0.8–21%).[2] Tape erosion into the urethra or vagina and bowel injuries are uncommon.

Complex Procedures

Colposuspension

The main continence operation in this group, which is still the gold standard for continence surgery, is the Burch colposuspension, first described by Burch in 1961.[5] Many papers have since been published about the technique, complications, and success rates.

The colposuspension can be performed either open or laparoscopically – most traditional clinicians prefer the former, which is more likely to be accompanied by posterior compartmental prolapse and voiding difficulties.[6] In a multicenter prospective random controlled trial of open and laparoscopic colposuspensions by Smith et al.,[7] the subjective and objective cure rates, mean length of hospital stay, and complications were similar for both operations. More pain was experienced by the open colposuspension.

The colposuspension works by elevating the bladder neck and may lead to some outflow obstruction. Intraoperative complications include venous hemorrhage, bladder injury, and ureteric injury, and, postoperatively, voiding difficulty and posterior compartmental prolapse.

At a mean follow-up of 12.1 years Alcalay et al. reported an objective cure rate of 94%.[8]

Sling

Sling procedures are now largely used as secondary continence operations – when other continence operations have failed and intrinsic sphincter defect is present. First described in 1907 by von Giordano, and later modified by Goebell (1910), Frangenheim (1914) and Stoeckel (1917), they were more popular in Central Europe until modified by Aldridge (1942) and Studdiford (1944) in the USA[9] and Moir (1968) in the United Kingdom.[10]

The earliest slings were of autologous tissue (rectus sheath fascia and fascia lata) then synthetic tissue (polyethylene and polypropylene), allograft tissue (cadaveric dura, fascia lata, etc.), and, more recently, xenograft tissue (porcine dermis, SIS, etc.); all require a combined abdominovaginal approach. The sling is inserted at the bladder neck (not midurethra) without tension. Cure rates vary enormously, with a quoted mean of 85%. Complications of all slings include voiding difficulties and erosion. With xenografts there is the added potential risk of transmission of viruses and DNA.[2]

Artificial Urinary Sphincter

In 1973, Scott, Bradley, and Timm described the artificial urinary sphincter (made by American Medical Systems), which could be used in both men and women.[11] In women, it is usually reserved for recurrent incontinence after failed conventional continence procedures. The device consists of a fluid-filled silastic-coated system with a cuff to be placed around the bladder neck, a pressure-regulating balloon in the retropubic space, and an activating device or pump in the left labium majus.

The sphincter allows controlled voiding via the urethra. The patient must be carefully selected – she must be mentally alert and aware of the complexity and complications of the artificial urinary sphincter and have adequate hand movements and coordination to operate the sphincter. She should have sterile urine, normal upper urinary tract, and a bladder capacity of >200 ml with minimal residual urine. Urodynamic testing should confirm USI without an overactive bladder or voiding difficulty. The sphincter is implanted through a lower abdominal incision and activated 6 weeks later to allow tissue healing. The success rate for social continence may be as high as 92%.[2] Complications include mechanical failure, cuff erosion, and infection. Careful follow-up is necessary.

Because of their complexity, sphincters should only be implanted in units (urological or gynecological) with specialized experience and capable of long-term follow-up.

The sphincter is the option for the woman with intractable incontinence who wishes to be dry, to void through her urethra, and to avoid urinary diversion. Most patients agree that it improves the quality of life.

Urethral Closure

This is a relatively simple procedure that is combined with suprapubic catheterization. Closure may be carried out vaginally (with a Martius fat pad) or suprapubically (with an omental interposition); the latter is usually more successful, but has a higher morbidity. Recanalization and fistula formation are known complications of both.[12]

Urinary Diversion

Urinary diversion is the final and most complex option for persistent urinary incontinence. Diversion may be carried out by several means. The most common is the conduit, in which a loop of small ileum is isolated and the ureters are reimplanted into one end, while the opposite end opens out through a stoma onto the anterior abdominal wall, where urine collects in a bag. The disadvantages include upper tract reflux and recurrent urinary tract infection with renal deterioration (caused by ureteric ileal stenosis) and conduit obstruction at the skin surface.

Alternatively, the ureters can be implanted into a detubularized rectosigmoid pouch (Mainz pouch), allowing the patient to void per rectum. The complications include the potential of metabolic acidosis and malignant change. Annual sigmoidoscopy is necessary to detect the latter.

Continent urinary diversion is achieved by a Mitrofanoff procedure. The bladder continues to be used as reservoir, but the bladder neck is closed off and a new "urethra" is constructed (using the appendix, ureter, or an isolated portion of ileum; Monti procedure), one end of which is anastomosed to the bladder and the other end opens out as a small stoma on the anterior abdominal wall. The bladder is emptied by frequent catheterization via the stoma. Complications include revision (as high as 25%) because of stenosis, urinary infection, stone formation, and failure to catheterize.[12]

Conclusion

There are many surgical options, and experienced advice from a urologist or urogynecologist is necessary when considering any of them.

References

1. Monga A, Robinson D, Stanton SL. Periurethral collagen injections for genuine stress incontinence: a two-year follow up. Brit J Urol. 1995;76:156–60.
2. Smith A. Surgery for urinary incontinence in women. In: Abrams P, Cardozo L, Khoury S, Wein A, editors. Incontinence – third international consultation. Plymouth, UK: Health Publications Ltd; 2005:1297–1370.
3. Petros P, Ulmsten U. An integral theory of female urinary incontinence. Experimental and clinical considerations. Acta Obstet Gynecol Scand. 1990;153:7–31.
4. Ulmsten U, Falconer C, Johnson P, et al. A multicentre study of TVT for surgical treatment of stress urinary incontinence. Int Urogynecol J. 1998:210–215.
5. Burch J. Urethro-vaginal fixation to Cooper's ligament for correction of stress incontinence,

cystocele and prolapse. Am J Obstet Gynecol. 1961; 81:281–290.
6. Alcalay M, Stanton SL. Retropubic suspensions – open retropubic suspensions. In: Stanton SL, Zimmern P, editors. Female pelvic reconstructive surgery. London: Springer-Verlag; 2002:93–101.
7. Smith A, Kitchener H, Dundee G, et al. A prospective randomised controlled trial of open and laparoscopic colposuspension. Neuro Urol Urodyn. 2005;24:422–423.
8. Alcalay M, Monga A, Stanton SL. Burch colposuspension: a ten-twenty year follow-up. Brit J Obstet Gynaecol. 1995;105:740–745.
9. Hohenfellner R, Petri E. Sling procedures. In: Stanton SL, Tanagho E, editors. Surgery for female incontinence. 2nd edition. Berlin: Springer-Verlag; 1986:105–113.
10. Moir JC. The gauze hammock operation. Br J Obstet Gynaecol. 1968;75:1–9.
11. Scott F, Bradley W, Timm G. Treatment of urinary incontinence by implantable prosthetic sphincter. Urol. 1973;1:252–259.
12. Venn S, Mundy T. Diversion and bladder neck closure. In: Stanton SL, Zimmern P, editors. Female pelvic reconstructive surgery. London: Springer-Verlag; 2002:261–269.

5.2
Genital Prolapse: Surgery for Failed Conservative Treatment

Stuart L. Stanton

Key Messages

Surgical treatment of prolapse can be deferred until symptoms become intrusive. Conditions that raise the intraabdominal pressure should be reduced. Sometimes a decision may have to be made between a vaginal or an abdominal approach, and whether to use a supportive biological or synthetic tissue. More properly controlled trials with a realistic follow-up are needed to help decision-making.

Introduction

Management of prolapse depends on the symptoms and their interference with quality of life, the extent of physical signs of prolapse, and associated conditions. If the symptoms are mild, the prolapse is stage I, and pelvic surgery is not required for any other condition (e.g. urinary incontinence, hysterectomy, or fibroids), surgical treatment may be deferred.

However, to prevent worsening of the prolapse, the following things should be minimized or avoided: weight gain, unnecessary heavy lifting, chronic cough, straining at stool, or anything else that raises the intraabdominal pressure.

Surgery is appropriate when the prolapse is a nuisance and conservative measures have failed or, rarely, when a stage 4 cystocele obstructs the ureters, leading to renal failure. Any decision to proceed to surgery should take the following into account:

1. Continued childbearing. As with urinary stress incontinence, it is wisest to defer surgery until childbearing is complete, as subsequent pregnancies and vaginal delivery are likely to disrupt any reconstructive surgery and lead to recurrence of symptoms with a decreased rate of success for further surgery.

2. Lifestyle. A job entailing heavy lifting or an active sporting life will need discussion, and even some curtailment, if a successful outcome is to be achieved. Advice will be needed to reduce excess or unnecessary physical activity.

3. Coincident pelvic conditions. An overall plan needs to be adopted if surgery is required for pelvic conditions, such as hysterectomy for fibroids, oophorectomy for ovarian pathology, or urinary or fecal incontinence.

4. Coitus. Care needs to be taken to avoid vaginal narrowing or shortening so that coitus is not compromised.

The patient's choice needs to be considered when informed consent is taken. Some women with multiple prolapses may want only one attended to and need to be warned that an incomplete cure of prolapse may result with continuation of prolapse symptoms.

Finally, the choice of operation for any prolapse may be influenced by previous prolapse surgery or other pelvic operations (e.g. fistula repair), the expertise of the surgeon, and the fitness of the patient (e.g. vaginal surgery may be less traumatic to the patient, but may not have the same success rate as an abdominal operation).

Role of Mesh

First time surgery for prolapse may fail in 30% of women, and up to a further 10% may require repeat surgery.[1] Some clinicians contest these figures and say that they are too high and that many women with prolapse may remain asymptomatic. Nevertheless, there is broad agreement that the success rate of prolapse surgery needs to be improved.

Over the last 5 years, synthetic meshes, autologous fascia, cadaveric fascia, and pig's small intestine have been used increasingly to strengthen prolapse repair. The most commonly used synthetic material is polypropylene, and of the varieties available monofilament is usually preferred because of the lower risk of infection caused by bacteria being trapped within its interstices. Whether synthetic or biological, all meshes and grafts rely on host fibroblasts growing into them to form a firm barrier and support. There are complications – the most common being rejection with infection and erosion. Because of the concern about the durability of biological grafts and the risk of infection caused by transmission of prions, synthetic meshes have become more popular, although many commercial products may have insufficient animal and human trial evidence to support the safety or efficacy of the mesh and the procedure, and these will need careful evaluation and discussion with the patient.

The main indications that meshes and grafts should be used are (a) previous failed prolapse surgery; (b) raised intraabdominal pressure, e.g. chronic obstructive airways disease, obesity, unavoidable heavy lifting; and (c) poor wound healing (e.g. coincident steroid therapy). Previous pelvic irradiation may be a contraindication to synthetic meshes.

Classification of Prolapse

The practical and anatomical classification of prolapse is:

1. Enterocele: anterior vaginal wall
   Cystocele: bladder only
   Cystourethrocele: bladder and urethra
2. Cervix and uterine prolapse: descent of the uterus
3. Vault prolapse: posthysterectomy prolapse in which the top of the vagina descends and may be indistinguishable from an enterocele. Both can contain small bowel.
4. Posterior vaginal wall: rectocele

The staging of severity or extent is fully explained in Chapter 1.6.

Surgery

Anterior Vaginal Wall Prolapse

The anterior repair, which was initially described simultaneously in 1888 by Donald in Manchester and Olshausen and Schroeder in Berlin, is the conventional procedure, with success rates varying between 31–85%; with mesh or other support this may rise to 40–100%.[3] The procedure involves a vertical incision of the vaginal wall to expose the pubocervical fascia. Several nonpermanent sutures are then used to draw the pubocervical fascia together from both sides to the midline, to buttress the urethra and bladder. Usually, some excess vaginal mucosa is excised and the remaining vaginal tissue is sutured in the midline. Care is exercised to avoid taking too much mucosa, which may lead to vaginal contraction. Catheter drainage is used afterward.

An alternative technique, the paravaginal repair, is based on the principle that a cystocele can occur when the pubocervical fascia is detached laterally from the white line (arcus tendineus). In this repair, which can be carried out vaginally, abdominally, or laparoscopically, several permanent sutures are inserted into the white line and then into the pubocervical fascia. The success rate for this procedure varies between 61% and 97%, but, to date, no randomized controlled trial has evaluated the paravaginal repair in isolation.[3,4]

Uterine Prolapse

Uterine prolapse can be managed either by a sacrohysteropexy, which leaves the uterus in place, or a hysterectomy, which can be carried out either vaginally or abdominally.

The sacrohysteropexy has a success rate of 92–100%.[3] Either an abdominal or laparoscopic approach may be used. Polypropylene or Vypro mesh is attached to the junction of the cervix and uterus and then to the anterior longitudinal ligament over the first sacral vertebra. Because mesh may shrink by up to 20%, the mesh is left quite loose and follows the curve of the sacrum. It is usually not peritonealized.

A hysterectomy is usually carried out vaginally when the main indication is prolapse and the uterus is not larger than 16 weeks. If there is concern about the difficulty of the hysterectomy, the risk of intraperitoneal adhesions, or a large uterus, it is wisest to carry out an abdominal hysterectomy. Either way, the vault has to be secured to the remnants of the broad and uterosacral ligaments to prevent subsequent vault prolapse, which may occur in 10% of patients.

Vault prolapse may be managed by many operations, including vaginal sacrospinous fixation and bilateral iliococcygeal fixation or abdominal sacrocolpopexy (either open or laparoscopic).

Sacrospinous fixation is performed by attaching the vault to one of the sacrospinous ligaments using nonabsorbable sutures; care has to be taken to avoid trauma to the pudendal nerve and vessels. This technique has the disadvantage of deviating the vagina to one side. Alternatively, a bilateral iliococcygeal fixation can be carried out, which involves securing the vault to the fascia over the left and right iliococcygeal muscles at the level of the ischial spine. There may be some shortening of the vagina in this operation. Maher and colleagues[5] have shown a similar objective success rate between the sacrospinous and iliococcygeal fixation of 67% and 53%, respectively.

The sacrocolpopexy involves suturing either Vypro or polypropylene mesh to the vault of the vagina using slowly absorbable sutures (PDS) and then to the anterior longitudinal ligament of the first sacral vertebrum. The mesh may be peritonealized and again is left loose to avoid the potential shortening due to contraction. If there is a concurrent rectocele the mesh can be extended down the posterior wall of the vagina and attached to the perineal body. The cure rate for sacrocolpopexy, either open or laparoscopic, varies between 90% and 100%.[3] The main complications of sacrocolpopexy are venous hemorrhage from the venous plexus over the sacrum and mesh erosion into the vagina.

Rectocele

The posterior colporrhaphy is the standard procedure for rectocele correction and to "tighten" the vagina in those women who complain of vaginal laxity. The conventional approach is a posterior wall vaginal incision that displays the fascia over the levator ani muscles, which are then approximated in the midline by absorbable sutures. Alternatively, a discrete fascial plication that looks for breaks in the fascia can be used, and then these breaks are sutured. The success rate for levator muscle plication is approximately 70% and for discrete fascial plication it is between 68% and 95%.[3] However, the follow-up for fascial plication at present does not exceed 18 months, so some judgment must be reserved about this procedure. Posterior colporrhaphy complications include dyspareunia and constipation. Kohli and Miklos[6] used dermal grafts and had a 93% surgical cure, whereas Sand et al[7] found no difference in cure rate when comparing a standard posterior repair with a repair using Polyglactin 910 mesh reinforcement.

Alternatively, colorectal surgeons prefer a transanal repair, where the patient is placed prone on the operating table and the anterior rectal mucosa is incised transversely, proximal to the dentate line, and dissected free of the underlying circular muscle. The circular muscle is then plicated longitudinally with 3 or 4 polypropylene permanent sutures, the excess mucosa is excised, and the defect then closed with a polydioxanone suture. Complications include prolonged wound healing of the perineum caused by infection and failure to correct any associated enterocele. The success rate of the transanal repair varies between 23% and 70%.[8]

Conclusion

Many of the trials were neither randomized nor controlled, nor did they have a follow-up of more than 2 years. Thus, there is insufficient reliable data to make a positive recommendation for one operation over another, or for a specific mesh reinforcement. Surgeons have to be guided by the patient's requirement and characteristics, and their own expertise, success, and complication rates.

References

1. Olsen A, Smith UG, Bergstrom J. Epidemiology of surgically managed pelvic organ prolapse and urinary incontinence. Obstet Gynecol. 1997;89:501–506.
2. Shaw F. Plastic vaginal surgery. In: Kerr JM, Johnstone R, Phillips M, editors. Historical review of British obstetrics and gynaecology. Edinburgh: E.S. Livingstone Ltd; 1954:370–381.
3. Brubaker L, Bump R, Fynes M, et al. Surgery for pelvic organ prolapse. In: Abrams P, Cardozo L, Khoury S, et al, editors. Incontinence. Paris: Health Publication Ltd; 2005:1373–1401.
4. Bent A, Yee A. Vaginal and abdominal paravaginal repair. In: Stanton SL, Zimmern P, editors. Female pelvic reconstructive surgery. London: Springer-Verlag; 2002:169–178.
5. Maher CF, Murray CJ, Carey MP, et al. Iliococcygeal or sacrospinous fixation for vaginal vault prolapse. Obstet Gynecol. 2001;98:40–44.
6. Kohli V, Miklos J. Dermal graft augmented rectocele repair. Int Urogynecol J. 2003;14:146–149.
7. Sand P, Koduri S, Lobel R, et al. Prospective randomised trial of Polyglactin 910 mesh to prevent recurrence of cystocele and rectocele. Amer J Obstet Gynecol. 2001;184:1357–1364.
8. Kahn M. Posterior compartment: rectocele reconstruction. In: Stanton SL, Zimmern P, editors. Female pelvic reconstructive surgery. London: Springer-Verlag; 2002:219–226.

5.3
The Anal Sphincter

Klaus E. Matzel, Manuel Besendörfer, and Stefanie Kuschel

Key Messages

We review the treatment of fecal incontinence, starting with biofeedback to teach awareness and recruitment of residual function. Should that fail, sacral nerve stimulation (SNS) is suggested. After a percutaneous test stimulation, a neurostimulatory device is implanted. The indications for this are a defect in the internal or external anal sphincter or neurogenic incontinence. Reconstructive techniques involve surgical repair of the anal sphincter by several techniques, with approximately 66% of patients showing improvement. Should this fail, sphincter replacement using either autologous tissues (e.g. gracilis transposition), the artificial bowel sphincter (developed by American Medical Systems), or stomal diversion can be considered. Outcome measurements for all of these techniques is disappointingly rare.

Introduction

Biofeedback is the first choice for functional rehabilitation. Based on the principle of operant conditioning, visual or acoustic signals are used to teach the patient awareness and use of specific physiologic functions and, thus, to recruit residual function. Success ranges widely, from 38% to 100%.[1] Retrograde irrigation is intended to improve rectal reservoir function (by distension and improved perception through a defined stimulus) and to establish a rhythm for sufficient bowel emptying (to ensure time intervals free of fecal loss). Only if these conservative therapies fail to improve symptoms should surgical intervention be considered.

Sacral Nerve Stimulation

Sacral nerve stimulation is based on the concept of recruiting residual function of the continent organ by stimulation of its peripheral nerve supply.[2] Various physiological functions contributing to continence are activated by low-frequency electrostimulation of 1 or more sacral spinal nerves by a fully implantable neurostimulation device that looks like a pacemaker.[3] A percutaneous test stimulation is usually done before surgery. The results of the test stimulation have a highly predictive value. Implantation of the final permanent neurostimulation device will only take place when a >50% reduction in incontinent episodes or in days with incontinence is achieved.[4]

With the help of test stimulation, the spectrum of indications for SNS has been continuously expanded to patients suffering from fecal incontinence, which is attributable to a wide variety of causes: weakness of the external anal sphincter, with concomitant urinary incontinence or a defect and/or deficit of the smooth-muscle internal anal sphincter; status postrectal resection; limited structural defects of the external anal sphincter combined with limited defects of the internal anal sphincter; and neurogenic incontinence.

The short-and long-term effects of SNS have been demonstrated in multiple single and multicenter trials. With chronic SNS the frequency of involuntary loss of bowel content is reduced, both the ability to postpone defecation and quality of life is improved,[5] and a substantial percentage of patients gain full continence. Complications are rare. In only less than 5% of patients has device removal become necessary, mostly because of pain or infection. After removal because of infection, reimplantation can be performed successfully at a later date.

The physiological mode of action of SNS is not yet clearly understood. Its effect is complex and multifactorial, involving somatomotor, somatosory, and autonomic functions of the anorectal continence organ.

Reconstructive Techniques

Morphological reconstruction is indicated if a defined, functionally relevant, sphincteric defect is diagnosed. Sphincter repair aims to reestablish function by reconstructing the morphological defect: a muscular gap is closed by adaptation of the dehiscent muscle. Several techniques, such as direct overlapping sphincter repair, postanal repair, and total pelvic floor repair, have been advocated in the past, and sphincter repair is now generally accepted as first-line treatment for incontinence caused by sphincteric defects. The results of sphincter repair are not reported uniformly and, thus, it is difficult to evaluate series and to compare the outcome of this technique with that of other procedures. Moreover, prospective outcome recording is rare; most reported results are based on patients' recall and are limited to functional issues without addressing quality of life. Approximately two thirds of patients report a significant improvement in continence. However, the long-term therapeutic effect of sphincter repair has recently been questioned, as several studies have reported a deterioration in function over time.

If sphincter repair – despite reestablishment of morphological integrity – fails to achieve success, or if function deteriorates over time, patients can be considered for functional rehabilitation, such as biofeedback, irrigation, and sacral nerve stimulation.

Sphincter Replacement

Sphincter replacement procedures are indicated if functional rehabilitation is not successful, if incontinence is the result of a substantial muscular defect that is not suitable for sphincter repair, or if a neurological defect is present. Two techniques have gained broad acceptance: dynamic graciloplasty (DGP)[6] and the artificial bowel sphincter (ABS).[7] The indications for both procedures are similar; end-stage incontinence in patients with a substantial muscular and/or neural defect of the anal sphincter complex. Both procedures represent an alternative to the creation of a stoma.

Dynamic Graciloplasty

Dynamic graciloplasty is a modification of the transposition of the gracilis muscle around the anus to function as a neosphincter, which was first described in the early 1950s.[8] The aim of this transposition is to encircle the anal canal completely with muscle tissue. Thus, the configuration of the muscle sling – alpha, gamma, or epsilon configuration – is determined by the length of the muscle and its tendon. This passive muscle wrap is rendered dynamic by the implantation of a neurostimulation device that is placed subcutaneously. Therefore, the innervation of the gracilis muscle must be intact. To adapt the muscle to prolonged contraction, the periods of stimulation are increased in stepwise fashion. The stimulator may be deactivated by an external magnet. Thus, bowel emptying becomes a voluntary act.

Artificial Bowel Sphincter

The ABS (Acticon neosphincter; American Medical Systems) consists of three components: an inflatable Silastic cuff placed around the anus via perianal tunnels; a liquid-filled, pressure-regulating balloon positioned in the preperitoneal fat; and a manual pump connecting these components, which is placed in either the labia majora or the scrotum. The anal canal is closed as the cuff fills with liquid. At the time for defecation the device is deactivated via the manual pump; the cuff empties and the anus opens to pass stool. The cuff is refilled and the anus is closed after a few minutes.

As with dynamic graciloplasty, opening of the ABS becomes a voluntary act and closure of the anal canal is maintained without conscious effort, mimicking the initiation of defecation in the healthy. Compared with DGP, there is a higher risk of infection with this implanted artificial material, especially if the Silastic cuff of the ABS cannot provide sufficient coverage.

Short- and long-term effects on function and quality of life have been published in several studies, both single- and multicenter.[9] Again, outcome measurement is inconsistent and data must be interpreted cautiously.

Both sphincter replacement procedures are associated with substantial morbidity in virtually all reports. In larger multicenter trials, the need for operative revision reached 42% for the DGP[10] and 46% for the ABS[11]; treatment had to be discontinued in 8% and 30%, respectively. The most severe complications were infections. Their occurrence is not surprising when one bears in mind that the operation is performed in a naturally contaminated area. In most cases, device removal is unavoidable. The functional complication most clinically relevant is outlet obstruction. This may be caused by a preexisting obstruction that is not identifiable because of incontinence or by "hypercontinence" subsequent to neosphincter creation. In most cases, this functional problem can be treated with regular enemas.

Stoma Creation

The creation of a diverting stoma should be considered an alternative to surgery for end-stage incontinence, although it doesn't address incontinence, per se, if comorbidity or intellectual or physical inability precludes the above-described sphincter replacements. Stoma creation carries its own risks, however, and patient counseling and performance of the procedure and postoperative management should be handled with great care.

Summary

The surgical options for fecal incontinence have increased during recent years, and a new treatment algorithm has evolved (Fig. 5.3.1). Symptoms and quality of life can be improved if patient

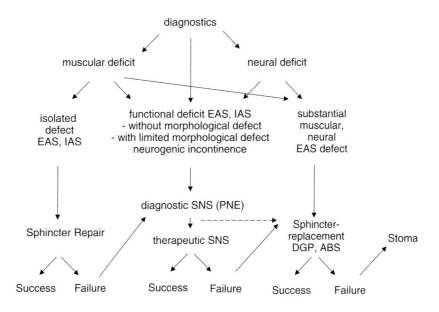

FIGURE 5.3.1. Surgery for fecal incontinence treatment alogrithm. EAS: external anal sphincter, IAS: internal anal sphincter.

selection is appropriate. Although these procedures carry some morbidity, they may offer an alternative to the creation of a diverting stoma.

References

1. Madoff RD, Parker SC, Varma MV, et al. Fecal incontinence in adults. Lancet. 2004;364:621–32.
2. Matzel KE, Schmidt RA, Tanagho EA. Neuroanatomy of the striated muscular anal continence mechanism: Implications for the use of neurostimulation. Dis Colon Rectum. 1990;33:666–673.
3. Matzel KE, Stadelmaier U, Hohenfellner M, et al. Electrical stimulation for the treatment of faecal incontinence. Lancet. 1995;346:1124–1127.
4. Matzel KE, Stadelmaier U, Hohenberger W. Innovations in fecal incontinence: Sacral nerve stimulation. Dis Colon Rectum. 2004;47;1720–1728.
5. Matzel KE, Kamm MA, Stösser M, et al. MDT 301 study group. Sacral nerve stimulation for fecal incontinence: a multicenter study. Lancet. 2004; 363:1270–1276.
6. Baeten C, Bailey RA, Bakka A, et al. Safety and efficacy of dynamic graciloplasty for fecal incontinence: Report of a prospective multicenter trial. Dis Colon Rectum. 2000;43:743–751.
7. Lehur PA, Roig J, Duinslaeger M. Artificial anal sphincter: Prospective clinical and manometric evaluation. Dis Colon Rectum. 2000;43:1213–1216.
8. Pickrell KL, Broadbent TR, Masters FW, et al. Construction of a rectal sphincter and restoration of anal continence by transplanting the gracilis muscle; a report of four cases in children. Ann Surg. 1952;135:853–862.
9. Chapmann AE, Geerdes B, Hewett P, et al Systematic review of dynamic graciloplasty in the treatment of faecal incontinence. Br J Surg. 2002;89: 138–153.
10. Matzel KE, Madoff R, LaFontaine LJ, et al and the Dynamic Graciloplasty Therapy Study Group. Complications of dynamic graciloplasty: Incidence, management and impact on outcome. Dis Colon Rectum. 2001;44:1427–1435.
11. Wong WD, Congliosi SM, Spencer MP. The safety and efficacy of the artificial bowel sphincter for fecal incontinence: results from a multicenter cohort study. Dis Colon Rectum. 2002;45:1139–1153.

Index

A

Abdominal/pelvic floor EMG, uroflow with, 114
Abdominal ultrasound, 137–138
Advanced uterovaginal prolapse, rectal prolapse and, 79
Afferent pathways, of PFMs, 27–28
Aging, muscle function and, 49–58
AI. *See* Anal incontinence
Alcock's canal, 12, 27
Alpha-agonists, for SUI, 225
Alternative methods, of pelvic floor muscle awareness/training, 208–211
Anal canal, high-pressure zone of, 124
Anal fissure, management pathway for, 244
Anal incontinence (AI), 75–82, 259–265
  anal sphincter laceration and, 41–42
  anatomy of, 75–76
  biofeedback for, 263–264
  cases of, 264–265
  exercises for, 263
  introduction to, 75, 259
  key message of, 75, 259
  neurogenic mechanism of, 81
  obstetric/maternal risk factors for, 42
  pathophysiology of, 77
  patient education and, 259
  physiology of, 75–76
Anal motility, electrical stimulation and, 264
Anal pressures, normal mean, 125

Anal probes, for electrical stimulation, 192
Anal/rectal sensation, 129
  balloon distension and, 129
Anal sphincter, 289–292
  anatomy of, 76
  complex, anatomy of, 18–19
  continuously firing tonic motor unit from, frequency histogram of, 24
  dynamic graciloplasty for, 290
  introduction to, 289
  key messages of, 289
  laceration, anal incontinence and, 41–42
  mechanism of, 76
  reconstructive techniques, 290
  replacement of, 290
  sacral nerve stimulation and, 289–290
Anal tear, management pathway for, 245
Anismus, 79
  functional causes of, 79
Anorectal constipation, 78–79
  functional causes, 79
  organic causes, 78–79
Anorectal function, neural control of, 30–31
Anorectal manometry, 124–126
  normal recordings and, 125–126
  perfusion systems and, 125
  resting anal pressure and, 125–126
Anorectal physiology, 124–132
  introduction to, 124
  key message of, 124
Anorectum, midcoronal section of, 18

Antepartum pelvic floor muscle training, to prevent postpartum urinary incontinence, 230–231
Antepartum/postpartum urinary incontinence prevention, PFMT and, during pregnancy, 231–232
Anterior rectocele, defecating proctogram of, 131
Anterior vaginal wall prolapse, surgery for, 286
Anterior wall support, urethra and, 14–16
Antidepressants, for OAB, 251
Antidiuretic agents, for OAB, 251
Antimuscarinic drugs, for OAB, 250–251
Arcus tendineus, pelvic attachment of, 15
Arcus tendineus levator ani (ATLA), 11
Arcus tendineus of fasciae pelvis (ATFP), 11
Artificial bowel sphincter, 290–291
Artificial urinary sphincter, for SUI, 282–283
ATFP. *See* Arcus tendineus of fasciae pelvis
ATLA. *See* Arcus tendineus levator ani

B

Balloon training, bowel routine and, 261–262
Behavioral/drug therapy combination, for OAB, 251

293

Behavioral treatment
  for OAB, 246
  with PFM rehabilitation, 248
  for PFMT, 215–219
    introduction to, 215
    key messages of, 215
Biofeedback
  for AI, 263–264
  for constipation, 263–264
  definition of, 184–185
  for detrusor overactivity control, 188
  with dynamic ultrasound, 188
  electromyographic, 187
  with inflated indwelling catheter, 186
  kinesthetic, 186–187
  manometric visual, 187–188
  motor learning and, 184
  PFM rehabilitation and, 185–189
  with sonography, 188
  SUI and, during sports/fitness activities, 268–269
  tactile/verbal, 185–186
  with vaginal cones, 187
  with vaginal EMG-electrode, 188
  with vaginal probe, 187
  visual, 186
Biomechanical stimulation, of pelvic floor, 211
Birth position, upright/lateral, 44
Bladder
  anatomy of, 5
  function of, 5
  innervation of, 5–6
  neck of, 5
  OAPF and, 83–84
Bladder base, MRI of, 147
Bladder diary, 215–217
  leakage frequency on, 166
  sample, 216
Bladder function, phases of, 62
Bladder neck, MRI of, T2 weighted transverse sections, 148
Bladder neck mobility, urethrovesical angle and, measurement of, 139
Bladder neck/perineum
  during pregnancy, 40
  after vaginal delivery, 40
Bladder training, 218

  guidelines for, 247
  for OAB, 246–248
Bonney's analogy, of vaginal prolapse, 9
Bowel
  management, urinary incontinence and, 219
  OAPF and, 84
  retraining, 263
  routine, 259–260
    balloon training and, 261–262
    diet and, 260
    fluids and, 260
    physical activity and, 260
    posture and, 261
  sphincter, artificial, 290–291
Bulking agents, for SUI, 281

C
Cantienica training, of pelvic floor, 210–211
Cellular mechanisms, of muscle repair/rehabilitation, 53–54
Central motor conduction time (CMCT), 28
Cesarean section, 44
Childbirth
  pelvic floor function and, 39–44
  perineal ultrasound findings on, 40
  urinary incontinence after, 41
  urodynamic findings on, 40
Circular smooth muscle (CSM), 7
Circumvaginal musculature rating scale, 98
Clam exercise, for pelvic floor, 209
CMCT. See Central motor conduction time
CMG. See Cystometrogram
CNE. See Concentric needle electrode
Coccygeus muscle, 10
Coccyx, pelvic floor muscles and, 10
Colonic constipation
  inorganic causes of, 77
  organic causes of, 77
Colonic function, measurement of, 129–130
Colonic manometry, 130
Colonic transit studies, 129–130
Colposuspension, for SUI, 282
Concentric needle electrode (CNE), 157

Constipation
  anorectal, 78–79
  biofeedback for, 263–264
  causes of, 77
  colonic, 77
  fecal incontinence and, 79–80
  pathogenesis of, 80
  pathophysiology of, 77
Continence
  mechanism
    deterioration of, 4
    levator ani muscle and, 12–13
  mechanism of, 77
  neural control of, 30
  with normal voiding, example of, 114–116
  physiology of, 76–77
  pregnancy and, 4
  promotion, with routine health care, 240–241
Continuity of care, 44
Contraceptive diaphragm, for SUI, 205
Coronal scan, 137–138
  with perineum ultrasound probe placement, 138
Coughing, sonographic evaluation of, 138–140
Cough stress profile, 115
  of continent woman, 115
  of stress-incontinent woman, 117
CSM. See Circular smooth muscle
Cystocele
  illustration of, 14
  left displacement, 15
Cystometrogram (CMG), 110

D
Darifenacin, for OAB, 250–251
Defecating proctogram, 130–131
  of anterior rectocele, 131
  of descending perineum, 131
  findings at, 131–132
Defecation
  mechanism of, 77
  OAPF and, 86
  physiology of, 76–77
  sonographic evaluation of, 140
Defecation scintigraphy, 130
Defecography, 130–131
Delayed voiding, 218
Denervation disorder, of skeletal muscle, 54

Index

Descending perineum, defecating proctogram of, 131
Descending perineum syndrome, 10, 73
Desmopressin, for OAB, 251
Detrusor overactivity control, biofeedback for, 188
Detrusor overactivity, pure, urodynamic findings and, 118–122
Devices for, for PFM rehabilitation, 201–206
Diet, bowel routine and, 260
Digital self-assessment, pelvic re-education regimen and, 97
Digital self examination, 91
Drydock boat concept, of POP, 73
Dynamic graciloplasty, for anal sphincter, 290
Dysfunctional voiding, 122–123
Dyspareunia, 253–254
  point-form treatment of, 256

E

Electrical neurostimulation
  evidence for use of, 193–194
  for fecal incontinence, 194
  for mixed incontinence, 193–194
  for OAPF, 194
  for stress urinary incontinence, 193
  for urge incontinence, 193–194
Electrical stimulation, 190–194
  action mechanisms of, 190–191
  anal motility and, 264
  anal probes for, 192
  application of, for pelvic dysfunction, 191
  introduction to, 190
  key messages of, 190
  methods of peripheral, 191–192
  in practice, 192–193
  vaginal probes for, 192
Electromagnetic induction therapy
  indications for, 197
  Neotonus chair for, 197
  neuromuscular tissue response to, 196–197
  pelvic floor rehabilitation and, 199
  principles of, 196
  RCTs of, 197–198
  treatment protocols for, 197–198
Electromyography (EMG), 128
  biofeedback, 187
  from pelvic floor, 157
  striated muscle activity and, 155–157
Electrophysiology, 155–161
  introduction to, 155
  key message of, 155
EMG. See Electromyography
Endoanal ultrasound, 132, 138
  abnormal, 132
  normal, 132
Endopelvic fascia
  of pelvic floor, 13–14
  of urethra, 6
Evacuation
  difficulties, 259–265
  normal, 131
  techniques, 261
    success rates for, 262
  training, 260–262
Evidence-based exercise parameters, for functional capacity improvement, of striated pelvic floor muscles, 238
Evoked potentials, of pelvic floor, 160–161
Exercise. See also Pelvic floor exercises
  active muscle use and, 224
  adherence/maintenance, 217–218
  initial program for, pelvic floor muscle training, 223–224
  for PFMs, 100–101, 184
  PFMT and, 179, 180, 216–217
  progression of, 224
  protocol adherence, for PFMT, 181–182
External palpation activations, of pelvic floor, 210
Extracorporeal magnetic stimulation, 196–199
  introduction to, 196
  key messages of, 196

F

Fecal incontinence (FI), 259
  constipation and, 79–80
  electrical neurostimulation for, 194
  medication for, 263
  studies on, 199
  surgical treatment algorithm for, 291
Feldenkrais approach, to pelvic floor, 209–210
Fembrace, for POP, 276
FI. See Fecal incontinence
Filling cystometry
  normal example of, 116
  in OAB female patient, 118
  pressure-flow studies and, 111–112
Fistulas, pessaries with, 275
Fluids
  bowel routine and, 260
  management of, 218
Frequency volume chart, leakage frequency on, 166

G

General muscle training
  overload and, 178
  principles of, 178
  reversibility and, 178
  specificity and, 178
Genital muscles, external, perineal membrane and, 16–17
Genital prolapse
  classification of, 286
  surgery for, 285–288
    introduction to, 285
    key messages of, 285
    role of mesh in, 286
Groutz Score, 165

H

High-pressure zone, of anal canal, 124
Hirschsprung's disease, functional causes of, 79
Hook wire electrodes, 157–158
  EMG recording with, 158
Hormone therapy, for SUI, 225–226
Hymenal ring, 17
Hypotone urethra, 116
Hysterectomy
  perineal descent/enterocele after, 153
  vaginal support levels after, 13

I

ICIQ-SF. *See* International Consultation on Incontinence Questionnaire-Short Form
ICS. *See* International Continence Society
Idiopathic lesions, PFMs and, 31–32
IIQ. *See* Incontinence Impact Questionnaire
Iliococcygeus muscle, 10
Imipramine, for OAB, 251
Incontinence. *See also* Anal incontinence; Fecal incontinence; Stress urinary incontinence; Urgency/urge incontinence; Urinary incontinence
   mixed, 67
      electrical neurostimulation, 193–194
      pathogenesis of, 80
      sexual dysfunction and, 254–255
      sports and, 4
      true mixed, 122
Incontinence Impact Questionnaire (IIQ), 170
Incontinence pessaries, 205
Incontinence Severity Index (ISI), 163
Ingelmann-Sundberg Scale, 163
International Consultation on Incontinence Questionnaire-Short Form (ICIQ-SF), 163, 164, 165
International Continence Society (ICS), 62
   urodynamic measurements of, 109
Intrapartum interventions, pelvic floor anatomy/function and, 42–43
Intraurethral ultrasound, 138
Intravaginal resistance devices, 201
   for pelvic organ prolapse, 202–204
   for SUI, 204–206
Introital ultrasound, 136–137
Intussusception, 79
Ischiopubic rami, perineal membrane and, 17
ISI. *See* Incontinence Severity Index
Isotope retention, mean segmental/total percentage of, 128

K

Kegel perineometer, 202
Kinesthetic biofeedback, 186–187
"The Knack," 13, 237

L

Large bowel, anatomy of, 75–76
Laxatives, 262–263
Leakage frequency
   on bladder diary, 166
   on frequency volume chart, 166
Leak point pressure, 114
Levator ani muscles, 3
   continence mechanism and, 12–13
   innervation of, 12
   macroscopic anatomy of, 10
   midurethral cross-section of, 6
   muscle fiber type of, 12
   muscle physiology of, 12
   permanent changes in, 147–149
   reconstructed, 56
   reversible changes in, 147–149
   stretching of, 186
Levator ani muscle thickness, MRI of, 146
Lifestyle changes, 218–219
Longitudinal smooth muscle (LSM), 7
Lower urinary tract (LUT)
   innervation of, 28
   neural control of, 29
LSM. *See* Longitudinal smooth muscle
LUT. *See* Lower urinary tract

M

Magnetic resonance imaging (MRI), 144–154
   axial section of, of middle urethra, 146
   of bladder base, 147
   dynamic, 150–154
      significance of, 153–154
      standard reference lines for, 150–151
   healthy anatomy and, in nulliparous women, 145–147
   imaging planes of, 145
   introduction to, 144
   key messages of, 144
   of levator ani muscle thickness, 146
   of middle urethra, 147
   of proximal urethra, 149
   static, 145–150
      anatomy and, 145
      normal variation and, 145
      pathology and, 145
      of stress urinary incontinence, 149–150
   T2 weighted images, of pelvic organs, 150, 151
   T2 weighted transverse sections, of bladder neck, 148
   technique of, 144–145
Manometric visual biofeedback, 187–188
Manual therapy, for OAPF, 256
Maximum urethral closure pressure (MUCP), 113
Maximum voluntary contraction (MVC), 97
Medication, for FI, 263
Mediolateral episiotomy, different rates of, 43
Micturition
   muscle mechanism of, 11
   neural control of, 28–29
   OAPF and, 85–86
   sonographic evaluation of, 140
Midline episiotomy, 42–43
Midurethral tape, for SUI, 282
Milex pessaries, 272
Mixed incontinence, 67
   electrical neurostimulation, 193–194
   true, 122
Modified Faecal Incontinence Score, 165
Motor learning, biofeedback and, 184
Motor nerve conduction studies, of sphincter function, 126–127
Motor unit, schematic of, 156

Index

Motor unit potential, component phases of, 156
MRI. *See* Magnetic resonance imaging
MUCP. *See* Maximum urethral closure pressure
Muscle. *See also* Levator ani muscles; Pelvic floor muscles; Skeletal muscle
  awareness, of PFMs, 25
  coordination assessment, of PFMs, 98
  dysfunction
    clinical impact of, 57
    future directions in, 58
  fibers, innervation of, 52
  function
    aging and, 49–58
    normal changes in, 57
  physiology, 49
  repair/rehabilitation, 53–57
    cellular mechanisms for, 53–54
    clinical aspects of, 54–56
    clinical assessment of, 54–56
    clinical relevance of, 56
    for OAPF, 256–257
  use, for accident prevention, 217
Musculoskeletal treatment, of OAPF, 255
MVC. *See* Maximum voluntary contraction

N

Neotonus chair, for electromagnetic induction therapy, 197
Nerve conduction studies, of pelvic floor, 159–160
Neurological lesions, PFMs and, 31–32
Neurologic conceptualization, of PFM physiotherapy, 33
Neuromuscular injury, to pelvic floor region, from vaginal delivery, 32–33
Nocturia, 219
  research on, 249
Normal recordings, of anorectal manometry, 125
Nulliparous woman, urethral anatomy of, 6

O

OAB. *See* Overactive bladder syndrome
OAPF. *See* Overactive pelvic floor
Observations in standing, PFMs and, 97
1-hour pad tests, 166–167
Onuf's nucleus, 64
Open pouch of Douglas, in female patient, 152
Orgasmic dysfunction, 254
Overactive bladder syndrome (OAB), 66–67, 246–251
  antidepressants for, 251
  antidiuretic agents for, 251
  antimuscarinic drugs for, 250–251
  behavioral/drug therapy combination for, 251
  behavioral interventions for, 246
  bladder training for, 246–248
  common causes for, 67
  darifenacin for, 250–251
  desmopressin for, 251
  imipramine for, 251
  introduction to, 246
  key messages of, 246
  oxybutynin for, 249
  patient example of, 117–119
  pharmacological therapy for, 249–251
  propiverine for, 250
  solifenacin for, 250
  studies on, 198–199
  tolterodine for, 250
  trospium for, 250
Overactive pelvic floor (OAPF), 83–86
  abnormal tissue areas of, 84
  bladder and, 83–84
  bowel and, 84
  causes of, 84–85
  conservative management of, 255–257
  defecation and, 86
  electrical neurostimulation for, 194
  evaluation of, 255
  findings of, 85
  introduction to, 83
  key message of, 83
  local pain of, 83
  manual therapy for, 256
  micturition and, 85–86
  muscle rehabilitation for, 256–257
  musculoskeletal treatment of, 255
  psychological issues with, 255
  referred pain of, 83
  sexual dysfunction and, 253–257
    introduction to, 253
    key messages of, 253
    symptoms/findings of, 253–255
  sexual function/sexual abuse and, 86
  symptoms of, 83
  urinary tract morbidity with, 85
  voiding patterns and, 86
Overload
  general muscle training and, 178
  PFMT and, 178–179
Oxford scale, 99
Oxybutynin, for OAB, 249

P

Pad tests, 166–168
Patient education, AI and, 259
Patient symptom scores, pelvic floor rehabilitation and, outcome measures in, 163–165
Pelvic attachment, of arcus tendineus, 15
Pelvic dysfunction, electrical stimulation for, application of, 191
Pelvic floor
  anatomic view of, 4
  anatomy of, 76
  awareness, enhancement of, 185
  biomechanical stimulation of, 211
  Cantienica training of, 210–211
  clam exercise for, 209
  connective tissue of, 13–14
  contraction, 151, 152
    sonographic evaluation of, 140
  deterioration of function, 4
  disorders, 3
    devices for, 202–206
  distension, during vaginal delivery, 37

Pelvic floor (*cont.*)
  dysfunction, identification of, 95
  electromyography activity from, 157
  endopelvic fascia of, 13–14
  evoked potentials of, 160–161
  external palpation activations of, 210
  Feldenkrais approach to, 209–210
  function
    childbirth and, 39–44
    deterioration of, 4
    loss reduction, 44
    sonographic evaluation of, 138–141
  functional anatomy, 3–19
  lateral view of, 16
  nerve conduction studies of, 159–160
  pain, pelvic floor muscle dysfunction and, 85
  physiotherapy, for POP, 275–276
  Pilates method for, 208–209
  POP and, 73–74
  postpartum management of, 235–241
  pudendal motor latency of, 160
  reflex studies of, 159
  rehabilitation of, 3
  at rest, 150, 151–152
  structures, lateral view of, 11
  during valsalva maneuver, 151, 152–153
  whole-body vibration therapy for, 211
  yoga for, 208–209
Pelvic floor exercises (PFX), 238, 240. *See also* Exercise
  program adherence to, 240
Pelvic floor muscle recruitment
  in stress incontinence, 180–181
  in urge incontinence, 181
Pelvic floor muscle re-education, 222–226
Pelvic floor muscles (PFMs), 3, 22–23, 91–103
  afferent pathways of, 27–28
  assessment of, 92–96
  clinical evaluation of, 91–103
    beginning of, 91–92
    introduction to, 91
    key message of, 91

coccyx and, 10
contraction
  assessment of, 96
  observations during, 96–97
denervation of, 22
digital evaluation of, clinical scale for, 101
dysfunction, pelvic floor pain and, 85
exercise for, 100–101, 184
function, visualization, 151
idiopathic lesions and, 31–32
innervation of, 26–28
lateral view of, digital palpation of, 94
layers of, 93
muscle awareness of, 25–26
muscle coordination assessment of, 98
neural control of, 22–33
  activity of, 22–23
  introduction to, 22
  key message of, 22
neurological lesions and, 31–32
observations in standing and, 97
palpation, 93–96
physiotherapy, neurologic conceptualization of, 33
rehabilitation/training of, 101–102
  behavioral training with, 248
  biofeedback and, 185–189
  devices for, 201–206
  instrumentation for, 102
  scales/assessment tools for, 98–100
schematic anatomy of, 92
somatic motor pathways of, 26–27
strength measurement of, 97–98
  other assessment methods for, 102–103
tenderness scale of, 95
tonic/phasic, 23–24
transvaginal palpation of, 94
two digital measurement of, 94
Pelvic floor muscle training (PFMT), 121
anatomy education and, 222
compared to no treatment, 228–229
devices for, 201–206

differing approaches to, 229–230
effectiveness of, 228–230
exercise and, 179–180, 216–217
exercise protocol adherence for, 181–182
initial exercise program for, 223–224
introduction to, 177–178
key messages of, 177
overload and, 178–179
PFM assessment and, 222–223
during pregnancy, antepartum/postpartum urinary incontinence prevention and, 231–232
principles of, 178–179
program descriptions of, 229, 231
reversibility and, 179
review/follow-up, 224–225
specificity and, 178–179
synergist coactivation, 179–180
voluntary contraction techniques for, 223
Pelvic floor rehabilitation
  electromagnetic stimulation and, 199
  outcome measures in, 162–171
    introduction to, 162
    patient symptom scores and, 163–165
    validation of, 162–163
  SUI and, during sports/fitness activities, 267–268
Pelvic muscle rating scale, 99
Pelvic organ prolapse (POP), 71–74, 105–108, 271–276
  clinical findings of, 71
  drydock boat concept of, 73
  etiology of, 71–72
  evaluation of, 272–276
  examination for, 72–73, 105
  fembrace for, 276
  historical perspective on, 271–272
  intravaginal resistance devices for, 202–204
  introduction to, 71
  key messages of, 71
  pelvic floor and, 73–74
  pelvic floor physiotherapy for, 275–276
  pessary treatment for, 271

Index

during pregnancy, 38–39
quantification system for, 72, 106–107
risk factors of, 71–72
space-occupying pessaries for, 203
support pessaries for, 203
types of, 272–276
Pelvic organs, support of, 8–10
Pelvic pain, point-form treatment of, 256
Pelvic re-education regimen
  digital self-assessment and, 97
  monitoring therapy and, 103
P.E.R.F. E. C. T. scale, 99
  examples of, 100
Perfusion systems, anorectal manometry and, 125
Periform vaginal electrode, with indicator, 224
Perineal clinic
  conditions treated in, 243–244
  experience in, 244
  introduction to, 242
  key messages of, 242
  referral reasons for, 244
  role of, 242–244
  structure of, 242–243
Perineal descent/enterocele, after hysterectomy, 153
Perineal massage, 44
Perineal membrane
  external genital muscles and, 16–17
  ischiopubic rami and, 17
Perineal pain, management pathway for, 244
Perineal ultrasound, 136–137
  findings on childbirth, 40
  of minimally filled bladder, 138
Pessaries
  complications with, 275
  cup/stem, 272
  fistulas with, 275
  fitting of, guidelines for, 273–274
  incontinence, 205
  milex, 272
  POP treatment with, 271
  success rates with, 274–275
  types of, 274
  vaginal discharge with, 275
  vaginal erosions with, 275
PFMs. *See* Pelvic floor muscles

PFMT. *See* Pelvic floor muscle training
PFX. *See* Pelvic floor exercises
Pharmacological therapy
  for OAB, 249–251
  for SUI, 225–226
Physical activity. *See also* Exercise
  bowel routine and, 260
Pilates method, for pelvic floor, 208–209
PMC. *See* Pontine micturition center
Point-form treatment
  of dyspareunia, 256
  of pelvic pain, 256
Pontine micturition center (PMC), 28–29
POP. *See* Pelvic organ prolapse
Posterior prolapse, 17
Posterior support, 17–18
Postpartum continence promotion program, components of, 239
Postpartum intervention, effectiveness of, 239–240
Postpartum management
  of pelvic floor, 235–241
    introduction to, 235
    key messages of, 235
  perineal, outline of, 236–237
Postpartum PFM rehabilitation, 238
Postpartum urinary incontinence prevention, with antepartum pelvic floor muscle training, 230–231
Posture, bowel routine and, 261
Pregnancy
  antepartum/postpartum urinary incontinence prevention and, PFMT during, 231–232
  bladder neck/perineum during, 40
  continence and, 4
  pelvic floor during
    examination for, 38–39
    findings of, 38–39
    imaging for, 39
    symptoms of, 38
    urodynamic studies for, 39
  pelvic floor dysfunction during, 37

  pelvic organ prolapse during, 38–39
  urodynamics during, 38–39
Pregnancy/childbirth effects
  hormonal changes/impact and, 37
  mechanical trauma and, 37–38
  on pelvic floor, 36–45
    introduction to, 36–37
    key messages on, 36
Pressure/flow recording, during filling/micturition, 112
Pressure-flow studies, filling cystometry and, 111–112
Pressure flow study, of patient, 118
Primiparous women, 38
Prolapse symptoms, illustrations of, 14
Propiverine, for OAB, 250
Psychological issues, with OAPF, 255
Pubic synphysis, as reference structure, 137
Pubococcygeus muscle, 10
  activation patterns of, 32, 33
Puborectalis muscle, 10
Pudendal electrode, disposable, 127
Pudendal latencies
  normal mean, 126
  normal recordings of, 127
Pudendal motor latency, of pelvic floor, 160
Pudendal nerve, from ventral rami, 26
Pudendal nerve terminal motor latency, 126–127
Pure detrusor overactivity, urodynamic findings and, 118–122
Pure stress incontinence, urodynamic findings and, 118–122

Q

Quality of life scores, urinary incontinence and, 170

R

Radioisotope colon transit study, 129
Randomized controlled trials (RCTs), of electromagnetic induction therapy, 197–198

RCTs. *See* Randomized controlled trials
Reconstructive techniques, for anal sphincter, 290
Rectal prolapse, 79
  advanced uterovaginal prolapse and, 79
Rectal sensation, showing balloon distension volumes, 127
Rectoanal reflex, 126
Rectocele, 79
  illustration of, 14
  surgery for, 287
Rectum, dynamic imaging of, 130–132
Reflex studies, of pelvic floor, 159
Resting anal pressure, anorectal manometry and, 125–126
Reversibility
  general muscle training and, 178
  PFMT and, 179
Routine mediolateral episiotomy, 42–43

S
Sacral nerve stimulation, anal sphincter and, 289–290
Sagittal scan/plan, 137
Schüssler diagram, 72
Selective serotonin-norepinephrine reuptake inhibitors, for SUI, 226
Sexual dysfunction
  incontinence and, 254–255
  OAPF and, 253–257
    introduction to, 253
    key messages of, 253
    symptoms/findings of, 253–255
Sexual function/sexual abuse, OAPF and, 86
Sexual response, neural control of, 31
Single-fiber electrode, superimposed on normal muscle, 128
Single-fiber electromyography, 128
Skeletal muscle
  adaptation of, within pelvic floor, 52–53
  cardiovascular damage of, 53

cross section of, 50
denervation disorder of, 54
gross anatomical aspects of, 49–50
mechanical/myopathic damage of, 53
micrograph of, 50
neural damage of, 53
normal adaptation of, 52
normal function of, 49
structure/function of, 50
three-dimensional schematic of, 51
weakness, mechanisms of, 53–57
Sling, for SUI, 282
Solifenacin, for OAB, 250
Somatic motor pathways, of PFMs, 26–27
Sonographic evaluation. *See also* Ultrasound imaging
  biofeedback with, 188
  of coughing, 138–140
  of defecation, 140
  of micturition, 140
  of pelvic floor contraction, 140
  of pelvic floor function, 138–141
  of straining, 138–140
  of valsalva, 138–140
Space-occupying pessaries, for pelvic organ prolapse, 203
Specificity
  general muscle training and, 178
  PFMT and, 178–179
Sphincter function
  electrophysiology of, 126
  motor nerve conduction studies of, 126–127
Sports/fitness activities
  incontinence and, 4
  SUI during, 267–269
Stamey Grading, of incontinence, 163
St. George Score, 165
Stoma creation, 291
Storage physiology, voiding phase and, 63–65
Storage reflex, 63
Straining, sonographic evaluation of, 138–140

Stress-incontinent woman, 116–117
  cough stress profile of, 117
  urethral closure pressure profile of, 116
Stress strategies, 217
Stress urinary incontinence (SUI), 65–66, 121, 221–226
  alpha-agonists for, 225
  artificial urinary sphincter for, 282–283
  assessment of, 221
  biofeedback techniques for, during sports/fitness activities, 268–269
  bulking agents for, 281
  colposuspension for, 282
  contraceptive diaphragm for, 205
  definition of, 221
  electrical neurostimulation for, 193
  hormone therapy for, 225–226
  intravaginal resistance devices for, 204–206
  introduction to, 221
  key messages of, 221
  midurethral tape for, 282
  MRI of, 149–150
  pelvic floor muscle recruitment in, 180–181
  pharmacological therapy for, 225–226
  pure, urodynamic findings and, 118–122
  selective serotonin-norepinephrine reuptake inhibitors for, 226
  sling for, 282
  during sports/fitness activities, 267–269
    etiology of, 267
    introduction to, 267
    key messages of, 267
    pelvic floor rehabilitation and, 267–268
  surgical choices for, 281–283
  tricyclic antidepressants for, 226
  urethral closure for, 283
  urinary diversion for, 283
Striated anal sphincter, 156
Striated muscle activity, electromyography and, 155–157

Index

Striated pelvic floor muscles, evidence-based exercise parameters of, for functional capacity improvement, 238
Striated urinary sphincter, 7
  of urethra, 7
SUI. *See* Stress urinary incontinence
Support pessaries, for pelvic organ prolapse, 203
Surface electrodes, 158–159
  EMG recording with, 159
  types of, 159
Surgery
  for anterior vaginal wall prolapse, 286
  for genital prolapse, 285–288
  for rectocele, 287
  for SUI, 281–283
    introduction to, 281
    key messages of, 281
    minimally invasive procedures for, 281–282
  for uterine prolapse, 286–287

T
Tactile/verbal feedback, 185–186
Tactile/visual biofeedback, during cystourethroscopy, 188–189
Tenderness scale, of PFMs, 95
Three-dimensional ultrasound, 138
Tolterodine, for OAB, 250
Tonic motor unit, continuously firing, falling leaf display of, 23
Tonic muscle activity, inhibition of, 24
TrA muscles. *See* Transversus abdominis muscles
Transmuscular hook electrodes, 157
Transvaginal palpation, of PFMs, 94
Transversus abdominis (TrA) muscles, 97
Tricyclic antidepressants, for SUI, 226
Trospium, for OAB, 250
True mixed incontinence, 122
24-hour pad tests, 167
  criterion validity of, 168
  repeatability of, 168
2-hour pad tests, 167
Two digital measurement, of PFMs, 94

U
UDI. *See* Urogenital Distress Inventory
Ultrasound imaging, 135–141. *See also* Sonographic evaluation
  applications of, 136–137
  as biofeedback tool, 140–141
  dynamic, biofeedback with, 188
  equipment for, 136–137
  key messages of, 135
  scope of, 135–136
UPP. *See* Urethral pressure profilometry
Urethra
  anatomy of, 6–8
  anterior wall support and, 14–16
  endopelvic fascia of, 6
  innervation of, 8
  middle, MRI of, 147
  palpation of, 96
  proximal, MRI of, 149
  support of, 8–10
Urethral closure
  pressure profile, 115
    of stress-incontinent woman, 116
  for SUI, 283
Urethral glands, 8
Urethral pressure measurement, 112–113
  schematic drawing of, 113
Urethral pressure profilometry (UPP), of urinary incontinence, 169
Urethral pressure transducer, 113
Urethral smooth muscle, 7
Urethral sphincter, external, MEPs of, 28
Urethral sphincter muscle, EMG recording of, 23
Urethral submucosal vasculature, 8
Urethrovesical angle, bladder neck mobility and, measurement of, 139
Urgency/frequency
  cycle, 247
  research on, 249

Urgency/urge incontinence
  electrical neurostimulation for, 193–194
  pelvic floor muscle recruitment in, 181
  studies of, 198
Urge suppression strategies, 217
  mechanism of, 248
  patient instructions for, 249
Urinary diversion, for SUI, 283
Urinary incontinence
  anatomical observations on, 168–171
  bowel management and, 219
  after childbirth, 41
  clinical treatment for, 56–57
  functional observations on, 168–171
  obstetric/maternal risk factors for, 41
  pad test classification of, 168
  prevention opportunities for, 56–57
  quality of life scores and, 170
  socioeconomic evaluation, 170
  symptom quantification for, 166–168
  symptoms of, 12, 13
  types of, 65–67
  UPP of, 169
  urodynamic assessment of, 168–169
  voiding dysfunction and, 62–69
  weight loss and, 218–219
Urinary sphincter
  artificial, for SUI, 282–283
  competent, 66
  striated, 5–6
Urinary tract, lower, anatomy of, 5–8
Urinary tract morbidity, with OAPF, 85
Urine flow curve, descriptive terminology and, 111
Urodynamic(s), 109–119
  indications of, 109
  key message of, 109
  during pregnancy, 38–39
Urodynamic assessment, of urinary incontinence, 168–169

Urodynamic findings
  in clinical practice, 118–123
  introduction to, 118
  key message of, 118
  pure detrusor overactivity and, 118–122
  pure stress incontinence and, 118–122
Urodynamic lab, with supine patient, 110
Urodynamic measurements, of ICS, 109
Urodynamic stress incontinence (USI), studies on, 198
Urodynamic studies, for pregnancy, pelvic floor during, 39
Urodynamic techniques
  definitions of, 109–110
  technical performance of, 109–110
Uroflow, with abdominal/pelvic floor EMG, 114
Uroflowmetry, 110
Urogenital diaphragm, 16–17
Urogenital Distress Inventory (UDI), 170
Urogenital hiatus, 10
U-shaped levator ani muscle, schematic of, 8
USI. See Urodynamic stress incontinence
Uterine prolapse
  illustration of, 14
  quantification of, 151
  surgery for, 286–287
Uterovaginal support, 16

V
Vaginal delivery
  bladder neck/perineum after, 40
  clinical examination findings after, 39
  neuromuscular injury to, pelvic floor region, 32–33
  neurophysiology of, 39–40
  pelvic floor distension during, 37
  pelvic floor dysfunction after, 37
  pelvic floor safety during, 44–45
  pelvic floor symptoms during, 37
Vaginal discharge, with pessaries, 275
Vaginal EMG-electrode, biofeedback with, 188
Vaginal erosions, with pessaries, 275
Vaginal probes, for electrical stimulation, 192
Vaginal prolapse, Bonney's analogy of, 9
Vaginal support
  diagrammatic display of, 9
  levels, after hysterectomy, 13
Vaginismus, 254
Valsalva, sonographic evaluation of, 138–140
Valsalva leak point pressures (VLPP), 169–170
Valsalva maneuver, pelvic floor during, 151, 152–153
Vesical pressure transducer, 113
Visual biofeedback, 186
VLPP. See Valsalva leak point pressures
Voiding dysfunction
  classification of, 68
  etiology of, 68
  patient example of, 117
  prevention of, 69
  symptoms of, 67–68
  treatment of, 69
  urinary incontinence and, 62–69
Voiding frequency
  decreasing, 218
  increasing, 218
Voiding habits/schedules, 218
Voiding patterns, OAPF and, 86
Voiding phase, storage physiology and, 63–65
Voiding reflex, 63
Voiding response, voluntary, 64
Voluntary contraction pressure, 126
Voluntary contraction techniques, for PFMT, 223

W
Weight loss, urinary incontinence and, 218–219
Wexner scale, 165
Whole-body vibration therapy, for pelvic floor, 211

Y
Yoga, for pelvic floor, 208–209